影印版说明

本手册由 Elsevier 授权影印出版。原版为全一册（共 19 章），考虑到内容的相对独立以及使用方便，影印版分为 6 册。

第 1 册 熔体·卷入·流动（第 1～3 章）。主要介绍金属熔体，包括熔体与周围环境可能发生的反应、熔体的流动性和可能存在的各种卷入性缺陷。

第 2 册 铸型和型芯·凝固组织·铸造合金（第 4～6 章）。第 4 章主要介绍铸型和型芯，包括铸型的性质、铸型和金属之间可能发生的反应。第 5、6 章主要介绍凝固组织的生长和可能存在的问题以及各种典型的铸造合金组织。

第 3 册 孔洞·热裂和冷裂·铸件的性能（第 7～9 章）。第 7、8 章主要介绍铸件中可能出现的缩孔和裂纹等缺陷。第 9 章主要介绍铸件的力学性能、抗氧化性能、耐腐蚀性能、密封性能以及表面处理。

第 4 册 获得优质铸件的 10 项准则（第 10 章）。

第 5 册 浇注系统设计基础、组成和设计实例（第 11～13 章）。主要介绍浇注系统的设计、组成，并给出浇注系统设计的实例分析。

第 6 册 熔炼·造型·铸造·凝固（第 14～19 章）。分别介绍铸造合金的熔炼、铸型材料和铸造成型方法的选择、凝固技术和铸件尺寸精度的控制、铸件的后续处理和探伤等。

本手册为从事铸造行业的科研人员和工程技术人员提供全面的指导和参考。目前国内尚没有如此完整介绍金属铸造工艺、冶金技术和设计、铸件质量控制和后续处理以及性能检测等方面的手册。

作者 John Campbell（约翰·坎贝尔）OBE，英国伯明翰大学（铸造专业）教授，铸件行业国际领先人物，Cosworth Casting 的创始人及 Baxi 铸造方法的发明人。

材料科学与工程图书工作室

联系电话 0451-86412421
　　　　　0451-86414559

邮　　箱 44739812@qq.com
　　　　　zhxh6414559@aliyun.com
　　　　　yh_bj@aliyun.com

JOHN CAMPBELL
COMPLETE CASTING HANDBOOK
Metal Casting Processes, Metallurgy, Techniques and Design
SECOND EDITION

影印版

铸造手册大全：金属铸造工艺、冶金技术和设计

第2册　铸型和型芯·凝固组织·铸造合金

黑版贸审字08-2018-075号

Elsevier (Singapore) Pte Ltd.
3 Killiney Road, #08-01 Winsland House I, Singapore 239519
Tel: (65) 6349-0200; Fax: (65) 6733-1817

Complete Casting Handbook, 2nd Edition
John Campbell
Copyright ©2015，2011 John Campbell. Published by Elsevier Ltd.All rights reserved.
ISBN-13: 9780444635099

This reprint of Complete Casting Handbook, 2nd Edition by John Campbell was undertaken by Harbin Institute of Technology Press and is published by arrangement with Elsevier (Singapore) Pte Ltd.

Complete Casting Handbook, 2nd Edition by John Campbell由哈尔滨工业大学出版社有限公司进行影印，并根据哈尔滨工业大学出版社有限公司与爱思唯尔（新加坡）私人有限公司的协议约定出版。

ISBN: 9787560373379

Copyright ©2018 by Elsevier (Singapore) Pte Ltd.

All rights reserved. No part of this publication may be reproduced or transmitted in any form or by any means, electronic or mechanical, including photocopying, recording, or any information storage and retrieval system, without permission in writing from Elsevier (Singapore) Pte Ltd. Details on how to seek permission, further information about the Elsevier's permissions policies and arrangements with organizations such as the Copyright Clearance Center and the Copyright Licensing Agency, can be found at our website: www.elsevier.com/permissions.

This book and the individual contributions contained in it are protected under copyright by Elsevier (Singapore) Pte Ltd. and Harbin Institute of Technology Press (other than as may be noted herein).
Online resources are not available with this reprint.
Printed in China by Harbin Institute of Technology Press under special arrangement with Elsevier (Singapore) Pte Ltd. This edition is authorized for sale in the People's Republic of China only, excluding Hong Kong SAR, Macau SAR and Taiwan. Unauthorized export of this edition is a violation of the contract.
本书英文影印版由Elsevier(Singapore) Pte Ltd.授权哈尔滨工业大学出版社有限公司在中华人民共和国境内（不包括香港特别行政区、澳门特别行政区以及台湾地区）出版与发行。未经许可之出口，视为违反著作权法，将受民事和刑事法律之制裁。
本书封底贴有Elsevier防伪标签，无标签者不得销售。

Notice

Knowledge and best practice in this field are constantly changing. As new research and experience broaden our understanding, changes in research methods, professional practices, or medical treatment may become necessary. Practitioners and researchers must always rely on their own experience and knowledge in evaluating and using any information, methods, compounds or experiments described herein. Because of rapid advances in the medical sciences, in particular, independent verification of diagnoses and drug dosages should be made. To the fullest extent of the law, no responsibility is assumed by Elsevier, authors, editors or contributors in relation to the adaptation or for any injury and/or damage to persons or property as a matter of products liability, negligence or otherwise, or from any use or operation of any methods, products, instructions, or ideas contained in the material herein.

图书在版编目（CIP）数据

铸造手册大全：金属铸造工艺、冶金技术和设计＝Complete Casting Handbook Metal Casting Processes，Metallurgy，Techniques and Design．第2册，铸型和型芯·凝固组织·铸造合金：英文／（英）约翰·坎贝尔（John Campbell）主编．—哈尔滨：哈尔滨工业大学出版社，2018.8
ISBN 978-7-5603-7337-9

Ⅰ.①铸… Ⅱ.①约… Ⅲ.①冶金－技术－手册－英文 Ⅳ.①TF1-62

中国版本图书馆CIP数据核字（2018）第094699号

责任编辑　杨　桦　许雅莹　张秀华
出版发行　哈尔滨工业大学出版社
社　　址　哈尔滨市南岗区复华四道街10号　邮编150006
传　　真　0451-86414749
网　　址　http://hitpress.hit.edu.cn
印　　刷　哈尔滨市工大节能印刷厂
开　　本　787mm×960mm　1/16　印张 19.75
版　　次　2018年8月第1版　2018年8月第1次印刷
书　　号　ISBN 978-7-5603-7337-9
定　　价　98.00元

（如因印刷质量问题影响阅读，我社负责调换）

Complete Casting Handbook
Metal Casting Processes, Metallurgy, Techniques and Design

Second Edition

John Campbell

*Emeritus Professor of Casting Technology,
Department of Metallurgy and Materials,
University of Birmingham, UK*

AMSTERDAM • BOSTON • HEIDELBERG • LONDON
NEW YORK • OXFORD • PARIS • SAN DIEGO
SAN FRANCISCO • SINGAPORE • SYDNEY • TOKYO

Butterworth-Heinemann is an imprint of Elsevier

目 录

前言 xxv

引言 xxvi

致谢 xxvii

铸造冶金学

第1册 熔体·卷入·流动（第1～3章）

第1章 熔体 3
1.1 熔体与环境的反应 5
1.2 气体在熔体中的传输 9
1.3 表面膜的形成 11
1.4 蒸发 14

第2章 卷入 17
2.1 卷入性缺陷 21
2.2 卷入过程 37
2.3 卷曲和展开 67
2.4 卷入膜的惰性化 77
2.5 可溶性瞬态膜 80
2.6 卷出过程 81
2.7 双层膜的证据 83
2.8 双层膜的重要性 88
2.9 四种常见的双层膜 89

第3章 流动 91
3.1 表面膜对填充的影响 91

3.2	最大流动性	95
3.3	延长流动性	128
3.4	连续流动性	131

第2册（本册） 铸型和型芯·凝固组织·铸造合金（第4~6章）

第4章 铸型和型芯 ... **135**
- 4.1 惰性铸型和活性铸型 ... 135
- 4.2 传质区 ... 136
- 4.3 蒸发区和凝聚区 ... 139
- 4.4 铸型气氛 ... 144
- 4.5 铸型表面反应 ... 147
- 4.6 金属表面反应 ... 155
- 4.7 铸型涂层 ... 159

第5章 凝固组织 ... **163**
- 5.1 传热 ... 163
- 5.2 基体组织的生长 ... 197
- 5.3 偏析 ... 213

第6章 铸造合金 ... **223**
- 6.1 锌合金 ... 223
- 6.2 镁合金 ... 227
- 6.3 铝合金 ... 236
- 6.4 铜合金 ... 269
- 6.5 铸铁 ... 275
- 6.6 钢 ... 314
- 6.7 镍基合金 ... 330
- 6.8 钛合金 ... 335

第3册 孔洞·热裂和冷裂·铸件的性能（第7~9章）

第7章 孔洞 ... **341**
- 7.1 收缩 ... 341
- 7.2 气孔 ... 387

| 7.3 孔洞的识别 | 413 |

第8章 热裂和冷裂 — 417
- 8.1 热裂 — 417
- 8.2 冷裂 — 442

第9章 铸件的性能 — 447
- 9.1 试棒 — 447
- 9.2 失效统计 — 450
- 9.3 缺陷的影响 — 463
- 9.4 拉伸性能 — 470
- 9.5 断裂韧性 — 493
- 9.6 疲劳性能 — 498
- 9.7 弹性（杨氏）模量和阻尼性能 — 507
- 9.8 残余应力 — 508
- 9.9 高温拉伸性能 — 509
- 9.10 抗氧化和耐腐蚀性能 — 511
- 9.11 密封性 — 519
- 9.12 表面处理 — 523
- 9.13 质量指数 — 526
- 9.14 无双层膜时性能 — 527

铸件制造（生产）

第4册 获得优质铸件的10项准则（第10章）
铸件制造业概论 — 529

第一部分 获得优质铸件的准则

第10章 获得优质铸件的10项准则 — 535
- 10.1 准则1：从高质量的熔体开始 — 536
- 10.2 准则2：避免液面湍流夹杂 — 542
- 10.3 准则3：避免液面层状夹杂 — 550

10.4	准则 4：避免裹气	557
10.5	准则 5：避免砂芯气孔	561
10.6	准则 6：避免缩孔	573
10.7	准则 7：避免对流	603
10.8	准则 8：减少偏析	610
10.9	准则 9：减少残余应力	613
10.10	准则 10：给定基准点	631

第 5 册　浇注系统设计基础、组成和设计实例（第 11 ~ 13 章）

第二部分　浇注系统设计

第 11 章　浇注系统设计基础 ... **641**
11.1	最大充型速度要求	642
11.2	重力浇注	642
11.3	降低和排除重力浇注问题	646
11.4	表面张力控制下的充型	651

第 12 章　浇注系统组成 ... **657**
12.1	浇口杯	657
12.2	直浇道	669
12.3	横浇道	683
12.4	内浇道	690
12.5	液流缓冲单元	716
12.6	防涡流单元	719
12.7	夹杂物的控制：过滤和集渣	724
12.8	过滤器	728

第 13 章　浇注系统设计实例 ... **747**
13.1	设计方案	748
13.2	浇注方案的确定	748
13.3	质量和体积的估计	748
13.4	封闭式和开放式	750
13.5	浇注时间的选择	755
13.6	薄壁件和缓慢充型	758

	13.7	充型速度	758
	13.8	浇口杯的设计	759
	13.9	直浇道的设计	759
	13.10	横浇道的设计	763
	13.11	内浇道的设计	765

第 6 册　熔炼·造型·铸造·凝固（第 14 ~ 19 章）

第三部分　生产过程（熔炼、造型、铸造、凝固）

第 14 章　熔　炼 … 769

　　14.1　间歇式熔炼 … 769
　　14.2　连续熔炼 … 772
　　14.3　保温、转运和分配 … 774
　　14.4　熔体处理 … 777
　　14.5　铸造材料 … 790
　　14.6　重熔 … 794

第 15 章　造　型 … 797

　　15.1　惰性铸型和铸芯 … 797
　　15.2　原砂 … 802
　　15.3　黏结剂 … 807
　　15.4　其他造型方法 … 813
　　15.5　橡胶铸型 … 814
　　15.6　原砂的回收和循环 … 814

第 16 章　铸　造 … 821

　　16.1　重力铸造 … 821
　　16.2　水平铸造 … 827
　　16.3　反重力铸造 … 838
　　16.4　离心铸造 … 856
　　16.5　压力辅助铸造 … 861
　　16.6　熔膜铸造和其他陶瓷型铸造工艺 … 871
　　16.7　消失模铸造 … 875

	16.8 真空铸造	878
	16.9 真空加压铸造	880
	16.10 真空熔炼和铸造	881

第 17 章 凝固控制技术 ... 883
17.1 常规凝固 ... 883
17.2 定向凝固 ... 883
17.3 单晶体凝固 ... 885
17.4 快速凝固 ... 887

第 18 章 尺寸精度 ... 893
18.1 净成型概念 ... 897
18.2 模样设计 ... 900
18.3 模样精度 ... 904
18.4 加工精度 ... 909
18.5 铸造精度 ... 910
18.6 测量技术 ... 924

第 19 章 铸件处理 ... 927
19.1 表面清理 ... 927
19.2 热处理 ... 929
19.3 热等静压 ... 935
19.4 加工 ... 938
19.5 上漆 ... 940
19.6 塑性加工（锻、轧、挤压）... 941
19.7 浸渗 ... 942
19.8 无损检测 ... 942

附录 I ... **947**

附录 II ... **951**

附录 III ... **955**

附录 IV ... **957**

参考文献 ... **959**

索引 ... **993**

Contents

Preface xxv
Introduction xxvi
Acknowledgements xxvii

CASTING METALLURGY

第1册　熔体·卷入·流动（第1~3章）

CHAPTER 1　The Melt 3
 1.1 Reactions of the Melt with its Environment 5
 1.2 Transport of Gases in Melts 9
 1.3 Surface Film Formation 11
 1.4 Vaporisation 14

CHAPTER 2　Entrainment 17
 2.1 Entrainment Defects 21
   2.1.1 Bifilms 21
   2.1.2 Bubbles 25
   2.1.3 Extrinsic Inclusions 30
 2.2 Entrainment Processes 37
   2.2.1 Surface Turbulence 37
   2.2.2 Oxide Skins from Melt Charge Materials 46
   2.2.3 Pouring 47
   2.2.4 The Oxide Lap Defect I: Surface Flooding 50
   2.2.5 Oxide Lap Defect II: The Confluence Weld 52
   2.2.6 The Oxide Flow Tube 56
   2.2.7 Microjetting 57
   2.2.8 Bubble Trails 58
 2.3 Furling and Unfurling 67
 2.4 Deactivation of Entrained Films 77
 2.5 Soluble, Transient Films 80
 2.6 Detrainment 81
 2.7 Evidence for Bifilms 83
 2.8 The Importance of Bifilms 88
 2.9 The Four Common Populations of Bifilms 89

CHAPTER 3　Flow 91
 3.1 Effect of Surface Films on Filling 91
   3.1.1 Effective Surface Tension 91

3.1.2 The Rolling Wave 92
3.1.3 The Unzipping Wave 92
3.2 Maximum Fluidity (The Science of Unrestricted Flow) 95
Fluidity Definition 96
3.2.1 Mode of Solidification 97
3.2.2 Effect of Velocity 110
3.2.3 Effect of Viscosity (Including Effect of Entrained Bifilms) 110
3.2.4 Effect of Solidification Time t_f 112
3.2.5 Effect of Surface Tension 121
3.2.6 Effect of an Unstable Substrate 124
3.2.7 Comparison of Fluidity Tests 125
3.2.8 Effect of Vibration 128
3.3 Extended Fluidity 128
3.4 Continuous Fluidity 131

第2册（本册） 铸型和型芯·凝固组织·铸造合金（第4～6章）

CHAPTER 4 Moulds and Cores 135
4.1 Moulds: Inert or Reactive 135
4.2 Transformation Zones 136
4.3 Evaporation and Condensation Zones 139
4.4 Mould Atmosphere 144
4.4.1 Composition 144
4.4.2 Mould Gas Explosions 145
4.5 Mould Surface Reactions 147
4.5.1 Pyrolysis 147
4.5.2 Lustrous Carbon Film 148
4.5.3 Sand Reactions 149
4.5.4 Mould Contamination 149
4.5.5 Mould Penetration 151
4.6 Metal Surface Reactions 155
4.6.1 Oxidation 155
4.6.2 Carbon (Pickup and Loss) 155
4.6.3 Nitrogen 156
4.6.4 Sulphur 157
4.6.5 Phosphorus 157
4.6.6 Surface Alloying 158
4.6.7 Grain Refinement 158
4.6.8 Miscellaneous 159
4.7 Mould Coatings 159
4.7.1 Aggregate Moulds 159
4.7.2 Permanent Moulds and Metal Chills 160
4.7.3 Dry Coatings 161

CHAPTER 5 Solidification Structure 163
5.1 Heat Transfer 163
5.1.1 Resistances to Heat Transfer 163
5.1.2 Increased Heat Transfer 178
5.1.3 Convection 193
5.1.4 Remelting 193
5.1.5 Flow Channel Structure 194
5.2 Development of Matrix Structure 197
5.2.1 General 197
5.2.2 Nucleation of the Solid 198
5.2.3 Growth of the Solid 201
5.2.4 Disintegration of the Solid (Grain Multiplication) 209
5.3 Segregation 213
5.3.1 Planar Front Segregation 213
5.3.2 Microsegregation 216
5.3.3 Dendritic Segregation 218
5.3.4 Gravity Segregation 219

CHAPTER 6 Casting Alloys 223
6.1 Zinc Alloys 223
6.2 Magnesium Alloys 227
6.2.1 Films on Liquid Mg Alloys + Protective Atmospheres 228
6.2.2 Strengthening Mg Alloys 230
6.2.3 Microstructure 234
6.2.4 Inclusions 235
6.3 Aluminium 236
6.3.1 Oxide Films on Al Alloys 237
6.3.2 Entrained Inclusions 239
6.3.3 Grain Refinement (Nucleation and Growth of the Solid) 239
6.3.4 Dendrite Arm Spacing (DAS) and Grain Size 244
6.3.5 Modification of Eutectic Si in Al-Si Alloys 244
6.3.6 Iron-Rich Intermetallics 259
6.3.7 Other Intermetallics 262
6.3.8 Thermal Analysis of Al Alloys 264
6.3.9 Hydrogen in Al Alloys 267
6.4 Copper Alloys 269
6.4.1 Surface Films 270
6.4.2 Gases in Copper-Based Alloys 270
6.4.3 Grain Refinement 274

CONTENTS

- **6.5** Cast Iron 275
  - 6.5.1 Reactions with Gases 275
  - 6.5.2 Surface Films on Liquid Cast Irons 277
  - 6.5.3 Cast Iron Microstructures 288
  - 6.5.4 Flake Graphite Iron and Inoculation 290
  - 6.5.5 Nucleation and Growth of the Austenite Matrix 299
  - 6.5.6 Coupled Eutectic Growth of Graphite and Austenite 300
  - 6.5.7 Spheroidal Graphite Iron (Ductile Iron) 302
  - 6.5.8 Compacted Graphite Iron 308
  - 6.5.9 Chunky Graphite 310
  - 6.5.10 White Iron (Iron Carbide) 312
  - 6.5.11 General 313
  - 6.5.12 Summary of Structure Hypothesis 314
- **6.6** Steels 314
  - 6.6.1 Carbon Steels 315
  - 6.6.2 Stainless Steels 316
  - 6.6.3 Inclusions in Carbon and Low-Alloy Steels: General Background 317
  - 6.6.4 Entrained Inclusions 319
  - 6.6.5 Primary Inclusions 322
  - 6.6.6 Secondary Inclusions and Second Phases 324
  - 6.6.7 Nucleation and Growth of the Solid 326
  - 6.6.8 Structure Development in the Solid 329
- **6.7** Nickel-Base Alloys 330
  - 6.7.1 Air Melting and Casting 331
  - 6.7.2 Vacuum Melting and Casting 332
- **6.8** Titanium 335
  - 6.8.1 Ti Alloys 336
  - 6.8.2 Melting and Casting Ti Alloys 336
  - 6.8.3 Surface Films on Ti Alloys 339

第3册 孔洞·热裂和冷裂·铸件的性能（第7～9章）

CHAPTER 7 Porosity 341
- **7.1** Shrinkage Porosity 341
  - 7.1.1 General Shrinkage Behaviour 341
  - 7.1.2 Solidification Shrinkage 342
  - 7.1.3 Feeding Criteria 348
  - 7.1.4 Feeding: The Five Mechanisms 352
  - 7.1.5 Initiation of Shrinkage Porosity 365
  - 7.1.6 Growth of Shrinkage Pores 381
  - 7.1.7 Shrinkage Pore Structure 382

xvi CONTENTS

    7.2 Gas Porosity 387
        7.2.1 Entrained Pores (Air Bubbles) 388
        7.2.2 Blow Holes 389
        7.2.3 Gas Porosity Initiated In Situ 398
    7.3 Porosity Diagnosis 413
        7.3.1 Gas Porosity 413
        7.3.2 Shrinkage Porosity 414

CHAPTER 8 Cracks and Tears 417
    8.1 Hot Tearing 417
        8.1.1 General 417
        8.1.2 Grain Boundary Wetting by the Liquid 420
        8.1.3 Pre-Tear Extension 421
        8.1.4 Strain Concentration 423
        8.1.5 Stress Concentration 424
        8.1.6 Tear Initiation 425
        8.1.7 Tear Growth 426
        8.1.8 Prediction of Hot Tearing Susceptibility 429
        8.1.9 Methods of Testing 433
        8.1.10 Methods of Control 436
        8.1.11 Summary of the Conditions for Hot Tearing and Porosity 440
        8.1.12 Hot Tearing in Stainless Steels 442
        8.1.13 Predictive Techniques 442
    8.2 Cold Cracking 442
        8.2.1 General 442
        8.2.2 Crack Initiation 444
        8.2.3 Crack Growth 444

CHAPTER 9 Properties of Castings 447
    9.1 Test Bars 447
    9.2 The Statistics of Failure 450
        9.2.1 Background of Using Weibull Analysis 453
        9.2.2 Procedure for Two-Parameter Weibull Analysis 455
        9.2.3 Three-Parameter Weibull Analysis 459
        9.2.4 bi-Weibull Distributions 459
        9.2.5 Limits of Accuracy 461
        9.2.6 Extreme Value Distributions 461
    9.3 Effect of Defects 463
        9.3.1 Inclusion Types 464
        9.3.2 Gas Porosity 465
        9.3.3 Shrinkage Porosity 466
        9.3.4 Tears, Cracks and Bifilms 467

CONTENTS

9.4 Tensile Properties 470
    9.4.1 Micro-structural Failure 470
    9.4.2 Ductility 473
    9.4.3 Yield Strength 477
    9.4.4 Tensile Strength 490
9.5 Fracture Toughness 493
9.6 Fatigue 498
    9.6.1 High Cycle Fatigue 498
    9.6.2 Low Cycle, High Strain and Thermal Fatigue 506
9.7 Elastic (Young's) Modulus and Damping Capacity 507
9.8 Residual Stress 508
9.9 High Temperature Tensile Properties 509
9.10 Oxidation and Corrosion Resistance 511
    9.10.1 Internal Oxidation 512
    9.10.2 Corrosion 514
    9.10.3 Pitting Corrosion 515
    9.10.4 Filiform Corrosion 517
    9.10.5 Inter-granular Corrosion 518
    9.10.6 Stress Corrosion Cracking 519
9.11 Leak-Tightness 519
9.12 Surface Finish 523
    9.12.1 Effect of Surface Tension 523
    9.12.2 Effects of a Solid Surface Film 525
9.13 Quality Indices 526
9.14 Bifilm-Free Properties 527

CASTING MANUFACTURE

第 4 册　获得优质铸件的 10 项准则（第 10 章）

Introduction to the Casting Manufacturing Industry 529

SECTION 1 RULES FOR GOOD CASTINGS

CHAPTER 10 The 10 Rules for Good Castings 535
10.1 Rule 1: Use a Good-Quality Melt 536
    10.1.1 Background 536
10.2 Rule 2: Avoid Turbulent Entrainment (The Critical Velocity Requirement) 542
    10.2.1 Introduction 542
    10.2.2 Maximum Velocity Requirement 543
    10.2.3 The No-Fall Requirement 547

- 10.3 Rule 3: Avoid Laminar Entrainment of the Surface Film (The Non-Stopping, Non-Reversing Condition) 550
  - 10.3.1 Continuous Expansion of the Meniscus 550
  - 10.3.2 Arrest of Forward Motion of the Meniscus 551
  - 10.3.3 Waterfall Flow: The Oxide Flow Tube 552
  - 10.3.4 Horizontal Stream Flow 553
  - 10.3.5 Hesitation and Reversal 555
  - 10.3.6 Oxide Lap Defects 556
- 10.4 Rule 4: Avoid Bubble Damage 557
  - 10.4.1 The Bubble Trail 557
  - 10.4.2 Bubble Damage 557
  - 10.4.3 Bubble Damage in Gravity Filling Systems 559
  - 10.4.4 Bubble Damage in Counter-gravity Systems 560
- 10.5 Rule 5: Avoid Core Blows 561
  - 10.5.1 Background 561
  - 10.5.2 Outgassing Pressure in Cores 564
  - 10.5.3 Core Blow Model Study 569
  - 10.5.4 Prevention of Blows 569
  - 10.5.5 Summary of Blow Prevention 573
- 10.6 Rule 6: Avoid Shrinkage Damage 573
  - 10.6.1 Definitions and Background 573
  - 10.6.2 Feeding to Avoid Shrinkage Problems 574
  - 10.6.3 The Seven Feeding Rules 575
  - 10.6.4 The New Feeding Logic 596
  - 10.6.5 Freezing Systems Design (Chills, Fins and Pins) 600
  - 10.6.6 Feeding: The Five Mechanisms 600
  - 10.6.7 Computer Modelling of Feeding 601
  - 10.6.8 Random Perturbations to Feeding 602
  - 10.6.9 The Non-Feeding Roles of Feeders 602
- 10.7 Rule 7: Avoid Convection Damage 603
  - 10.7.1 The Academic Background 603
  - 10.7.2 The Engineering Imperatives 603
  - 10.7.3 Convection Damage and Casting Section Thickness 608
  - 10.7.4 Countering Convection 610
- 10.8 Rule 8: Reduce Segregation Damage 610
- 10.9 Rule 9: Reduce Residual Stress 613
  - 10.9.1 Introduction 613
  - 10.9.2 Residual Stress from Casting 613
  - 10.9.3 Residual Stress from Quenching 617
  - 10.9.4 Controlled Quenching Using Polymer and Other Quenchants 620
  - 10.9.5 Controlled Quenching Using Air 623
  - 10.9.6 Strength Reduction by Heat Treatment 624

	10.9.7 Distortion	626
	10.9.8 Heat Treatment Developments	626
	10.9.9 Beneficial Residual Stress	628
	10.9.10 Stress Relief	628
	10.9.11 Epilogue	631
10.10	Rule 10: Provide Location Points	631
	10.10.1 Datums	631
	10.10.2 Location Points	633
	10.10.3 Location Jigs	637
	10.10.4 Clamping Points	637
	10.10.5 Potential for Integrated Manufacture	638

第 5 册　浇注系统设计基础、组成和设计实例（第 11 ~ 13 章）

SECTION 2 FILLING SYSTEM DESIGN

CHAPTER 11　Filling System Design Fundamentals 641
11.1 The Maximum Velocity Requirement 642
11.2 Gravity Pouring: The 'No-Fall' Conflict 642
11.3 Reduction or Elimination of Gravity Problems 646
11.4 Surface Tension Controlled Filling 651

CHAPTER 12　Filling System Components 657
12.1 Pouring Basin 657
    12.1.1 The Conical Basin 657
    12.1.2 Inert Gas Shroud 660
    12.1.3 Contact Pouring 660
    12.1.4 The Offset Basin 661
    12.1.5 The Offset Step (Weir) Basin 662
    12.1.6 The Sharp-Cornered or Undercut Basin 667
    12.1.7 Stopper 669
12.2 Sprue (Down-Runner) 669
    12.2.1 Multiple Sprues 675
    12.2.2 Division of Sprues 677
    12.2.3 Sprue Base 678
    12.2.4 The Well 679
    12.2.5 The Radial Choke 681
    12.2.6 The Radius of the Sprue–Runner Junction 681
12.3 Runner 683
    12.3.1 The Tapered Runner 687
    12.3.2 The Expanding Runner 688
12.4 Gates 690
    12.4.1 Siting 690

12.4.2 Direct and Indirect 690
12.4.3 Total Area of Gate(s) 691
12.4.4 Gating Ratio 691
12.4.5 Multiple Gates 692
12.4.6 Premature Filling Problem via Early Gates 692
12.4.7 Horizontal Velocity in the Mould 695
12.4.8 Junction Effect 696
12.4.9 The Touch Gate 700
12.4.10 Knife Gate 701
12.4.11 The Pencil Gate 702
12.4.12 The Horn Gate 703
12.4.13 Vertical End Gate 703
12.4.14 Direct Gating Into the Mould Cavity 706
12.4.15 Flow Channel Structure 706
12.4.16 Indirect Gating (Into an Up-Runner/Riser) 707
12.4.17 Central versus External Systems 710
12.4.18 Sequential Gating 711
12.4.19 Priming Techniques 712
12.4.20 Tangential Filter Gate 714
12.4.21 Trident Gate 715

12.5 Surge Control Systems 716
12.6 Vortex Systems 719
12.6.1 Vortex Sprue 720
12.6.2 Vortex Well 721
12.6.3 Vortex Runner (The Offset Runner) 722
12.6.4 Vortex Gate 723

12.7 Inclusion Control: Filters and Traps 724
12.7.1 Dross Trap (or Slag Trap) 724
12.7.2 Slag Pockets 725
12.7.3 Swirl Trap 726

12.8 Filters 728
12.8.1 Strainers 729
12.8.2 Woven Cloth or Mesh 729
12.8.3 Ceramic Block Filters 733
12.8.4 Leakage Control 735
12.8.5 Filters in Running Systems 739
12.8.6 Tangential Placement 741
12.8.7 Direct Pour 742
12.8.8 Sundry Aspects 744
12.8.9 Summary 745

CHAPTER 13　Filling System Design Practice 747
13.1　Background to the Methoding Approach 748
13.2　Selection of a Layout 748
13.3　Weight and Volume Estimates 748
13.4　Pressurised Versus Unpressurised 750
13.5　Selection of a Pouring Time 755
13.6　Thin Sections and Slow Filling 758
13.7　Fill Rate 758
13.8　Pouring Basin Design 759
13.9　Sprue (Down-Runner) Design 759
13.10　Runner Design 763
13.11　Gate Design 765

第6册　熔炼・造型・铸造・凝固（第14～19章）

SECTION 3　PROCESSING (MELTING, MOULDING, CASTING, SOLIDIFYING)

CHAPTER 14　Melting 769
14.1　Batch Melting 769
14.1.1　Liquid Metal Delivery 769
14.1.2　Reverberatory Furnaces 770
14.1.3　Crucible Melting (Electric Resistance or Gas Heated) 771
14.1.4　Induction Melting 772
14.1.5　Automatic Bottom Pouring 772
14.2　Continuous Melting 772
14.2.1　Tower (Shaft) Furnaces 772
14.2.2　Dry Hearth Furnaces for Non-Ferrous Metals 773
14.3　Holding, Transfer and Distribution 774
14.3.1　Holder Failure Modes 776
14.3.2　Transfer and Distribution Systems 776
14.4　Melt Treatments 777
14.4.1　Degassing 777
14.4.2　Detrainment (Cleaning) 783
14.4.3　Additions 787
14.4.4　Pouring 788
14.5　Cast Material 790
14.5.1　Liquid Metal 790
14.5.2　Partially Solid Mixtures 791
14.6　Re-Melting Processes 794
14.6.1　Electro-Slag Re-Melting 794
14.6.2　Vacuum Arc Re-Melting 795

CHAPTER 15 Moulding 797
15.1 Inert Moulds and Cores 797
- 15.1.1 Permanent Metal Moulds (Dies) 797
- 15.1.2 Salt Cores 799
- 15.1.3 Ceramic Moulds and Cores 800
- 15.1.4 Magnetic Moulding 802

15.2 Aggregate Moulding Materials 802
- 15.2.1 Silica Sand 803
- 15.2.2 Chromite Sand 804
- 15.2.3 Olivine Sand 804
- 15.2.4 Zircon Sand 805
- 15.2.5 Other Minerals 805
- 15.2.6 Carbon 806
- 15.2.7 Synthetic Aggregates 806

15.3 Binders 807
- 15.3.1 Greensand (Clay + Water) 807
- 15.3.2 Dry Sand 808
- 15.3.3 Chemical Binders 808
- 15.3.4 Effset Process (Ice Binder) 812
- 15.3.5 Loam 812
- 15.3.6 Cement Binders 813
- 15.3.7 Fluid (Castable) Sand 813

15.4 Other Aggregate Mould Processes 813
- 15.4.1 Precision Core Assembly 813
- 15.4.2 Machined-to-Form 813
- 15.4.3 Unbonded Aggregate Moulds 814

15.5 Rubber Moulds 814
15.6 Reclamation and Re-Cycling of Aggregates 814
- 15.6.1 Aggregate Reclamation in an Al Foundry 814
- 15.6.2 Aggregate Reclamation in a Ductile Iron Foundry 816
- 15.6.3 Aggregate Reclamation with Soluble Inorganic Binders 817
- 15.6.4 Facing and Backing Sands 818

CHAPTER 16 Casting 821
16.1 Gravity Casting 821
- 16.1.1 Gravity Pouring of Open Moulds 821
- 16.1.2 Gravity Pouring of Closed Moulds 822
- 16.1.3 Two-Stage Filling (Priming Techniques) 825
- 16.1.4 Vertical Stack Moulding 826
- 16.1.5 Horizontal Stack Moulding (H Process) 826
- 16.1.6 Postscript to Gravity Filling 827

	16.2 Horizontal Transfer Casting	827
	16.2.1 Level Pour (Side Pour)	828
	16.2.2 Controlled Tilt Casting	829
	16.2.3 Roll-over as a Casting Process	835
	16.2.4 Roll-over After Casting (Sometimes Called Inversion Casting)	836
	16.3 Counter-Gravity	838
	16.3.1 Low-Pressure Casting	842
	16.3.2 Liquid Metal Pumps	846
	16.3.3 Direct Vertical Injection	853
	16.3.4 Programmable Control	855
	16.3.5 Feedback Control	855
	16.3.6 Failure Modes of Low-Pressure Casting	856
	16.4 Centrifugal Casting	856
	16.5 Pressure-Assisted Casting	861
	16.5.1 High-Pressure Die Casting	862
	16.5.2 Squeeze Casting	868
	16.6 Lost Wax and Other Ceramic Mould Casting Processes	871
	16.7 Lost Foam Casting	875
	16.8 Vacuum Moulding (V Process)	878
	16.9 Vacuum-Assisted Casting	880
	16.10 Vacuum Melting and Casting	881
CHAPTER 17	**Controlled Solidification Techniques**	**883**
	17.1 Conventional Shaped Castings	883
	17.2 Directional Solidification	883
	17.3 Single Crystal Solidification	885
	17.4 Rapid Solidification Casting	887
CHAPTER 18	**Dimensional Accuracy**	**893**
	18.1 The Concept of Net Shape	897
	18.1.1 Effect of the Casting Process	898
	18.1.2 Effect of Expansion and Contraction	899
	18.1.3 Effect of Cast Alloy	900
	18.2 Mould Design	900
	18.2.1 General Issues	900
	18.2.2 Assembly Methods	902
	18.3 Mould Accuracy	904
	18.3.1 Aggregate Moulds	904
	18.3.2 Ceramic Moulds	907
	18.3.3 Metal Moulds (Dies)	907
	18.4 Tooling Accuracy	909
	18.5 Casting Accuracy	910
	18.5.1 Uniform Contraction	910

	18.5.2 Nonuniform Contraction (Distortion) 915
	18.5.3 Process Comparison 923
	18.5.4 General Summary 924
18.6	Metrology... 924

CHAPTER 19 Post-Casting Processing 927

19.1 Surface Cleaning.. 927
19.2 Heat Treatment 929
    19.2.1 Homogenisation and Solution Treatments... 929
    19.2.2 Heat Treatment Reduction and/or Elimination... 930
    19.2.3 Blister Formation... 932
    19.2.4 Incipient Melting... 933
    19.2.5 Fluid Beds.. 934
    19.2.6 Quenching... 935
19.3 Hot Isostatic Pressing 935
19.4 Machining... 938
19.5 Painting 940
19.6 Plastic Working (Forging, Rolling, Extrusion) 941
19.7 Impregnation 942
19.8 Non-Destructive Testing 942
    19.8.1 X-ray Radiography 942
    19.8.2 Dye Penetrant Inspection 943
    19.8.3 Leak Testing 943
    19.8.4 Resonant Frequency Testing 944

Appendix I..947
Appendix II 951
Appendix III 955
Appendix IV...957
References 959
Index..993

Preface

In this first update of the *Handbook*, the major revisions are probably those relating to running system design in which the vestiges of filling defects have finally been eliminated from castings.

Thus, the powerful benefits of contact pouring (in which the universal conical trumpet decorating all traditional filling systems is now eliminated) is finally shown to have been hugely underestimated by a number of foundries. Contact pouring has probably been the most important (and the most simple and zero-cost) initiative to revolutionise quality in castings. In addition, the adoption of various forms of tangential filter designs to gates has finally eliminated the problem of the entrainment of priming bubbles. These residual bubbles have long impaired the benefits of previous filling systems.

Gravity pouring has now advanced to the point at which I find myself having to admit that it starts to threaten my cherished and favoured casting production system: countergravity.

This is seen to be especially true for those low-pressure systems which use a refractory lining for the pressurised furnace. I only recently discovered the hugely damaging emission of bubbles from these linings during depressurisation of the furnace. This problem has clearly been a major source of impaired castings in the low-pressure casting industry and has hampered this industry since its beginnings.

The use of my pneumatic pump is described for the first time. It would lower costs and solve most of the problems of this industry. Thus, I continue to stand by countergravity as the optimum casting system where it can be used. My hope is that it will be teamed up with a good melting and metal handling system. Only careful foundry design will minimise bifilm populations in metals. Only when castings can be produced substantially free from bifilms will we enjoy the full benefits of castings, and metals in general, resistant to hot tearing, cracking, blisters, corrosion pitting and attack of grain boundaries, plus the benefits of extraordinary mechanical properties, potentially eliminating future failure by fracture or fatigue.

These are heady predictions. However, early results in foundries are already indicating that beautiful defect-free castings with revolutionary metallurgical benefits appear to be routinely attainable. Despite challenges from the undoubtedly unique benefits of such new processes as additive manufacture, my hope for the future for castings is based on the adoption of simple principles which could not only secure the future of our casting industry, but improve the welfare and environment of all of us whose lives depend on it.

JC
Ledbury, Herefordshire, England
02 April 2015

Introduction

CASTINGS HANDBOOK, 2ND EDITION, 2015

When *Castings Handbook* first appeared in 2011, I had not expected to revise it so soon. However, the latest findings require publicising as quickly as possible—there is still a long way for the industry to go! The message of the book, summarised in the section "Bifilm-free Properties", is that a quality improvement of astonishing scale is possible now. When I first started to experiment with novel filling systems for castings, there were naturally many disappointments. However, those days are long gone. The concepts of entrainment and bifilm creation laid out in the book are now proven. Some foundries are already being designed to take advantage of a unique and easily affordable quality revolution and scrap reduction. More need to follow. The risks are minimal and the rewards are great.

With regard to improved casting techniques which can reduce or even eliminate the usual vast populations of bifilms in our metals, I have always been aware of the potential benefit of contact pouring, but had completely underestimated its effects. It achieves miraculous improvements to castings by eliminating the 50% air mixing step. Contact pouring is strongly recommended in this volume as a major but low-cost step forward.

The ultimate step forward is countergravity casting which should always be targeted if possible. However, although I discuss the traditional pumping techniques, I present here for the first time my new pneumatic pump. It is another low-cost, reliable technique which enjoys uniquely low turbulence and might allow a rapid takeup of this unique technology.

Turning to the seriousness of the current position in casting, and in metallurgical engineering as a whole, the fact that most current metals can fail by cracking should alert us to the glaring inconsistency in our metallurgical thinking because many of our metals and alloys are ductile, so failure by cracking should be impossible. In a ductile metal an attempt to propagate a crack should merely result in the crack tip blunting, preventing propagation.

In the absence of any other viable alternative mechanism, that metals do crack is a strong indication that cracks pre-exist in metals in the form of bifilms formed in the liquid state during pouring of the metal to make a casting. The poor practice is almost universally associated with pouring methods which ensure that the molten metal is mixed and emulsified with at least 50 volume % of air during its journey into the mould. I defy anyone to make a respectable casting from such a disgracefully inappropriate and inept technique. This book presents the case that this need not be so; metals need not contain bifilms, and thus need not contain those Griffith cracks which can initiate failure by cracking. To achieve this, we simply have to improve our casting technology.

It remains the case that bifilms are still lamentably researched, so that this book has to resort to sifting through inconclusive and fragmentary evidence from researchers who were not looking for bifilms. Unfortunately, researchers up until now have not been aware of their presence, and certainly did not suspect their overwhelming influence on their results. Although a welcome start is being made by a few workers, I remain impatient for more definitive research to be carried out.

In the meantime, while researchers slowly get around to proving the background theory, founders need not wait for answers. Practical low bifilm casting techniques have already been developed and are described here. They promise the quality improvement and cost reduction that the casting industry so badly needs.

Acknowledgements

It is a pleasure to acknowledge the significant help and encouragement I have received from many good friends. John Grassi has been my close friend and associate at Alotech, the company promoting the new, exciting ablation castings process. Ken Harris has been an inexhaustible source of knowledge on silicate binders, aggregates and recycling. His assistance is clear in Chapter 15. Clearly, the casting industry needs more chemists like him. Bob Puhakka has been the first regular user of my casting recommendations for the production of large aluminium and steel castings, which has provided me with inspirational confirmation of the soundness of the technology described in this book. In addition, the practical feedback and warm friendship over several years from those at the UK steel foundry Furniss and White is a pleasure to record. Murat Tiryakioglu has been a loyal supporter and critic, and provides the elegantly written publications that have provided welcome scientific underpinning. He has provided generous and invaluable help with the important section *The Statistics of Failure*. Naturally, many other acknowledgements are deserved among friends and students whose research has been a privilege to supervise. I do not take these for granted. Even if not listed here, they are not forgotten.

The American Foundry Society is thanked for the use of a number of illustrations from *Transactions*.

CHAPTER 4

MOULDS AND CORES

As in so much technology, the production of moulds and cores is a complex subject that we can only touch upon here. Although this text is about the metallurgy of the casting, the mould can be profoundly influential. The mould and the casting co-exist for sufficiently long, and at a sufficiently high temperature, that the two cannot fail to have an important mutual impact. Essentially, this section is about the *science* of mould and casting interactions.

In the later part of the book, *Casting Manufacture*, we concentrate on the *technology* of moulds and castings, outlining the main benefits and problems of the various types of moulds and cores to allow a user to make a more informed choice of manufacturing route.

4.1 MOULDS: INERT OR REACTIVE

Very few moulds and cores are really inert towards the material being cast into them. However, some are very nearly so, especially at lower temperatures. This short section lists those types of moulds and cores that in general do not react with their cast metals.

The usefulness of a relatively inert mould is emphasised by the work of Stolarczyk (1960), who suffered 4.5% porosity in gunmetal test bars cast in greensand moulds compared with 0.5% porosity for identical castings into steel-lined moulds. This simple experiment confirms one of the most important advantages of metal moulds: they are impressively inert towards their liquid metal and cast product. Thus all of the gravity and low pressure dies (permanent moulds) are reassuringly inert.

Unfortunately, the inertness of high pressure die casting dies and some squeeze casting dies is compromised by die coolants and lubricants which definitely impair the casting to varying degrees.

Greensand moulds that have been dried in an oven (i.e. dry sand moulds) have been found to be largely inert, as shown by Locke and Ashbrook (1950).

This behaviour contrasts with that of the original greensand in its various forms. The water and hydrocarbon contents of greensand mixes lead to masses of outgassing and very rapid attack, leading to oxidation or deposit of carbon on the liquid metal front, plus the generation of copious amounts of lower hydrocarbons (such as methane) and even more hydrogen. At first sight, it is perhaps amazing that any useful casting could be produced by such a reactive system. However, greensand remains, justifiably, the most important volume producer of good castings worldwide. Its many benefits, particularly its unbeatable moulding speed, are discussed in more detail in Chapter 15.

Whereas the various resin binders used for sand moulds do not outgas as impressively as greensand, the quantities and nature of the various volatiles that are attempting to escape during the dramatic early seconds and minutes of attack by the melt are seen in Figures 4.1 and 4.2. Clearly, there is plenty of chemistry involved, and the high temperatures ensure that this is energetic.

In recent years, several synthetic aggregates have become available in tonnage quantities. The hardness, wear and fracture resistance of these new moulding materials has meant that their high cost can sometimes be justified because of the very high efficiency with which they can be recycled because of low losses. The materials are mainly based on highly stable, high melting point oxides such as alumina and silica. A mixture of these two oxides to give a crystal composition such as mullite has the great benefit of avoiding the alpha/beta quartz transition, and so has the potential to improve the accuracy of castings. The synthetic grains so far available are produced in two ways, both of which result in beautifully spherical grains which have excellent flowability. However, they can have quite different results for some castings.

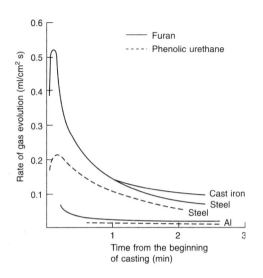

FIGURE 4.1

Measured gas evolution rates from castings of aluminium, iron and steel in chemically bonded sand moulds (Bates and Monroe, 1981).

1. Aggregates can be produced in the solid state, aggregating powders with an organic binder, by rolling around on oscillating trays. The weakly bonded spherical grains are then sintered, eliminating the temporary binder, and forming strong grains but containing several percent of porosity, some closed and some continuously linked through the grain. My experience with this variety of aggregate is that the mould binder tends to enter the pores of the grains, unfortunately boiling and exploding when contacted with liquid aluminium alloy, thus punching vapour pores into the surface of the casting. It is not known whether such 'microblows' occur with other metals and alloys. Such microblows will, of course, occur in addition to the chemical reactions normally to be expected between the binder and the liquid metal.
2. Aggregates can be produced by a melting route, one of which employs a jet of liquid oxide which is atomised by a blast of air. Such grains tend to be perfectly sound and can still be very respectably spherical. Moulds made from such aggregates would be expected to react only by reason of the presence of the aggregate binder.

4.2 TRANSFORMATION ZONES

As the hot metal is poured into a greensand mould, the blast of heat from the melt at temperature greatly exceeding the boiling point of water, very rapidly heats the surface of the mould, boiling off the water (and other volatiles). As the heat continues to advance into the mould, the moisture continues to migrate away, only to condense again in the deeper, cooler parts of the mould. As the heat continues to diffuse in, the water evaporates again and migrates further. This is of course a continuous process. Dry and wet zones travel through the mould like weather systems in the atmosphere. The evaporation of water in greensand moulds has been the subject of much research.

Looking at these in detail, four zones can be distinguished, as shown in Figure 4.3.

1. The dry zone is where the temperature is high and all moisture has been evaporated from the binder. It is noteworthy that this very high temperature region will continue to retain a relatively stagnant atmosphere composed of nearly 100% water vapour as a hot dry gas. However, of course, some of this superheated and very dry steam will be reacting at the casting surface to produce oxide and free hydrogen.

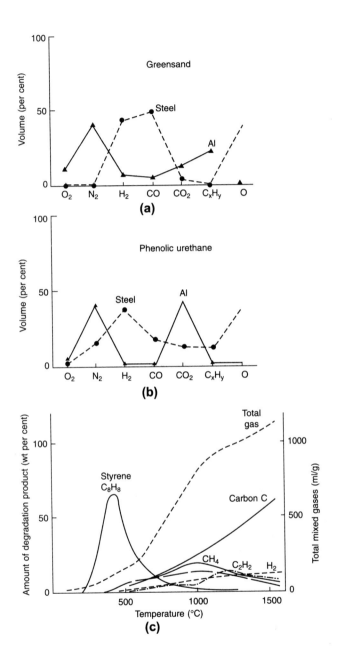

FIGURE 4.2

Composition of mould gases (a) from greensand (Chechulin, 1965); (b) phenolic urethane (Bates and Monroe, 1981) and (c) thermally decomposing expanded polystyrene (Goria et al., 1986).

138 CHAPTER 4 MOULDS AND CORES

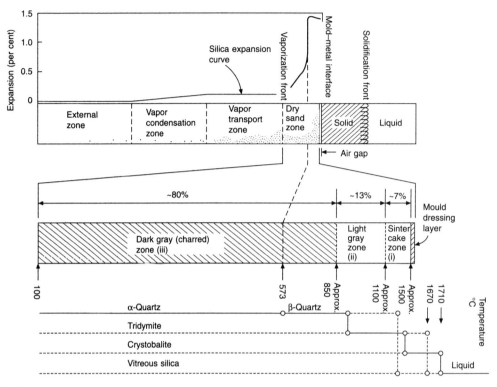

FIGURE 4.3

Structure of the heated surface of a greensand mould against a steel casting and the forms of silica (after Sosman, 1927), with solid lines denoting stable states and broken lines denoting unstable states.

2. The vapour transport zone, essentially at a uniform temperature of 100°C, and at a roughly constant content of water, in which steam is migrating away from the casting.
3. The condensation zone, where the steam re-condenses. This zone was for many years the subject of some controversy as to whether it was a narrow zone or whether it was better defined as a front. The definitive theoretical model by Kubo and Pehlke (1986) has provided an answer where direct measurement has proved difficult; it is in fact a zone, confirming the early measurements by Berry, Kondic and Martin (1959). This zone gets particularly wet. The raised water content usually greatly reduces the strength of greensand moulds, so that mechanical failure is most common in this zone.
4. The external zone where the temperature and water content of the mould remain as yet unchanged.

It is worth taking some space to describe the structure of the dry sand zone. When casting light alloys and other low-temperature materials, the dry sand layer has little discernible structure.

However, when casting steel it becomes differentiated into various layers that have been detailed from time to time (e.g. Polodurov, 1965; Owusu and Draper, 1978). These are, counting the mould coating as number zero:

1. Dressing layer of usually no more than 0.5 mm thickness, and having a dark metallic lustre as a result of its high content of metal oxides.
2. Sinter cake zone, characterised by a dark brown or black colour. It is mechanically strong, being bonded with up to 20% fayalite, the reaction product of iron oxide and silica sand. The remaining silica exists as shattered quartz

grains partially transformed to tridymite and cristobalite, which is visible as glittering crystals (explaining the origin of the name cristobalite). This layer is largely absent when casting grey iron at ordinary casting temperatures.
3. Light-grey zone, with few cracked quartz grains and little cristobalite. What iron oxides are present are not alloyed with the silica grains. This zone is only weakly bonded, and disintegrates on touch.
4. Charred zone, of dark-grey colour, of intermediate strength, containing unchanged quartz grains but significant levels of iron oxide. Polodurov speculates that this must have been blown into position by mould gases.

The pattern of these zones is further complicated by convection effects inside the mould or core for a relatively long period after casting, carrying carbonaceous vapours back into the heated zones, chemically 'cracking' the compounds to release hydrogen and depositing carbon. The inner layers therefore become black, with the sand grains seen under the microscope to be coated with a kind of fibrous, furry layer of graphite. Churches and Rundman (1995) studying a phenolic urethane mould found this reaction to occur between 15 min and 3 h, and at temperatures down to about 540°C for their grey iron castings. Highly heated cores had lost up to 50% of their carbon after finally cooling to room temperature.

The changes in form of the silica sand during heating are complicated. An attempt to illustrate these relations graphically is included in Figure 4.3. This complexity, and particularly the expansion accompanying the phase change from alpha to beta quartz, has prompted a number of foundries to abandon silica sand in favour of more predictable moulding aggregates. This advantageous move is expected to become more widespread in future especially as interesting new synthetic aggregates become available.

4.3 EVAPORATION AND CONDENSATION ZONES

As the heat diffuses from the solidifying casting into the mould (Figure 4.4), the transformation zones migrate deeper into the mould. We can follow the progress of the advance of the zones by considering the distance d that a particular isotherm reaches as a function of time t. The solution to this simple one-dimensional heat-flow problem is

$$d = (Dt)^{1/2} \tag{4.1}$$

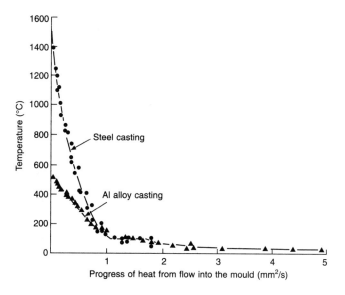

FIGURE 4.4

Temperature distribution in a greensand mould on casting an aluminium alloy (Ruddle and Mincher, 1949, 1950) and a steel (Chvorinov, 1940).

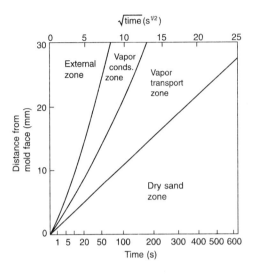

FIGURE 4.5

Position of the vapour zones after the casting aluminium in a greensand mould.

Data from Kubo and Pehlke (1986).

where D is the coefficient of diffusion. This solution is of course equivalent to the solute diffusion given earlier (Eqn (1.5)).

In the case of the evaporation front, the isotherm of interest is that at 100°C. We can see from Figures 4.4 and 4.5 that the value of D is close to 1 mm^{2}s^{-1}. This means that the evaporation front at 1 s has travelled 1 mm, at 100 s has travelled 10 mm, and requires 10,000 s (nearly 3 h!) to travel 100 mm. It is clear that the same is true for aluminium, as well as steel. (This is because we are considering a phenomenon that relies only on the rate of heat flow in the mould—the metal and its temperature are not involved.)

For the condensation zone the corresponding value of D is approximately 3 mm^{2}s^{-1}, so that the position of the front at 1, 100, and 10,000 s is 1.7, 17, and 170 mm, respectively.

These figures are substantiated to within 10 or 20% by the theoretical model by Tsai et al. (1988). This work adds interesting details such as that the rate of advance of the evaporation front depends on the amount of water present in the mould, higher water contents making slower progress. This is to be expected, because more heat will be required to move the front, and this extra heat will require extra time to arrive. The extra ability of the mould to absorb heat is also reflected in the faster cooling rates of castings made in moulds with high water content. Measurements of the thermal conductivity of various moulding sands by Yan et al. (1989) have confirmed that the apparent thermal conductivity of the moisture-condensation zone is about three or four times as great as that of the dry sand zone.

An earlier computer model by Cappy et al. (1974) also indicates interesting data that would be difficult to measure experimentally. They found that the velocity of the vapour was in the range of 10–100 mms^{-1} over the conditions they investigated. Their result for the composition and movement of the zones is given in Figure 4.6. Kubo and Pehlke calculate flow rates of 20 mms^{-1}. These authors go on to show that moisture vapourises not only at the evaporation front, but also in the transportation and condensation zones. Even in the condensation zone a proportion of the water vapourises again at temperatures below 100°C (Figure 4.5).

The pressure of water vapour at the evaporation front will only be slightly above atmospheric pressure in a normal greensand mould. However, because the pressure must be the same everywhere in the region between the mould-metal interface and the evaporation front, it follows that the dry sand zone must contain practically 100% water vapour. This is

4.3 EVAPORATION AND CONDENSATION ZONES

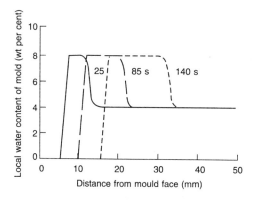

FIGURE 4.6

Water content of the vapour transport zone with time and position (smoothed computed results of Capp et al., 1974).

surprising at first sight. However, a moment's reflection will show that there is no paradox here. The water vapour is very dry and hot, reaching close to the temperature of the mould-metal interface. At such high temperatures, water vapour is highly oxidising. There is no need to invoke theories of additional mechanisms to get oxygen to this point to oxidise the metal—there is already an abundance of highly oxidising water vapour present (the breakdown of the water vapour also providing a high-hydrogen environment, of course, to enter the metal, and to increase the rate of heat transfer in the dry sand zone).

Kubo and Pehlke (1986) confirm that gas in the dry sand and transportation zones consists of nearly 100% water vapour. In the condensation zone, the percentage of air increases, until it reaches nearly 100% air in the external zone (water vapour would be expected to be present at its equilibrium vapour pressure, 32 mmHg or 42 mb at 30°C).

It is found that similar evaporation and condensation zones are present for other volatiles in the greensand mould mixture. Marek and Keskar (1968) have measured the movement of the vapour transport zone for benzene and xylene. The evaporation and condensation fronts of these more volatile materials travel somewhat faster than those of water. When such additional volatiles are present they will, of course, contribute to the 1 atm of gas pressure in the dry zone, helping to dilute the oxidising effect of water vapour and helping to explain part of the beneficial effect of such additives. In the following section, we will see that many organics decompose at these high temperatures, providing a deposit of carbon, which further assists, in the case of such metals as cast iron, in preventing oxidation and providing a non-wetting metal surface of carbon which contains the liquid iron and helps to prevent its contact with the sand mould.

It is to be expected that vapour transport zones will also be present to various degrees in chemically bonded sands. The zones will be expected to have traces of water mixed with other volatiles such as organic solvents. Little work appears to have been carried out for such binder systems, so it is not easy to conclude how important the effects are, if any. In general, however, the volatiles in such dry sand systems usually total less than 10% of the total volatiles in greensand, so that the associated condensation zones will be expected to be less than one-tenth of those occurring in greensand. It may be, therefore, that they will be unimportant. However, at the time of writing we cannot be sure; it would be nice to know.

What is certain is that silicate binders appear to have a high chilling effect in ferrous castings because of their water content. The loss of the water will create evaporation and condensation zones that will carry heat away from the casting.

All of these considerations on the rate of advance of the moisture assume no other flows of gases through the mould. This is probably fairly accurate in the case of the drag mould, where the flow of the liquid metal over the surface of the mould effectively seals the surface against any further ingress of gases. A certain amount of convection is expected in the mould, but this will probably not affect the conditions in the drag significantly.

In the vertical walls of the mould, however, convection may be significant. Close to the hot metal, hot gases are likely to diffuse upwards and out of the top of the mould, their place being taken by cold air being drawn in from the surroundings at the base of the mould, or the outer regions of the cope.

General conditions in the cope, however, are likely to be more complicated. It was Hofmann in 1962 that first emphasised the different conditions experienced during the heating up and outgassing of the cope. He pointed out that the radiated heat from the rising melt would cause the cope surface of the mould to start to dry out before the moment of contact with the melt. During this pre-contact period two different situations can arise:

1. If the mould is open, as the cope surface heats up, the water vapour can easily escape through the mould cavity and out via the opening (Figure 4.7). The rush of water vapour through an open feeder can easily be demonstrated by holding a piece of cold metal above the opening. It quickly becomes covered with condensate. The water vapour starts its life at a temperature of only 100°C. It is therefore a relatively cool gas, and is thus most effective in cooling the surface of the mould as it travels out through the surface of the cope on its escape route.
2. If the mould is closed, the situation is quite different. The air being displaced and expanded by the melt will force its way through the mould, carrying away the vapour from the interface (Figure 4.7). The rate of flow of air is typically in the 10–100 $ls^{-1}m^{-2}$ range (the reader is encouraged to confirm this for typical castings and casting rates). This is in the same range of flow rate as the transport of vapour given in computer models. Thus if the casting rate is relatively low, then the vapour transport zone is likely to be relatively unaffected, although perhaps a little accelerated in its progress. When the casting rate is relatively high, then the vapour transport zone will be effectively blown away, diluted with the gale of air so that no condensation can occur.

Because the water vapour is driven away from the surface and into the interior of the mould, its beneficial cooling effect at the surface is not felt, with the result that the surface reaches much higher temperatures as a result of the direct

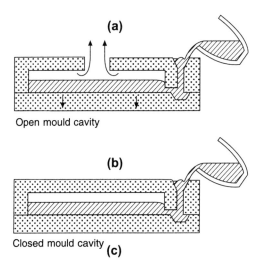

FIGURE 4.7

Three conditions of vapour transport in moulds: (a) unrestricted free evaporation from the cope; (b) evaporation from the cope constrained to occur via the cope mould; (c) evaporation from the drag confined by the cover of metal, and possibly confined by the substrate on which the mould sits, and possibly at its sides by a moulding box, leaving no option but for bubbling through the metal.

Partly from Hofman (1962).

radiation of heat from the melt, as is seen in Figure 4.8. The prospect of the failure of the cope surface by expansion and spalling of the sand is therefore much enhanced.

However, the rate of heating of the surface by radiation from the melt may be reduced by a white mould coat, such as a zircon- or alumina-based mould wash, now widely applied for large castings of iron and steel. Boenish (1967) confirmed that light-coloured mould coats resist scabbing for up to 400% longer than the jet-black graphitic surfaces commonly used for cast iron.

One final aspect of vapour transport in the mould is worth noting. There has been much discussion over the years about the contribution of the thermal transpiration effect to the flow of gases in moulds.

Although it appears to have been widely disputed, the effect is certainly real. It follows from the kinetic theory of gases and essentially is the effect of heated gases diffusing away from the source of heat, allowing cooler gases to diffuse up the temperature gradient. In this way it has been argued that oxygen from the air can arrive continuously at the casting to oxidise the surface to a greater degree than would normally have been expected.

Williams (1970) described an experiment that demonstrated this effect. He took a sample of clay approximately 50 mm long in a standard 25 mm diameter sand sampling tube. When one end was heated to 1000°C and the other was at room temperature, he measured a pressure difference of 10 mmHg if one end was closed, or a flow rate of 20 ml per min if both ends were open. If these results are typical of those that we might expect in a sand mould, then we can make a comparison as follows. The rate of thermal transpiration is easily shown to convert to $0.53 \, \text{ls}^{-1}\text{m}^2$ for the conditions of temperature gradient and thickness of sample used in the experiment. From the model of Cappy et al. (1974), we obtain an estimate of the rate of transport of vapour of $100 \, \text{ls}^{-1}\text{m}^2$ at approximately this same temperature gradient through a similar thickness of mould. Thus thermal transpiration is seen to be less than 1% of the rate of vapour transport.

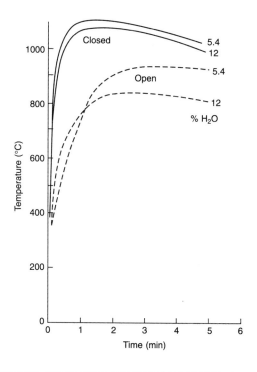

FIGURE 4.8

Temperature in the cope surface seen to be significantly lowered by open moulds and high moisture levels.

Data from Hofman (1962).

144 CHAPTER 4 MOULDS AND CORES

Additional flows such as the rate of volume displacement during casting and the rate of thermal convection in the mould will further help to swamp thermal transpiration.

Thermal transpiration does seem to be a small contributor to gas flow in moulds. It is possible that it may be more important in other circumstances. More work is required to reinstate it to its proper place, or lay it to rest as an interesting but unimportant detail.

4.4 MOULD ATMOSPHERE
4.4.1 COMPOSITION

On the arrival of liquid iron or steel in the mould, a rich soup of gases boils from the surface of the mould and the cores. The air originally present in the cavity dilutes the first gases given off, but this is quickly expelled through vents or feeders or may diffuse out through the cope. After the air is forced out, the composition of the mould gas is relatively constant. The speed at which the original gas in the mould is expelled by the arrival of the metal is the reason why attempts to fill the mould with inert gases to avoid oxidising reactions are practically always a waste of effort. The casting of Al alloys is rather less dramatic, and the original air is expelled only gradually as seen from experiments described in the following section.

In the case of steel being cast into greensand moulds, the mould gas mixture has been found to contain up to 50% hydrogen (Figure 4.2). The content of hydrogen depends almost exactly on the percentage water in the sand binder, with dry sand moulds having practically no hydrogen as found by Locke and Ashbrook (1950, 1972). Other changes brought about by increased moisture in the sand were a decrease in oxygen, an increase in the CO/CO_2 ratio and the appearance of a few% of paraffins (it is not clear whether these originated from the use of lubricant sprayed on to the pattern). The presence of cereals in the binder was found to provide some oxygen, even though the concentration of oxygen in the atmosphere fell because of dilution with other gases (Locke and Ashbrook, 1950). Chechulin (1965) describes the results for greensand when aluminium alloys, cast irons and steels are cast into them. His results are given in Figure 4.2(a). Irons and steels produce rather similar mould atmospheres, so only his results for steel are presented.

The high oxygen and nitrogen content of the atmosphere in the case of moulds filled with aluminium simply reflects the high component of residual air (originally, of course, at approximately 20% oxygen and 80% nitrogen). The low temperature of the incoming metal is insufficient to generate enough gas and expand it to drive out the original atmosphere. This effective replacement of the atmosphere is only achieved in the case of iron and steel castings.

The atmosphere generated when ferrous alloys are cast into chemically bonded sand moulds is, perhaps rather surprisingly, not so different from that generated in the case of greensand (Figure 4.2(b)). The mixture consists mainly of hydrogen and carbon monoxide.

The kinetics of gas evolution have been studied by a number of researchers. Jones and Grim (1959) found some clays evolved significant water at just over 100°C, particularly Western bentonite, although Southern bentonite results went wildly off scale. A second peak of evolution at around 550°C was relatively unimportant. Scott and Bates (1975) found that hydrogen evolution peaked within 4–5 min for most chemical binders. However, for the sodium silicate binder, a rapid burst of hydrogen was observed which peaked in less than 1 min.

Scarber et al. (2006) measured 200 ml gas from about 100 g (they do not give the exact weight of their samples) of a core bonded with some resin binders which rose to more than 500 ml if the core was coated with a water-based wash even though the core had been thoroughly dried. Even this paled into insignificance when compared with the outgassing of greensand which they found to be nearly 10 times that from resin-bonded sands. In addition, they found multiple peaks associated with solvent evaporation, and, probably, water evaporation. The process seemed to be complicated by the re-condensation of these volatiles in the cooler centres of cores, and their final re-evaporation (a re-enactment of the condensation and evaporation zones previously discussed in Section 4.3). Yamamoto et al. (1980) described the variation in output they observed as (1) peak I attributed to the expansion of air plus the evaporation of free moisture and other volatiles in the core when the core was first covered with liquid metal; (2) peak II after about 30 s identified as the release of combined water (water of crystallisation) in the binder and/or aggregate; and finally (3) a broad smooth peak III, constituting a nearly constant pressure output period for about 50–80 s, attributed to the general breakdown of the organics in the binder. They confirmed that peak II was much higher for sodium silicate/CO_2 cores.

Lost-foam casting, where the mould cavity is filled with polystyrene foam (the 'full mould' process), is a special case. Here it is the foam that is the source of gases as it is vapourised by the molten metal. At aluminium casting temperatures the polymerised styrene merely breaks down into styrene, but little else happens, as is seen in Figure 4.2(c). It seems that the liquid styrene soaks into the ceramic surface coating on the foam, so that the coating will temporarily become completely impermeable. This unhelpful behaviour probably accounts for many of the problems suffered by aluminium alloy castings made by the lost foam process.

At the casting temperatures appropriate for cast iron, more complete breakdown occurs, with the generation of hydrogen and considerable quantities of free carbon. Rao and Lee (1984) show how even methane is largely decomposed at these temperatures, forming less than 1% of a $C-H_2-O_2$ mixture at 1 atm. The carbon deposits on the advancing metal front as a pyrolytic form of carbon widely known as 'lustrous carbon'. Once formed, the layer is rather stable at iron-casting temperatures, and can therefore lead to serious defects if entrained in the metal. The problem has impeded the successful introduction of lost foam technology. For steel casting the temperature is sufficiently high to cause the carbon to be taken into solution. Steel castings of low or intermediate carbon content are therefore contaminated by pockets of high-carbon alloy. This problem has prevented lost-foam technology in the form of the full mould process being used for low- and medium-carbon steel castings.

Lost-foam iron castings are not the only type of ferrous castings to suffer from lustrous-carbon defects. The defect is also experienced in cast iron made in phenolic urethane-bonded moulds, and at times can be a serious headache. The absence of carbon is therefore a regrettable omission from the work reported in Figure 4.2(a). At the time Chechulin carried out this study, the problem would not have been known. Later, Gloria (1986) does report carbon as a product of decomposition of polystyrene foam.

It seems reasonable to expect that carbon may also be produced from the pyrolysis of other binder systems. More work is required to check this important point.

4.4.2 MOULD GAS EXPLOSIONS

The various reactions of the molten metal with the volatile constituents of the mould, particularly the water in many moulding materials, would lead to explosive reactions if it were not for the fact that the reactions are dampened by the presence of masses of sand in mould materials or cores. Thus although the reactions in the mould are fierce, and not to be underestimated, in general they are not of explosive violence because the 90% or more of the materials involved are inert (simply sand and nitrogen) and have considerable thermal inertia. Taleyarkhan (1998) draws attention to the important role of non-condensable gases (e.g. nitrogen) in the suppression of explosive reactions. Outgassing reactions are therefore rather steady and sustained.

When carrying out a series of experiments in about 1992, liquid aluminium was poured while the interior of a polyurethane-bonded mould was recorded by video though a glass window, Helvaee was surprised to note that in every case, on arrival of the metal in the mould, a flicker of flame licked across the mould cavity, sometimes more energetically than others. On no occasion was there an explosion. Much earlier, in 1948, Johnson and Baker recorded similar flashes of light when recording the filling of open greensand moulds by molten steel at 1650°C. It seems such phenomena might be common or even normal.

It is noteworthy that the gases from the outgassing of an aggregate mould may contain a number of potentially flammable or explosive gases. These include a number of vapours such as hydrocarbons such as methane, other organics such as alcohols and a number of reaction products such as hydrogen and carbon monoxide.

Because of the presence of these gases, explosions sometimes occur and sometimes not. The reasons have never been properly investigated. This is an unsatisfactory situation because the explosion of a mould during casting can be an unpleasant event. The author has witnessed explosions a number of times in furan-bonded boxless moulds when casting an aluminium alloy casting weighing more than 50 kg: there was a muffled explosion, and large parts of the sand mould together with liquid metal flew apart in all directions. After several repeat performances, the operators developed ways of pouring this component at the end of long-handled ladles to keep as far away as possible, never knowing whether the mould would explode. The cause always remained a mystery. Everyone was relieved when the job came to an end.

Explosions in and around moulds containing iron or steel castings are relatively common. One of the most common is from under the mould, between the mould and its base plate, after the casting has solidified, so that there is less danger either to personnel or casting. The sound of muffled explosions from the mix of CO and air under moulds is common in many greensand foundries.

With subsequent experience, and in the absence of any other suggestions, the following is suggested as a possible cause of the problem in the case of the light alloy casting.

Explosions can, of course, only happen when the flammable components of the gas mix with an oxidising component such as oxygen from the air. The mixing has to be efficient, which suggests that turbulence is important. Also, the mix often has to be within close compositional limits, otherwise either no reaction occurs, or only slow burning takes place. The limits for the carbon monoxide, oxygen and inert (carbon dioxide and nitrogen) gas mixtures are shown in Figure 4.9.

In the first edition of this book the mixing with air, which is essential for explosions, was thought to occur only under certain conditions including feeder heads open to the air and the use of two sprues which would be not easily synchronised. Subsequent experience has modified these conclusions, even though definitive reasons are still not known. It is now clear that at least 50% of the mixture arriving at the base of the sprue is air, scrambled in with the hot metal, as a result of the pumping action of the conical trumpet entrance to the sprue, and the fact that most sprues are oversized and therefore allow the ingress and entrainment of air. Thus bubbles of air, together with hot splashes of metal, mixing turbulently with the mould gases are a recipe for explosive conditions.

We can surmise that contact pouring, together with a slim 'naturally pressurised' filling system design would eliminate explosions as a result of the melt displacing the air in the filling system, then filling the mould uphill, with minimal mixing of the hot mould gases in contact with the melt; melt and gases will both tend to rise steadily through the mould as progressive layers. In a system filling quiescently from the bottom upwards, the outgassing of the mould and cores will provide a spreading blanket of gas over the liquid. There will be almost no air in this cover, so no burning or explosion can occur. The air will be displaced ahead and will diffuse out of the upper parts of the mould. Where the flammable gas blanket meets the air, it is expected to be cool, well away from the liquid metal. Thus any slight mixture that will occur at the interface between these layers of gases is not likely to ignite to cause an explosion.

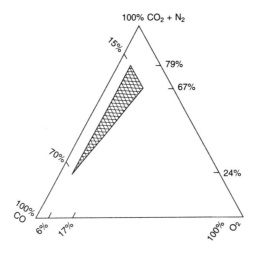

FIGURE 4.9

Shaded region defines the explosive regime for mixtures of CO, O_2, and a mixture of $CO_2 + N_2$ (Ellison and Wechselblatt, 1966).

Occasionally, Al-Si alloys which have been usually docile during pouring have been known to explode after the addition of Sr as a modifier. This is thought to be the enhanced reactivity of the Sr, converting moisture to hydrogen gas. The elimination of the trumpet intake and the provision of a properly calculated filling system should both help, as would the reduction or elimination of the Sr addition.

4.5 MOULD SURFACE REACTIONS
4.5.1 PYROLYSIS

When the metal has filled a sand mould, the mould becomes hot. A common misconception is to assume that the sand binder then burns; however, this is not true. It simply becomes hot. There is insufficient oxygen to allow any significant burning. What little oxygen is available is consumed in a minor, transient oxidation which quickly comes to a stop. What happens then to the binder is not burning, but pyrolysis.

Pyrolysis is the decomposition of compounds, usually organic compounds, simply by the action of heat. Oxygen is absent, so that no burning (i.e. high temperature oxidation) takes place. Pyrolysis of various kinds of organic binder components to produce carbon is one of the more important reactions that take place in the mould surface. Carbon is poorly wetted by many liquid metals, so the formation of carbon on the grains of sand, as a pyrolysed residue of the sand binder, produces a non-wetted mould surface, which can lead to an improved surface finish to the casting (although, as will become clear, the effect on the surface film on the liquid metal is probably more important).

This non-wetting feature of residual carbon on sand grains is at first sight curious, because carbon is soluble in many metals, and so should react, and should therefore wet. Cast iron would be a prime candidate for this behaviour. Why does this not happen?

In the case of ductile iron, sand cores do not need a core coat because the solid magnesium oxide rich film (probably a magnesium silicate) on the surface is a mechanical barrier that prevents penetration of the metal into the sand.

In the case of grey cast iron in a greensand mould the atmosphere may be oxidising, causing the melt surface to grow a film of liquid silicate. This is highly wetting to sand grains, so that the application of a core coating such as a refractory wash may be necessary to prevent penetration of the metal into the core.

However, in the case of grey iron cast in a mould rich in hydrocarbons (i.e. greensand with sufficient additions of coal dust or certain resin-bonded sands), metal penetration is prevented when the hydrocarbons in the atmosphere of the mould pyrolyse on contact with the surface of the hot liquid metal to deposit a thin film of solid carbon on the liquid. Thus the reason for the robust non-wetting behaviour is that a solid pyrolysed carbon film on the liquid contacts a solid, pyrolysed carbon layer on the sand grains of the mould surface. This carbon-against-carbon behaviour explains the excellent surface finish that grey irons can achieve if bottom-gated and filled in a laminar fashion (Figure 6.34).

For the casting of iron, powdered coal additions or coal substitutes are usually added to greensands to improve surface finish in this way, providing a carbon layer to both the sand grains and the liquid surface. The reactions in the pyrolysis of coal were originally described by Kolorz and Lohborg (1963):

1. The volatiles are driven out of the coal to form a reducing atmosphere.
2. Gaseous hydrocarbons break down on the surface of the liquid metal. (Kolorz and Lohborg originally thought the hydrocarbons broke down on the sand grains to form a thin skin of graphite, but this is now known to be not true.)
3. The coal swells, and on account of its large expansion, is driven into the pores of the sand. This plastic phase of the coal addition appears to plasticise the binder temporarily and thereby eases the problems associated with the expansion of the silica, allowing its expansion to be accommodated without the fracture of the surface. As the temperature increases further and the final volatiles are lost, the mass becomes rigid, converting to a semi-coke. The liquid metal is prevented from contacting and penetrating the sand by this in-filling of carbon that acts as a non-wetted mechanical barrier.

Kolorz and Lohborg recommend synthetically formulated coal dusts with a high tendency to form anthracitic carbon, of good coking capacity and with good softening properties. They recommend that the volatile content be near 30% and sulphur less than 0.8% if no sulphur contamination of the surface is allowable. If some slight sulphurisation is permissible, then 1.0–1.2% sulphur could be allowed. Peterson and Blanke (1980) emphasise the bituminous nature and low ash content of desirable coals.

In the case of phenolic urethane and similar organic chemical binders based on resin systems, the thermal breakdown of the binder assists the formation of a good surface finish to cast irons and other metals largely in the manner described previously. The binder usually goes through its plastic stage prior to rigidising into a coke-like layer. The much smaller volume fraction of binder, however, does not provide for the swelling of the organic phase to seal the pores between grains. So in principal, the sand remains somewhat vulnerable to penetration by the liquid metal. The second aspect of non-wettability is discussed next.

The founder should be aware of binder pyrolysis reactions that can have adverse effects on the surface quality of the casting. These are more likely to occur with binders that contain nitrogen (phenol urethanes and some furans) which can be converted to cyanide and amine gases. Different reactions occur in the case of furans, where problems may arise from the use of the sulphonic acid or phosphoric acid catalysts which become cumulative, being least when new sand is used and progressively more concentrated as the content of mechanically reclaimed sand in the mix increases. Sulphonic acids react with iron and iron oxide residues in the sand, with feldspars, with limestone often present in poor quality sand, and with certain types of coatings, to form the corresponding sulphonates. These will in turn be reduced to sulphides during casting and these may then cause sulphide damage to cast parts.

Phosphoric acid damage is caused by a different mechanism because phosphates are stable under casting conditions. It seems that the damage caused is primarily from the reaction of its vapour with iron^{II} oxide (or chromite) dust, leading to the formation of iron^{II} phosphate which coats the sand grains. Both the phosphoric acid vapour and the iron^{II} phosphate can interact with components in the (ferrous) metal being cast.

4.5.2 LUSTROUS CARBON FILM

The carbonaceous gases evolved from the binder complete their breakdown at the white hot surface of the advancing front of liquid metal, giving up carbon and hydrogen to the advancing liquid front. For steel, the carbon dissolves quickly and usually causes relatively little problem. For cast iron, the carbon dissolves hardly at all, because the temperature is lower, and the metal is already nearly saturated with carbon; thus the carbon cannot dissolve away into the liquid iron as quickly as it arrives, so that it accumulates on the surface as a film. The time for dissolution seems to be about the same as the time for mould filling and solidification; thus the film has a life sufficiently long to affect the flow of the liquid for good or ill, depending on the circumstances.

If the metal is rising nicely in the mould, the surface film migrates across the liquid surface as it splits and reforms and then thickens with time on its approach to the side walls. The rising metal causes the liquid iron to rolls out on the surface of the mould, rolling out and laying down its surface film as the liquid advances. This laying down of the surface film can be likened to the laying down of the tracks of a track-laying vehicle. The film confers a non-wetting behaviour on the liquid itself because the liquid is effectively sealed in a non-wetting, dry solid skin. The skin forms a mechanical barrier between the liquid and the mould. This barrier is laid down as the liquid progresses because of friction between the liquid and the mould, the friction effectively stretching the film and tearing it near the centre of the meniscus where it immediately re-forms to continue the process (Figures 2.2 and 6.34). The strength and rigidity of the carbon film helps the liquid surface to bridge unsupported regions between sand grains or other imperfections in the mould surface. By this mechanism the surface of the casting is smoothed to a significantly finer finish that could be achieved by surface tension alone attempting to bridge between the sand grains.

The benefits of laying out a solid film as the liquid iron rolls out on the mould surface is only achieved if the liquid is encouraged to fill progressively, as a rolling action up against the mould surface. The mechanism for the improvement of surface finish can only operate effectively if the progress of the meniscus is steady and controlled, i.e. in the absence of surface turbulence. The casting surface can then take on an attractive, smooth glossy shine. It makes sense to fill the casting nicely using a properly designed bottom-gated technique. However, if the melt is top-poured, splashing and hammering the mould, no benefit is obtained; in fact, the liquid penetration can be severe giving a casting surface so spiky, sharp and rough to take the skin off your hands.

In addition to surface finish problems, a further problem arises if surface turbulence causes the carbon film to become entrained in the liquid. The film will necessarily become entrained as a double film, a carbon bifilm (because single films cannot be entrained) constituting a serious crack-like defect. Fortunately, in heavy-section castings, the entrained double film defect has time to dissolve, and never seems to be a problem. Light-section castings should only be attempted using binder systems that produce little lustrous carbon because defects will not have time to dissolve and will therefore be effectively permanent. Alternatively, a preferable strategy would of course be the use of a running system that could guarantee the absence of surface turbulence.

The lustrous carbon film, in contrast to all other surface films on other metals that I know, appears to detach easily from the iron during cooling. This curious behaviour means that after breaking the casting out of the mould, the film is sometimes seen attached to the surface of the mould, covering the surface as a smooth, glossy sheet, leading many observers to the mistaken conclusion that the film had initially formed on the mould surface (Naro, 2004). This is clearly not possible, of course, because pyrolysis reactions can only occur in high temperature conditions, such as on the melt surface, in contrast to the sand grains which are relatively cold.

4.5.3 SAND REACTIONS

Other reactions in the mould surface occur with the sand grains themselves. The most common of sand reactions is the reaction between silica (SiO_2) and iron oxide (wustite, FeO) to produce various iron silicates, of which the most commonly quoted (but not necessarily the most common nor the most important) is fayalite (Fe_2SiO_4). This happens frequently at the high temperatures required for the casting of irons and steels. It causes the grains to fuse and collapse as they melt into each other, because the melting point of some of the silicates is below the casting temperature. (The much quoted melting point of fayalite at only 1205°C may be an error because fayalite may have a significantly higher melting temperature. A more likely candidate to melt is grunerite, $FeSiO_3$). The reacted grains adhere to the surface of the casting because of the presence of the low-melting-point liquid 'glue'. This is known as burn-on.

The common method of dealing with this problem is to prevent the iron oxidising to form FeO in the first place. This is usually achieved by adding reducing agents to the mould material, such as powdered coal to greensand or aluminium powder additions to mould washes and the like. The problem is also reduced in other sands that contain less silica, such as chromite sand. However, the small amounts of silica present in chromite can still give trouble in steel castings, where the extreme temperature causes the residual silica to fuse with the clay. At these temperatures, the chromite may break down, releasing FeO or even droplets of Fe on the surface of the chromite sand grains. Metal penetration usually follows as the grains melt into each other, and the mould surface generally collapses. The molten, fused mass is sometimes known as 'chromite glaze'. It is a kind of burn-on, and is difficult to remove from steel castings (Petro and Flinn, 1978). Again, carbon compounds added to the moulding material are useful in countering this problem (Dietert et al., 1970).

Peterson and Blanke (1980) draw attention to the potential role of hydrogen released during the pyrolysis of coal dust in greensand, proposing that hydrogen plays an important part in avoiding the formation of oxides and silicates, and which may explain its contribution to good surface finish and the avoidance of burn-on. If hydrogen is actually valuable in this way, it clearly cannot arise from the main carbon constituent of coal, but must be generated from its minor hydrocarbon content, strongly suggesting that coal might be eliminated in favour of a more targeted binder addition.

4.5.4 MOULD CONTAMINATION

There are a few metallic impurities that find their way into moulding sands as a result of interaction between the cast metal and the mould. We are not thinking for the moment of the odd spanner or tonnes of iron filings from the steady wearing away of sand plant. Such ferrous contamination is retrieved in most sand plants by a powerful magnet located at some convenient point in the recirculating sand system. (The foundry maintenance crew always have interesting stories to tell of items found from time to time attached to the magnet.) Nor are we thinking of the pieces of tramp metal such as flash and other foundry returns. Our concern is with the microscopic traces of metallic impurities that lead to a number of problems, particularly because of the need to protect the environment from contamination.

Foundries that cast brasses find that the grains of their moulding sand become coated with a zinc-rich layer, with lead-rich nodules on the surface of the zinc (Mondloch et al., 1987). The metals are almost certainly lost from the casting by evaporation from the surface after casting. The vapour condenses among the cool sand grains in the mould as either particles of metallic alloy, or reacts with the clay present, particularly if this is bentonite, to produce Pb-Al silicates. If there is no clay present, as in chemical binder systems such as furan resins, then no reaction is observed so that metallic lead remains (Ostrom et al., 1982). Thus ways of reducing this problem are: (1) the complete move, where possible in simple castings, to metal moulds; (2) the complete move, where possible, from lead-containing alloys; or (3) the use of chemical binders, together with the total recycling of sand in-house. This policy will contain the problem, but the sand will have a fair degree of toxicity. If the metallic lead can be separated from the sand in the sand recycling plant, the proceeds might provide a modest economic return, and the sand toxicity could be limited.

There has been a suggestion that iron can evaporate from the surface of a ferrous casting in the form of iron carbonyl $Fe(CO)_5$. This suggestion appears to have been eliminated on thermodynamic grounds; Svoboda and Geiger (1969) show that the compound is not stable at normal pressures at the temperature of liquid iron. Similar arguments eliminated the carbonyls of nickel, chromium and molybdenum. These authors carry out a useful survey of the existing knowledge of the vapour pressures of the metal hydroxides and various sub-oxides but find conclusions difficult because the data is sketchy and contradictory. Nevertheless, they do produce evidence that indicates vapour transport of iron and manganese occurs by the formation of the sub-oxides $(FeO)_2$ and $(MnO)_2$. The gradual transfer of the metal by a vapour phase, and its possible reduction back to the metal on arrival on the sand grains coated in carbon might explain some of the features of metal penetration of the mould, which is often observed to be delayed, and then occur suddenly. More work is required to test such a mechanism.

The evaporation of manganese from the surface of castings of manganese steel is an important factor in the production of castings. The surface depletion of manganese seriously reduces the surface properties of the steel. In a study of this problem, Holtzer (1990) found that the surface concentration of manganese in the casting was depleted to an impressive non-trivial depth of 8 mm and the concentration of manganese silicates in the surface of the moulding sand was increased.

The process of Mn evaporation can be observed to occur from a drop of liquid steel placed on a water-cooled copper substrate. A 'halo' is seen to develop around the drop, indicating mainly Mn condensation, although Cr and Fe can also contribute (Nolli and Cramb, 2008).

Figure 6.26 confirms that the vapour pressure of manganese is significant at the casting temperature of steel. However, the depth of the depleted surface layer is nearly an order of magnitude larger than can be explained by diffusion alone. It seems necessary to assume, therefore, that the transfer occurs mainly while the steel is liquid, and that some mixing of the steel is occurring in the vicinity of the cooling surface.

It is interesting that a layer of zircon wash on the surface of the mould reduces the manganese loss by about half. This seems likely to be the result of the thin zircon layer heating up rapidly, thereby reducing the condensation of the vapour. In addition, it will form a barrier to the progress of the manganese vapour, keeping the concentration of vapour near the equilibrium value close to the casting surface. Both mechanisms will help to reduce the rate of loss. If, however, the protective wash is applied after the moulding sand has already become significantly contaminated with Fe and Mn oxides in the recycling process, the underlying sand may partially melt and collapse (Kruse, 2006). This instability of the underlying sand will cause mechanical penetration of the zircon wash, and extensive permeation of the metal into the underlying partially melted sand. The carryover of such contamination should be eliminated by careful control of the recycling process or revised selection of moulding aggregate (see the section on mould aggregates).

Gravity die casters that use sand cores (semi-permanent moulds) will be all too aware of the serious contamination of their moulds from the condensation of volatiles from the breakdown of resins in the cores. The buildup of these products can be so severe as to cause the breakage of cores, and the blocking of vents. Both lead to the scrapping of castings. The blocking of vents by tar-like deposits in permanent moulds is the factor that controls the length of a production run prior to the mould being taken out of service for cleaning. On carousels of dies in Al alloy cylinder head production, a die may need to be taken out of service every 10th or 15th casting. The absence of such problems in sand moulds is a natural advantage, aiding the already high productivity of sand moulding that is usually overlooked.

4.5.5 MOULD PENETRATION

Svoboda (1994) reviews the wide field of mould penetration by melts, and concludes that (1) mechanical balance between the driving pressure and capillary effects lead to liquid state penetration in 75% of cases, (2) chemical reactions cause 20% of cases of penetration, and (3) vapour-state reactions may control 5% of penetration problems. We shall examine his categories in detail next. In addition, the time dependence of penetration is an interesting behaviour that requires explanation.

4.5.5.1 Liquid penetration: effect of surface tension and pressure

Any liquid will immediately start to impregnate a porous solid if the solid is wetted by the liquid. The effect is the result of capillary attraction. For this reason, moulding materials are selected for their non-wetting behaviour, so that penetration by liquid metals is resisted; the resistance is known as capillary repulsion.

It was the French workers Portevin and Bastien (1936) that first proposed that capillary forces should be significant for the penetration of liquid metals into sand. However, this prediction was neglected until the arrival of Hoar and Atterton (1950) who first set out the basic physics, creating a quantitative model in which the surface tension γ of the melt would hold back the penetration of a liquid subjected to gradually increasing pressure against the surface of a porous, non-wetted aggregate. The liquid interface bridging the inter-particle spaces would gradually swell out, its bulging action progressively confined to a steadily smaller radius until it became hemispherical (radius r in Figure 4.11). Up to this critical value, the melt could be held back, but beyond this condition, further advance of the meniscus would cause the radius of curvature to increase as indicated by the formula that follows, lowering the resistance. The balance condition at the critical pressure for penetration is that quoted in Eqn (3.11), $P = 2\gamma/r$.

If we substitute $\rho g H$ for pressure from depth H in a metal of density ρ, we have the critical depth of metal for penetration

$$H = 2\gamma/r\rho g$$

Thus, beyond this critical point, penetration would instantly occur as a runaway effect; Figure 4.10 shows the meniscus expanding rapidly away after the narrowest point between sand grains is passed. Penetration might subsequently only be stopped by the advancing front losing its heat and freezing. Such elementary physics is helpful to understand why finer grain size of the aggregate or the provision of any extremely fine-grained surface coating are both helpful to reduce penetration for any fixed value of H by this simple mechanical model.

Resistance to penetration until some critical pressure is achieved has been demonstrated many times. For instance, Draper and Gaindhar (1975) find a metal depth of approximately 300 mm is required for steel to penetrate consistently into greensand moulds in their experiment. This critical depth was influenced to some degree by mould compaction, oxidising conditions and the temperature of mould hot-spots. If we assume a density of 7000 kg/m^3, surface tension approximately 2 N/m and acceleration from gravity of about 10 m/s^2, we find the radius r is approximately 0.02 mm, indicating the interparticle diameter to be roughly 0.04 mm (40 µm) in their moulds which seems a reasonable value.

All of the discussion so far has assumed perfect non-wetting between the metal and the mould (i.e. a contact angle of 180°). Hayes et al. (1998) go to some lengths to emphasise the effect of conditions intermediate between wetting and non-wetting, taking account of the contact angle between the melt and its substrate. They find a good correlation between the contact angle and the penetration of liquid steel into silica sand moulds. Petterson (1951) suggests that the penetration of steel into sand moulds would be aided by a gradual fall in the contact angle as is commonly seen between liquids and substrates. There may be some truth in this in some situations, particularly for some alloys where vapour transport or chemical reactions are occurring. However, Hayes and colleagues measure the change in contact angle in their work and find the change insignificant. In their careful study the penetration appeared to be a simple mechanical process involving varying degrees of capillary repulsion.

Levelink and Berg have investigated and described conditions in which they suggested dynamic conditions were important (Figure 4.11). They claimed that iron castings in greensand moulds were subject to a problem that they suggested was a water explosion. This led to a severe but highly localised form of mould penetration by the metal. However, careful evaluation of their work indicates that it seems most likely that they were observing a simple

152 CHAPTER 4 MOULDS AND CORES

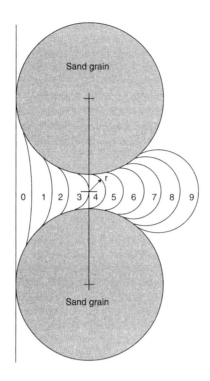

FIGURE 4.10

Metal penetration between two cylinders to simulate two sand grains. Surface tension resists the applied pressure up to the minimum radius r. Beyond this curvature penetration is a run-away instability.

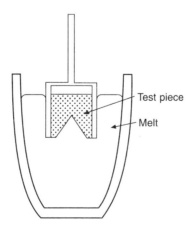

FIGURE 4.11

Water hammer (momentum effect) test piece (Levelink and Berg, 1971).

conservation of momentum effect. As the liquid metal fills their conical mould, it accelerates as the area of the cone decreases. As the melt nears the tip it reaches its maximum velocity, slamming the melt into the ever-decreasing space. The result is the generation of a high shock pressure, forcing metal into the sand. The effect is similar to a cavitation damage event associated with the collapse of a bubble against a ship's propeller. Although they cite the presence of bubbles as evidence of some kind of explosion, the oxides and bubbles present in many of their tests seem to be the result of entrainment in their rather poor filling system, rather than associated with any kind of explosion.

The impregnation of the mould with metal in last regions to fill is commonly observed in all metals in sand moulds. A pressure pulse generated by the filling of a boss in the cope will often also cause some penetration in the opposite drag surface of a thin-wall casting. The point discontinuity shown in Figure 2.27 will be a likely site for metal penetration into the mould. If the casting is thin-walled, the penetration on the front face will also be mirrored on its back face. Such surface defects in thin-walled aluminium alloy castings in sand moulds are highly unpopular, because the silvery surface of an aluminium alloy casting is spoiled by these dark spots of adhering black sand (sometimes called 'the black plague'!) and thus will require the extra expense of being removed by hand, or blasting with shot or grit.

Levelink and Berg (1968) report that the problem is increased in greensand by the use of high-pressure moulding. This may be the result of the general rigidity of the mould accentuating the concentration of momentum (weak moulds will yield more generally, and thus dissipate the pressure over a wider area). They list a number of ways in which this problem can be reduced:

1. Reduce mould moisture.
2. Reduce coal and organics.
3. Improve permeability or local venting; gentle filling of mould to reduce final filling shock.
4. Retard moisture evaporation at critical locations by local surface drying or the application of local oil spraying.

The reduction in the mechanical forces involved by reduced pouring rates or by local venting is understandable as reducing the final impact forces. Similarly, the use of a local application of oil will reduce permeability, causing the air to be compressed, acting as a cushion to decelerate the flow more gradually.

The other techniques in their list seem less clear in their effects, and raise the concern that they may possibly be counter-productive! It seems there is plenty of scope for additional studies to clarify these problems.

Work over a number of years at the University of Alabama, Tuscaloosa (Stefanescu et al., 1996), has clarified many of the special issues relating to the penetration of sand moulds by cast iron. Essentially, this work concludes that hot spots in the casting, corresponding to regions of isolated residual liquid, are localised regions in which high pressures can be generated in the residual liquid inside the casting by the expansion of graphite, forcing the liquid out via microscopic patches of unfrozen surface and causing local penetration of the mould. The pressure can be relieved by careful provision of 'feed paths' to allow the expanding liquid to be returned to the feeder. The so-called feed paths are, of course, allowing residual liquid to escape, working in reverse of normal feeding.

Naturally, any excess pressure inside the casting will assist in the process of mould penetration. Thus large steel castings are especially susceptible to mould penetration because of the high metallostatic pressure. This factor is in addition to the other potential high temperature reactions listed previously. This is reason for the widespread adoption in steel foundries of the complete coating of moulds with a ceramic wash as an artificial barrier to penetration.

4.5.5.2 Natural barriers to penetration

We have seen previously that in some conditions cast iron can develop a strong carbon film (the lustrous carbon film) on its advancing meniscus which is pulled down into the gap between the mould and the metal. Here it effectively separates the liquid from the aggregate, providing a mechanical barrier to prevent penetration, and retain a smooth shiny surface on the casting. Also, as we shall see, oxidising conditions can eliminate the carbon film replacing it with a liquid silicate. Penetration in oxidising conditions can then be practically guaranteed.

In copper-based foundries, the alloy aluminium bronze is renowned for its ability to resist penetration of the mould. The reason is also certainly the mechanical barrier presented by the strong alumina film; the high content of Al, in the 5–10% range, and the high temperature combine to create one of the most tenacious films in the casting world. In contrast, those bronzes not protected by a strong film in the liquid state, tin-, lead- and phosphor-bronze, all suffer penetration problems.

Similarly, Gonya and Ekey (1951) compared the behaviour of the common copper-base alloy 85-5-5 to 5-5 and the Al-5Si alloy, investigating a number of moulding variables. However, their most significant result related to pressure. They gradually increased pressure, finding penetration in the copper alloy at a critical pressure 17 kPa (2.5 psi), whereas the Al alloy continued to resist penetration up to the maximum they were able to provide in their experiment, about 30 kPa (4.5 psi). These pressures correspond to head of liquid metal of approximately 230 and 1240 mm, respectively. Clearly the low density of Al combined with the presence of the alumina film on the surface of the liquid alloy is a major benefit in avoiding mould penetration.

4.5.5.3 Temperature and time dependence of penetration

Clearly, the metal cannot penetrate the mould if the metal has solidified. Taking this impressively undeniable basic logic to heart, Brookes et al. (2007) draw attention to the importance of the mould temperature. By both computer model and experiment, they were able to demonstrate that as the mould surface temperature increased on contact with the melt and later decreased during cooling, penetration of liquid steel did not occur until the mould temperature exceeded a critical value. The penetration subsequently continued, causing the penetration effect to worsen while the temperature remained above the critical temperature. Penetration finally stopped when the temperature fell once again below the critical value. It is interesting that this relatively simple model appears to provide an explanation for the time dependence of penetration as has been known for many years (Jones, 1948; Shirey and Williams, 1968).

It should be noted, however, that the reduction of contact angle, i.e. progressively increasing effectiveness of wetting, also has been observed for many metal/mould combinations to be a function of time (for instance Wu et al., 1997; Shen et al., 2009). Also, of course, mould atmosphere greatly affects wetting ability; this too is a function of time. Thus a reducing or neutral atmosphere is useful to reduce penetration in low carbon steels in greensand moulds, whereas an oxidising atmosphere encouraged penetration as found by Draper and Gaindhar in 1975.

4.5.5.4 Chemical interactions

There seems little doubt that the contact angle between the mould particles and the melt can change with both time and temperature. Hayes et al. (1998) noted a reduction of the contact angle of liquid steel on silica sand from 110° to 93° over a 30 minute period. However, this lengthy period and relatively small change are not likely to greatly affect most steel castings because the majority will have frozen in this time.

To effect a major change in behaviour would require a major change in contact angle over a relatively short period such as a few minutes. The penetration of grey irons into silica sand moulds in oxidising condition is exactly such a candidate as has already been noted previously.

4.5.5.5 Effects of vapour transport

For a general overview of the knowledge of vapour transport, the reader is referred to the excellent review by Svoboda and Geiger (1969). It is salutary to reflect that it seems relatively little additional knowledge on this important subject has been gained since this early date.

On a microscale, the effects of vapour transport are likely to be complicated. Ahn and Berghezan (1991) studied the infiltration of liquid Sn, Pb and Cu inside metal capillaries using a scanning electron microscope. They found evidence of the deposition of metal vapours over a region up to 0.1 mm ahead of the advancing front. Clearly, the presence of this freshly deposited metal influenced the effect of the wetting of the liquid that followed closely behind.

Shen et al. (2009) drew similar conclusions from experiments with sessile drops in vacuum. They found that the contact angle of a zirconium/copper-based alloy on an alumina substrate fell from about 90° to 0° in about 10 min. They found Zr adsorption at the liquid/solid interface, followed by a Zr-Cu precursor film that accounted for the excellent wetting.

On a macroscale, vapour transport from metals into moulds is a common feature in foundries and will be referred to repeatedly elsewhere in this book. To give just two contrasting instances here:

a. Foundries casting magnesium into plaster moulds filled with vacuum assistance (Sin et al., 2006) find that Mg diffuses into the plaster, reacting with the SiO_2, taking the oxygen to form MgO, and reducing the Si to Mg_2Si according to the reaction

$$4Mg + SiO_2 = 2MgO + Mg_2Si$$

b. Bronze foundries casting classical tin-bronzes and lead-bronzes find that both alloys suffer mould penetration. However, only lead has any significant vapour pressure at the casting temperature, its pressure being between 100 and 1000 times greater than tin (see Figure 6.26). We can conclude therefore that in this particular case, vapour pressure is seen to be of minor, if any, importance; a high vapour pressure does not necessarily lead to enhanced penetration. The expected absence of any significant mould reactions or any strong surface films leaves only the relatively weak action of surface tension to inhibit penetration. Thus mould penetration for these alloys seems possible only by mechanical means: the provision of a fine mould aggregate, or a ceramic surface coat.

4.6 METAL SURFACE REACTIONS

Easily the most important metal/mould reaction is the reaction of the metal with water vapour to produce a surface oxide and hydrogen, as discussed in Chapter 1.

However, the importance of the release of hydrogen and other gases at the surface of the metal, leading to the growth of porosity in the casting, is to be dealt with in Chapter 6. Here we shall devote ourselves to the many remaining reactions. Some are reviewed by Bates and Scott (1977). These and others are listed briefly next.

4.6.1 OXIDATION

Oxidation of the casting skin is common for low carbon–equivalent cast irons and for most low-carbon steels. It is likely that the majority of the oxidation is the result of reaction with water vapour from the mould, and not from air which is expelled at an early stage of mould filling as shown earlier. Carbon additions to the mould help to reduce the problem.

The catastrophic oxidation of magnesium during casting, leading to the casting (and mould) being consumed by fire, is prevented by the addition of so-called inhibitors to the mould. These include sulphur, boric acid, and other compounds such as ammonium borofluoride. More recently, much use has been made of the oxidation-inhibiting gas, sulphur hexafluoride (SF_6) which is used diluted to about 1–2% in air or other gas such as CO_2 to prevent the burning of magnesium during melting and casting. However, since its identification as a powerful ozone-depleting agent, SF_6 is being discontinued for good environmental reasons. A return is being made to dilute mixtures of SO_2 in CO_2 and other more environmentally friendly atmospheres are now under development.

In any case, the burning of Mg alloys during pouring of the mould is almost certainly the result of surface turbulence. For instance, if liquid Mg can be introduced into a mould quiescently, so that it rises steadily with a substantially flat liquid surface in the mould, the huge heat of oxidation released on the surface is rapidly conveyed away into the bulk of the melt so that the surface temperature never rises to a dangerous level. Such a quiescent fill is therefore safe. Conversely, if the melt is jumping and splashing, the heat of oxidation will diffuse into the thin liquid splash from both sides of the splash surface, and quickly heats the small amount of metal in the splash. The ignition temperature of the melt is quickly exceeded, with disastrous results. There is a saying, which regrettably probably reflects some truth, that the few Mg foundries in existence nowadays are the result of most of them having burned down. True or not, the turbulent handling of Mg alloys is clearly dangerous but mostly avoidable. The use of an offset step basin and stopper, and a naturally pressurised filling system would make a huge difference to Mg foundries.

Titanium and its alloys are also highly reactive. Despite being cast under vacuum into moulds of highly stable ceramics such as zircon, alumina or yttria, the metal reacts to reduce the oxides of the mould, contaminating the surface of the casting with oxygen, thereby stabilising the alpha phase of the alloy. The 'alpha case' usually has to be removed by chemical machining.

4.6.2 CARBON (PICKUP AND LOSS)

Mention has already been made of the problem of casting titanium alloy castings in carbon-based moulds. The carburisation of the surface, again results in the stabilisation of the alpha phase, and requires to be subsequently removed.

The difficulty is found with stainless steel of carbon content less than 0.3% cast in Croning resin-bonded shell moulds (McGrath, 1973). The relatively high resin binder content of these moulds, generally in the region of 2.5–3.0 wt%, causes steels to suffer a carburised layer between 1 and 2 mm deep. Tordoff (1996) also found significant carbon pickup in stainless steels from phenolic urethane binders which rarely exceed 1.0–1.2 wt% based on sand. The carburisation, of course, appears more severe the lower the carbon content of the steel. Naturally, the problem is worse on drag than on cope faces as a result of the casting sitting hard down on the drag face of the mould, and contracting away from the upper surfaces. Tordoff also found that iron oxide in the moulding sand reduced the effect, whereas the use of a furan resin eliminated pickup altogether. The author could not find a record of the effect of silicate-based binders, although their usually low content of organics would be expected to give minimal problems.

Carbon pickup is the principal reason why low carbon steel castings are not produced by the lost-foam process. The atmosphere of styrene vapour, which is created in the mould as the polystyrene decomposes, causes the steel to absorb carbon (and presumably hydrogen). The carbon-rich regions of the casting are easily seen on an etched cross-section as swathes of pearlite in an otherwise ferritic matrix.

In controlled tests of the rate of carburisation of low carbon steel in hydrocarbon/nitrogen mixtures at 925°C (Kaspersma, 1982), methane was the slowest and acetylene the fastest of the carburising agents tested, and hydrogen was found to enhance the rate, possibly by reducing adsorbed oxygen on the surface of the steel. At high ratios of H_2/CH_4 at this temperature, hydrogen decarburises steel by forming methane (CH_4). This may be the important reaction in the casting of steel in greensand sand moulds containing only low carbonaceous additions.

Decarburisation can also be a problem. For instance, surface decarburisation of steels is often noted with acid-catalysed furan resin binders (Naro and Wallace, 1992).

In the investment casting of steel in air, the decarburisation of the surface layer is particularly affected because atmospheric oxygen persists in the mould as a consequence of the inert character of the mould and its permeability to the surrounding environment. Doremus and Loper (1970) have measured the thickness of the decarburised layer on a low carbon steel investment casting and find that it increases mainly with mould temperature and casting modulus. The placing of the mould immediately after casting into a bin filled with charcoal helps to recarburise the surface. However, Doremus and Loper point out the evident danger that if the timing and extent of recarburisation is not correct, the decarburised layer will still exist below a recarburised layer!

In iron castings, the decarburisation of the surface gives a layer free from graphite. This adversely affects machinability, giving pronounced tool wear, especially in large castings such as the bases of machine tools. The decarburisation seems to be mainly the result of oxidation of the carbon by water vapour because dry moulds reduce the problem. An addition of 5–6% coal dust to the mould further reduces it. Rickards (1975) found that the reaction seems to start at about the freezing point of the eutectic, 1150°C, and proceeds little further after the casting has cooled to 1050°C (Figure 4.12). Stefanescu et al. (2009) find that the depth of the casting skin in ductile iron is controlled by diffusion to only about 0.5 mm if the liquid in the casting is relatively quiescent, whereas the strong convection currents during solidification of larger castings causes mixing, increasing the skin depth to nearly 3 mm.

4.6.3 NITROGEN

Nitrogen pickup in grey cast irons appears to be directly related to the nitrogen content of mould binders (Graham et al., 1987). Ammonia is released during the pyrolysis of urea and amines contained in hot box and Croning shell systems when they become recycled into a greensand system. Ammonia appears to be reversibly absorbed by the bentonite clays and is released on heating. The pyrolysis of ammonia releases nascent nitrogen and nascent hydrogen by the simple decomposition

$$NH_3 = N + 3H$$

Graham and coworkers confirmed that subsurface porosity and fissures in irons do not correlate well with the total nitrogen content of the sand, but were closely related to the total ammonia content. Lee (1987) confirms the usefulness of an ammoniacal nitrogen test which in his work pointed to wood flour as a major contributor of ammonia in his greensand system.

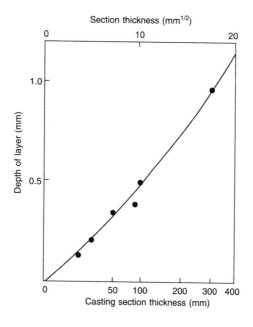

FIGURE 4.12

Depth of decarburisation in grey iron plates cast in greensand.

Data from Rickards (1975).

The link of ammonia and the so-called nitrogen fissures in iron castings suggests the formation of nitride bifilms which might be opened, becoming visible, by inflation with the copious amounts of hydrogen released by the decomposition of ammonia.

At the other end of the casting value system, vacuum-cast Ni-based superalloys suffer severely from nitrogen pickup from the moment the door of the vacuum furnace is opened, allowing air to rush in and react while the casting is still hot. Casting returns in such foundries are known to contaminate the new melts, although the contamination would be expected not to be nitrogen in solution, but nitride bifilms in suspension. These conjectures require more research to clarify the situation.

4.6.4 SULPHUR

The use of moulds bonded with furane resin catalysed with sulphuric and/or sulphonic acid causes problems for ferrous castings because of the pickup of sulphur in the surface of the casting (Naro and Wallace, 1992). This is especially serious for ductile iron castings, because the graphite reverts from spheroidal back to flake form in the high-sulphur surface layer. This has a serious impact on the fatigue resistance of the casting.

4.6.5 PHOSPHORUS

The use of moulds bonded with furane resin catalysed with phosphoric acid leads to the contamination of the surfaces of ferrous castings with phosphorus (Naro and Wallace, 1992). In grey iron, the presence of the hard phosphide phase in the surface causes machining difficulties because of rapid tool wear.

4.6.6 SURFACE ALLOYING

There has been some Russian (Fomin, 1965) and Japanese work (Uto, 1967) on the alloying of the surface of steel castings by the provision of materials such as ferrochromium or ferromanganese in the facing of the mould. Because the alloyed layers that have been produced have been up to 3 or 4 mm deep, it is clear once again that not only is diffusion involved but also some additional transport of added elements must be taking place by mixing in the liquid state. Omel'chenko further describes a technique to use higher-melting-point alloying additions such as titanium, molybdenum and tungsten by the use of exothermic mixes. Predictably enough, however, there appear to be difficulties with the poor surface finish and the presence of slag inclusions. Until this difficult problem is solved, the technique does not have much chance of attracting any widespread interest.

4.6.7 GRAIN REFINEMENT

The use of cobalt aluminate ($CoAl_2O_4$) in the primary mould coat for the grain refinement of nickel and cobalt alloy investment castings is now widespread as described in early reports by Rappoport (1964) and Watmough (1980). The mechanism of refinement is not yet understood. It seems unlikely that the aluminate as an oxide phase can wet and nucleate metallic grains. That the surface finish of grain refined castings is somewhat rougher than that of similar castings without the grain refiner indicates that some wetting action has occurred. This suggests that the particles of $CoAl_2O_4$ decompose to some metallic form, possibly CoAl. This phase has a melting point of 1628°C. It would therefore retain its solid state at the casting temperatures of Ni-based alloys. In addition it has an identical face-centred–cubic crystal structure to nickel. On being wetted by the liquid alloy, it would be expected to constitute an excellent substrate for the initiation of grains.

Watmough also investigates a number of other additives to coatings for Ni-base high temperature alloy castings including cobalt oxide, CoO. Because oxides are almost certainly not effective nuclei for solid formation and because CoO is not especially stable (Llewelyn and Ball, 1962), it is likely that the compound decomposes at casting temperature forming metallic cobalt which would be expected to be an effective nucleus for the alloy. Although its melting point is not significantly higher than the alloy, that the Co particles sit at the mould wall will ensure that they remain cool and therefore will resist melting and continue to act as nucleating particles.

It is to be expected that there can be no refinement if all the CoAl particles are either melted or dissolved, or if the newly nucleated grains are themselves re-melted. Unfortunately, such re-melting problems are common in investment castings as a result of (1) casting at too high a temperature and (2) convection problems that have so far been overlooked. It is regrettable that the beautifully fine grain structure produced by these techniques is often lost by subsequent convection in the casting, conveying so much heat from hotter regions that parts of the surface in the path of the flow re-melt, destroying the nuclei and the early grains, and replacing these with massive grains of uncontrolled size. In addition, the small depth of grain refinement (Watmough reports only 1.25 mm) is another probable consequence of uncontrolled convection. It is expected that control and suppression of convection in investment castings would greatly improve cast structures in polycrystalline high temperature alloy castings. These issues are dealt with in *Casting Manufacture* in Casting Rule 7 'Avoiding Convection'.

Cobalt addition to a mould coat is also reported to grain-refine malleable cast iron (Bryant, 1971), presumably by a similar mechanism to that enjoyed by the vacuum-cast Ni-base alloys.

The use of zinc in a mould coat by Bryant and Moore to achieve a similar aim in iron castings must involve a quite different mechanism, because the temperature of liquid iron (1200–1450°C) greatly exceeds not only the melting point, but even the boiling point (907°C) of zinc! It may be that the action of the zinc boiling at the surface of the solidifying casting disrupts the formation of dendrites, detaching them from the surface so that they become freely floating nuclei within the melt. Thus the grain refining mechanism in this case is grain multiplication rather than nucleation.

The zinc-containing coating and others listed next all appear to induce grain multiplication as a result of the rafts of dendrites attempting to grow on an unstable, moving and collapsing substrate. The effect seems analogous to that described in Section 3.2.6 for the enhancement of the fluidity of Al alloy castings by coatings of acetylene black or hexachloroethane on moulds.

Nazar et al. (1979) report the use of hexachlorethane-containing coatings for the grain refinement of Al alloys.

4.6.8 MISCELLANEOUS

Boron has been picked up in the surfaces of stainless steel castings from furane bonded moulds that contain boric acid as an accelerator (McGrath, 1973).

Tellurium is sometimes deliberately added as a mould wash to selected areas of a grey iron casting. This element is a strong carbide former, and will locally convert the structure of the casting from grey to a fully carbidic white iron. Chen et al. (1989) describe how TeCo surface alloying is useful to produce wear-resistant castings.

In other work, the carbide-promoting action of Te is said to reduce local internal shrinkage problems, although its role in this respect seems difficult to understand. It has been suggested that a solid skin is formed rapidly, equivalent to a thermal chill (Vandenbos, 1985). The effect needs to be used with caution: tellurium and its fumes are toxic, and the chilled region causes difficulties in those parts requiring to be machined.

The effect of tellurium converting grey to white irons is used to good purpose in the small cups used for the thermal analysis of cast irons. Tellurium is added as a wash on the inside of the cup. During the pouring of the iron, it seems to be well distributed into the bulk of the sample, not just the surface, so that the whole test piece is converted from grey to white iron. This simplifies the interpretation of the cooling curve, allowing the composition of the iron to be deduced.

4.7 MOULD COATINGS
4.7.1 AGGREGATE MOULDS

Although we have dealt at length with reactions that can occur between the metal and the mould, the purpose of a mould coating is to prevent such happenings by keeping the two apart. It has to be admitted that for many formulations, these attempts are of only limited success. Some useful reviews of coatings from which the author has drawn are given by Vingas (1986), Wile et al. (1988) and Beeley (2001). Vingas in particular describes techniques for measuring the thickness of coatings in the liquid and solid states. All describe the various ways in which coatings can be applied by dipping, swabbing, brushing and flow-over. These issues will not be dealt with here.

With regard to the definition of a coating, it is a creamy mixture made up from of a number of constituents:

1. *Refractory*, usually oxides of many kinds ground to a particle size of 50 μm or less, but for cast irons carbonaceous material is common in the form of graphite or ground coke etc.
2. *Carrier* nowadays becoming more commonly water (alcohols and chlorinated hydrocarbons have environmental problems). This has to be removed after the coating has been applied, usually requiring the application of heat.
3. *Suspension agent* required to maintain a uniform dispersion of the refractory. Commonly used materials include sodium bentonite clay or cellulose compounds. Beeley (2001) describes a problem with the clay addition, as a result of the water soaking into the mould surface, but being hindered during the subsequent drying because of the presence of the impervious clay coat. The development of a coating without the clay addition doubled the rate of drying.
4. *Binder* which is often the same as that used for bonding moulds. Thus, organic resins, colloidal silica and sodium silicate-based binders are common. Interesting incompatibilities are sometime experienced, such as the fact that magnesite coatings for steel castings may be difficult to use with sand bonded with furan resins because of the acid content of the resin, whereas magnesite coatings perform excellently when used in conjunction with silicate-bonded moulds and cores.
5. *Sundry chemicals* including surfactants to improve wettability, antifoaming agents and bactericides.

The advantages of coatings include the following

1. Reduction of cleaning costs because of improved surface finish (finer surface, reduced or eliminated veining, reduced penetration and/or burn-on and reduced reactions between metal and mould, for instance between (a) manganese steel and silica sand, and (b) binder gases such as sulphur compounds from furan binders)

2. Improved shake-out because of improved sand peel.
3. Reduced machining time and tool wear.

However, an important aspect to bear in mind, as with the provision of feeders, the best action (if possible) is to avoid coatings. There are many reasons for avoiding coatings. Disadvantages include the following

1. Cost of materials, especially those based on expensive minerals such as zircon.
2. Potential loss of accuracy because of difficulty of controlling coat thickness and coating penetration.
3. Possibility of cosmetic defects from runs and drops.
4. Floor space for coating station.
5. Energy cost to dry and possible capital cost and floor space for drying ovens and extraction ducting and fans.
6. Floor space to dry if dried naturally (this might exceed moulding space because drying is slow).
7. Drying time can severely reduce productivity.
8. Cores appear to be never fully dried, despite all attempts, after the application of a coating, so that there is enhanced danger of blow defects if the core cannot be vented to the atmosphere.
9. Environmental problems: at this time many coatings are still alcohol (ethanol) based, and so burned off rather than dried. This reduces floor space and energy, but does add to the loading of volatile organic compounds in the environment. This approach is likely to be banned under future legislation, forcing the use of water-based coatings.

The water-based coatings pose a significantly increased drying problem. Puhakka (2009) describes the novel use of a remote infra-red camera to monitor the drying process; while water is still evaporating the surface will remain cool. Only when fully dry will the surface temperature of the mould rise to room temperature.

Coating costs for the foundry are usually justified on (1) and (2) of the previous advantages and should be based on the dry refractory deposited. Naturally, costs are directly related to the surface area to volume ratio of the castings, and so will vary from foundry to foundry. Alternatively, surface area per ton of castings is a useful measure.

Coatings are not required in general for the lower melting point metals such as the zinc-based alloys and the light metals, Al- and Mg-based alloys, but are widely used for cast iron, copper-based alloys and steels.

The use of coatings on cores to prevent core blows is usually a mistake. Although the coating does reduce the permeability of the surface, assisting to keep in the expanding gases, the additional volatiles from the coat, which appear to be never completely removed by drying, are usually present in excess to overwhelm the coating barrier.

Another interesting mistake is the use of coatings to prevent sand erosion. The presence of sand defects in a casting is never, in my experience, the fault of the sand. It immediately signals that the filling system is faulty. A turbulent filling system that entrains air will splatter and hammer against the sand surface, oxidising away any binder, and so erode away cores and moulds. Although, of course, a coating will reduce the problem, it is often less costly and more effective to upgrade the filling system to a naturally pressurised design. In this way, the moulds and cores are at all times gently pressurised by the melt, causing the sand grains of the mould to be held safely in place. An additional bonus is a better casting, enjoying more reliable properties and freedom from other defects such as porosity and cracks.

4.7.2 PERMANENT MOULDS AND METAL CHILLS

Permanent moulds or chills in grey iron or steel are practically inert, so any mould coat is not required to prevent chemical reactions between the two. A coat will help to

1. protect the mould from thermal shock;
2. avoid the premature chilling of the metal that might result in cold lap defects and
3. confer some 'surface permeability' on the impermeable surface to allow the melt to flow better over the surface, and allowing the escape of any volatiles or condensates (particularly from the surface of chills).

It is often seen that two different mould coatings are used on a permanent mould. A thin smooth coating is applied to the mould cavity which forms the casting, whereas a thick, rough, insulating coating is applied to the running system in an effort to avoid temperature losses during mould filling. This is a mistake. The metal is in the running system for perhaps only 1 to 2 s (because its velocity is in the range of metres per second, and the distance involved is usually only a metre or less) that any loss of temperature is negligible. Furthermore, the roughness of this coat endangers the flow because of the consequential turbulence. It is far better, much less trouble, less time-consuming and less costly simply to coat the whole die with the same thin, smooth coat. The castings will probably be improved.

4.7.3 DRY COATINGS

Dry coatings are of course an attractive concept because no drying is involved and there is no danger of introducing additional volatiles into moulds or cores.

Greensand moulds have benefited somewhat from the use of dusting powders of various kinds and more recently by the electrostatic precipitation of dry zircon flour (precoated with an extremely thin layer of thermosetting resin, activated by the heat of the metal during casting). This development appears to have been resistant to extension to dry sand processes because of the generally poorer electrical conductivity of dry sand moulds particularly when bonded with a phenol-urethane resin. It would be interesting to know whether more success might be expected from moulds made with materials having higher electrical conductivity such as sand bonded with acid-catalysed resins, silicates or alkaline phenolics, or with chromite sand.

A successful technique for the application of dry coatings to dry sand moulds has therefore proved elusive. Simply dusting on dry powder, or attempting to apply it in a fluidised bed, suffers from most of the powder being blown off as clouds of dust when the mould is cleaned by blowing out with compressed air prior to closure.

A Japanese patent (Kokai, 1985) describes how fine refractory powder in a fluid bed is drawn into the surface pores of an aggregate mould or core by applying a vacuum via some far location on the mould or core. However, of course, despite probably being useful for simple geometries, this technique fails for complex shapes. Another Japanese patent describes how fine powder in a fluid bed is forced into the pores of a mould by a mechanical brushing action. A Canadian automotive foundry has developed a fluid bed containing a mixture of the fine coating powder together with relatively heavy zircon grains. The jostling of the zircon grains effectively hammers the powder into the mould pores.

These dry coating techniques deserve wider use. When correctly applied, the coating simply fills the interstices between the surface grains of the mould, not adding any thickness of deposit. Thus the technique has the advantage of preserving the accuracy of the mould while improving surface finish with no drying time penalty.

CHAPTER 5

SOLIDIFICATION STRUCTURE

In this section, we consider how the metal changes state from the liquid to the solid and develops its structure.

It is a widely accepted piece of dogma, often quoted, that the properties of the casting are controlled by its structure. This seems to me to be largely untrue. For instance, in meeting mechanical property specifications of a casting its solidification structure, its grain size or dendrite arm spacing (DAS) bears only a superficial relation to the properties. The feature really in control is the bifilm population. That the grain size and DAS appear to be important seems mainly in relation to the way in which they influence the unfurling of bifilms and the bifilm distribution. More of this later.

In a later chapter, we consider the problems of the volume deficit on solidification, and the so-called shrinkage problems that lead to a set of void phenomena, sometimes appearing as porosity, even though, as we shall see, such problems are actually relatively rare in castings (mainly because foundry people understand how to avoid shrinkage porosity). Most of the problems attributed to shrinkage are either not shrinkage problems at all, or are only secondarily shrinkage problems. Most problems that *appear* to be 'shrinkage' are actually oxides, actually bifilms together with their entrained layers of air and bubbles.

When we understand (1) how shrinkage voids can form, we are then in a better position to understand the development of (2) gas pores. Both of these sets of defects generally grow from bifilms.

These issues highlight the problem faced by the author. The problem is how to organise the descriptions of the complex but inter-related phenomena that occur during the solidification of a casting into a logical presentation necessary for a book. This book could be organised in many different ways. For instance, naturally, the gas and shrinkage contributions to the overall pore structure are complimentary and additive, and both rely on the presence and character of the pre-existing population of bifilms.

The reader is requested to be vigilant to see this integration. I am conscious that, although spelling out the detail in a didactic dissection of phenomena, emphasising the separate physical mechanisms, the holistic vision for the reader is easily lost. Take care not to lose the overall coherence of our fascinating field.

5.1 HEAT TRANSFER
5.1.1 RESISTANCES TO HEAT TRANSFER

The hot liquid metal takes time to lose its heat and solidify. The rate at which the heat escapes is controlled by a number of resistances described by Flemings (1974). We shall follow his clear treatment in this section.

The five main resistances to heat flow from the interior of the casting (starting at zero for convenience of numbering later) are:

0. The liquid.
1. The solidified metal.
2. The metal-mould interface.
3. The mould.
4. The surroundings of the mould.

All these resistances add, as though in series, as shown schematically in Figure 5.1.

164 CHAPTER 5 SOLIDIFICATION STRUCTURE

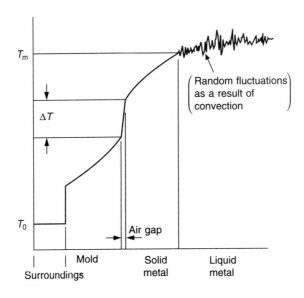

FIGURE 5.1

Temperature profile across a casting freezing in a mould, showing the effect of the addition of thermal resistances that control the rate of loss of heat.

As it happens, in nearly all cases of interest, resistance (0) is negligible as a result of stirring by forced convection during filling and thermal convection during cooling. The turbulent flow and mixing quickly transport heat and so smooth out temperature gradients. This happens quickly because the viscosity of the melt is usually low, so that the flow of the liquid is fast, and the heat is transported out of the centre of large ingots and castings in a time that is short compared to that required by the remaining resistances, whose rates are controlled by much slower diffusion processes.

In many instances, resistance (4) is also negligible in practice. For instance, for normal sand moulds the environment of the mould does not affect solidification because the mould becomes hardly warm on its outer surface by the time the casting has solidified inside.

However, there are, of course, a number of exceptions to this general rule, all of which relate to various kinds of thin-walled moulds that, because of the thinness of the mould shell, are somewhat sensitive to their environment. Iron castings made in Croning shell moulds (the Croning shell process is one in which the sand grains are coated with a thermosetting resin, which is cured against a hot pattern to produce a thin, biscuit-like mould) solidify faster when the shell is thicker, or when the shell is thin but backed up by steel shot. Conversely, the freezing of steel castings in investment shell moulds can be delayed by a thick backing of thermal insulation around the shell, all preheated to high temperature. Alternatively, without a backing, cooling is relatively fast, radiating heat away freely to cooler surroundings. Naturally, aluminium alloys in iron or steel dies can be cooled even faster when the dies are water cooled.

Nevertheless, despite such useful ploys for coaxing greater productivity, it remains essential to understand that in general the major fundamental resistances to heat flow from castings are items (1), (2) and (3). For convenience we shall call these resistances 1, 2 and 3.

The effects of all three simultaneously can nowadays be simulated with varying degrees of success by computer. However, the problem is both physically and mathematically complex, especially for castings of complex geometry.

There is therefore still much understanding and useful guidance to be obtained by a less ambitious approach, whereby we look at the effect of each resistance in isolation, considering only one dimension (i.e. uni-directional heat flow). In this way, we can define some valuable analytical solutions that are surprisingly good approximations to casting problems. We shall continue to follow the approach by Flemings.

Resistance 1: The casting

The rate of heat flow through the casting, helping to control the freezing time of the casting, applies in the case of such examples as Pb-Sb alloy cast into steel dies for the production of battery grids and terminals; or the casting of steel into metal moulds; or the casting of hot wax into metal dies as in the injection of wax patterns for investment casting. It would be of wide application in the plastics industry.

However, it has to be admitted that this type of freezing regime does not apply for many metal castings of high thermal conductivity such as the light alloys or Cu-based alloys.

For the unidirectional flow of heat from a metal poured exactly at its melting point T_m against a mould wall initially at temperature T_0, the transient heat flow problem is described by the partial differential equation, where α_s is the thermal diffusivity of the solid

$$\frac{\partial T}{\partial t} = \alpha_s \frac{\partial^2 T}{\partial x^2} \tag{5.1}$$

The boundary conditions are $x = 0$, $T = T_0$; at $x = S$, $T = T_m$, and at the solidification front the rate of heat evolution must balance the rate of conduction down the temperature gradient, i.e.

$$H\rho_s \left(\frac{\partial S}{\partial t}\right) = K_s \left(\frac{\partial T}{\partial x}\right)_{x=s} \tag{5.2}$$

where K_s is the thermal conductivity of the solid, H is the latent heat of solidification, and for which the solution is:

$$S = 2\gamma\sqrt{\alpha_s t} \tag{5.3}$$

The reader is referred to Flemings for the rather cumbersome relation for γ. The important result to note is the parabolic time law for the thickening of the solidified shell. This agrees well with experimental observations. For instance, the thickness S of steel solidifying against a cast iron ingot mould is found to be:

$$S = at^{1/2} - b \tag{5.4}$$

where the constants a and b are of the order of 3 and 25, respectively, when the units are millimetres and seconds. The result is seen in Figure 5.2.

The apparent delay in the beginning of solidification shown by the appearance of the constant b is a consequence of the following: (1) The turbulence of the liquid during and after pouring, resulting in the loss of superheat from the melt, and so slowing the start of freezing. (2) The finite interface resistance further slows the initial rate of heat loss. Initially, the solidification rate will be linear, as described in the next section (and hence giving the initial curve in Figure 5.2 because of this plot using the square root of time). Later, the resistance of the solidifying metal becomes dominant, giving the parabolic relation (shown, of course, as a straight line in Figure 5.2 because of using the square root plot of time).

Resistance 2: The metal-mould interface

In many important casting processes, heat flow is controlled to a significant extent by the resistance at the metal-mould interface. This occurs when both the metal and the mould have reasonably good rates of heat conductance, leaving the boundary between the two the dominant resistance. The interface becomes overriding in this way when an insulating mould coat is applied, or when the casting cools and shrinks away from the mould (and the mould heats up, expanding away from the metal), leaving an air gap separating the two. These circumstances are common in the die casting of light alloys.

For unidirectional heat flow, the rate of heat released during solidification of a solid of density ρ_s and latent heat of solidification H is simply:

$$q = -\rho_s HA \frac{\partial S}{\partial t} \tag{5.5}$$

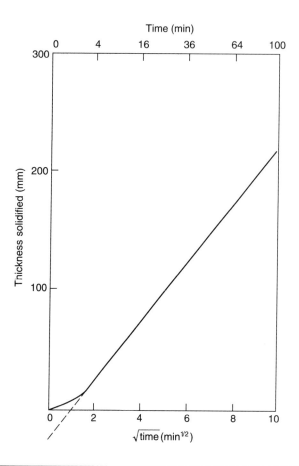

FIGURE 5.2

Unidirectional solidification of pure iron against a cast iron mould coated with a protective wash.

From Flemings (1974).

This released heat has to be transferred to the mould. The heat transfer coefficient h across the metal/mould interface is simply defined as the rate of transfer of energy q (usually measured in watts) across unit area (usually a square metre) of the interface, per unit temperature difference across the interface. This definition can be written:

$$q = -hA(T_m - T_0) \tag{5.6}$$

assuming the mould is sufficiently large and conductive not to allow its temperature to increase significantly above T_0, effectively giving a constant temperature difference $(T_m - T_0)$ across the interface. Hence equating (5.5) and (5.6) and integrating from $S = 0$ at $t = 0$ gives:

$$S = \frac{h(T_m - T_0)}{\rho_s H} \cdot t \tag{5.7}$$

It is immediately apparent that because shape is assumed not to alter the heat transfer across the interface, Eqn (5.7) may be generalised for simple shaped castings to calculate the solidification time t_f in terms of the volume V to cooling surface area A ratio (the geometrical modulus) of the casting:

$$t_f = \frac{\rho_s H}{h(T_m - T_0)} \cdot \frac{V}{A} \tag{5.8}$$

All of these calculations assume that h is a constant. As we shall see later, this is perhaps a tolerable approximation in the case of gravity die (permanent mould) casting of aluminium alloys where an insulating die coat has been applied. In most other situations h is highly variable, and is particularly dependent on the geometry of the casting.

The air gap

As the casting cools and the mould heats up, the two remain in good thermal contact while the casting is still liquid. When the casting starts to solidify, it rapidly gains strength and can contract away from the mould. In turn, as the mould surface increases in temperature it will expand. Assuming for a moment that this expansion is *homogeneous*, we can estimate the size of the gap d as a function of the diameter D of the casting:

$$d/D = \alpha_c\{T_f - T\} + \alpha_m\{T_{mi} - T_0\}$$

where α is the coefficient of thermal expansion, and subscripts c and m refer to the casting and mould respectively. The temperatures T are T_f the freezing point, T_{mi} the mould interface, and T_0 the original mould temperature.

The benefit of the gap equation is that it shows how straightforward the process of gap formation is. It is simply a thermal contraction-expansion problem, directly related to interfacial temperature. It indicates that for a steel casting ($\alpha_c = 14 \times 10^{-6}$ K^{-1}) of 1 m diameter that is allowed to cool to room temperature the gap would be expected to be of the order of 10 mm at each of the opposite sides simply from the contraction of the steel, neglecting any expansion of the mould. This is a substantial gap by any standards!

Despite the usefulness of the elementary formula in giving some order-of-magnitude guidance on the dimensions of the gap, there are a number of interesting reasons why this simple approach requires further sophistication.

In a thin-walled aluminium alloy casting of section only 2 mm the room temperature gap would be only 10 µm. This is only one-twentieth of the size of an average sand grain of 200 µm diameter. Thus the imagination has some problem in visualising such a small gap threading its way amid the jumble of boulders masquerading as sand grains. It really is not clear whether it makes sense to talk about a gap in this situation.

Woodbury and co-workers (2000) lend support to this view for thin wall castings. In horizontally sand cast aluminium alloy plates of 300 mm square and up to 25 mm thickness, they measured the rate of transfer of heat across the metal-mould interface. They confirmed that there appeared to be no evidence for an air gap. Our equation would have predicted a gap of 25 µm. This small distance could easily be closed by the slight inflation of the casting because of two factors. (1) The internal metallostatic pressure provided by the filling system (no feeders were used). (2) The precipitation of a small amount of gas; for instance, it can be quickly shown that 1% porosity would increase the thickness of the plate by at least 70 µm. Thus the plate would swell by creep with minimal difficulty under the combined internal pressure from head height and the growth of gas pores. The 25 µm movement from thermal contraction would be so comfortably overwhelmed that a gap would probably never have chance to form.

Our simple air gap formula assumes that the mould expands *homogeneously*. This may be a reasonable assumption for the surface of a greensand mould, which will expand into its surrounding cool bulk material with little resistance. A more rigid chemically bonded sand would be subject to rather more restraint, thus preventing the surface from expanding so freely. The surface of a metal die will, of course, be most constrained of all by the surrounding metal at lower temperature, but the higher conductivity of the mould will raise the temperature of the whole die more uniformly, giving a better approximation once again to homogeneous expansion.

Also, the sign of the mould movement for the second half of the equation is only positive if the mould wall is allowed to move outwards because of small mould restraint (i.e. a weak moulding material) or because the interface is concave. A rigid mould and/or a convex interface will tend to cause inward expansion, reducing the gap, as shown in Figure 5.3. It might be expected that a flat interface will often be unstable, buckling either way. However, Ling, Mampaey and co-workers (2000) found that both theory and experiment agreed that the walls of their cube-like mould poured with white cast iron distorted always outwards in the case of greensand moulds, but always inwards in the case of the more rigid chemically bonded moulds.

There are further powerful geometrical effects to upset our simple linear temperature relation. Figure 5.4 shows the effect of linear contraction during the cooling of a shaped casting. Clearly, anything in the way of the contraction of the straight lengths of the casting will cause the obstruction to be forced hard against the mould. This happens in the corners

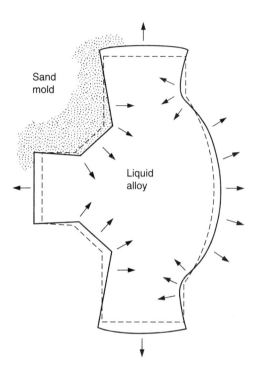

FIGURE 5.3

Movement of mould walls, illustrating the principle of inward expansion in convex regions and outward expansion in concave regions.

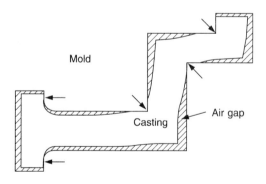

FIGURE 5.4

Variable air gap in a shaped casting: arrows denote the probable sites of zero gap.

at the ends of the straight sections. Gaps cannot form here. Similarly, gaps will not occur around cores that are surrounded with metal, and onto which the metal contracts during cooling. Conversely, large gaps open up elsewhere. The situation in shaped castings is complicated and is only just being tackled with some degree of success by computer models. Even so, Kron (2004) is sceptical of the computer models to date. Few attempt to allow for the mechanical factors influencing the air gap. Even where such allowance is attempted, there is no agreement on a suitable constitutive equation for mechanical deformation of solids at temperatures near their melting points. The swelling of the casting in the mould from the hydrostatic pressure in the liquid, and the friction of the rather soft casting against the mould are additional complicating factors. Large square section ingots of several tons in weight solidifying in ingot moulds are known to contract first from the mould corners, but under the high internal metallostatic pressure the casting balloons out to form a good contact over the central areas of the flat sides for much longer, causing these areas of the mould to suffer cracking by thermal fatigue.

Richmond and Tien (1971) and Tien and Richmond (1982) demonstrate via a theoretical model how the formation of the gap is influenced by the internal hydrostatic pressure in the casting and by the internal stresses that occur within the solidifying solid shell. Richmond (1990) goes on to develop his model further, showing that the development of the air gap is not uniform, but is patchy. He found that air gaps were found to initiate adjacent to regions of the solidified shell that were thin, because, as a result of stresses within the solidifying shell, the casting-mould interface pressure first dropped to zero at these points. Conversely, the casting-mould interface pressure was found to be raised under thicker regions of the solid shell, thereby enhancing the initial non-uniformity in the thickness of the solidifying shell. Growth becomes unstable, automatically moving away from uniform thickening. This rather counter-intuitive result may help to explain the large growth perturbations that are seen from time to time in the growth fronts of solidifying metals. Richmond reviews a considerable amount of experimental evidence to support this model. All the experimental data seem to relate to solidification in metal moulds. It is possible that the effect is less severe in sand moulds.

Lukens et al. (1991) confirmed that increased feeder height increased the heat transfer at the base of the casting. Furthermore, for a horizontal cylinder 91 mm diameter in a chromite greensand mould, as the cylinder contracted away from the mould, gravity caused the cylinder to sit at the bottom of the mould, therefore contacting the mould purely along the line of contact along its base. Thus for this rather larger casting than the 25 mm thick plate cast by Woodbury, it seems an air gap definitely occurs in a sand mould. This result confirms Lukens earlier result (1990) in which most heat appears to be extracted through the drag, and increased head pressure enhances metal/mould contact.

Attempts to measure the gap formation directly (Isaac et al., 1985; Majumdar and Raychaudhuri, 1981) are difficult to carry out accurately. Results averaged for aluminium cast into cast iron dies of various thickness reveal the early formation of the gap at the corners of the die where cooling is fastest, and the subsequent spread of the gap to the centre of the die face. A surprising result is the reduction of the gap if thick mould coats are applied. (The results in Figure 5.5 are plotted as straight lines. The apparent kinks in the early opening of the gap reported by these authors may be artefacts of their experimental method.)

It is not easy to see how the gap can be affected by the thickness of the coating. The effect may be the result of the creep of the solid shell under the internal hydrostatic pressure of the feeder. This is more likely to be favoured by thicker mould coats as a result of the increased time available and the increased temperature of the solidified skin of the casting. If this is true, then the effect is important because the hydrostatic head in these experiments was modest, only about 200 mm. Thus, for aluminium alloys that solidify with higher heads and times as long or longer than a minute or so, this mechanism for gap reduction will predominate. It seems possible, therefore, that in gravity die casting of aluminium the die coating will have the major influence on heat transfer, giving a large and stable resistance across the interface. The air gap will be a small and variable contributor. For computational purposes, therefore, it is attractive to consider the great simplification of neglecting the air gap in the special case of gravity die casting of aluminium.

It is probably helpful to draw attention to the fact that the name 'air gap' is perhaps a misnomer. The gap will contain almost everything except air. As we have seen previously, mould gases are often high in hydrogen, containing typically 50%. At room temperature, the thermal conductivity of hydrogen is approximately 5.9 times higher than that of air, and at 500°C the ratio rises to 7.7. Thus, the conductivity of a gap at the casting-mould interface containing a 50:50 mixture of air and hydrogen at 500°C can be estimated to be approximately a factor of four higher than that of air. In the past, therefore, most investigators in this field have probably chosen the wrong value for the conductivity of the gap, and by a substantial margin!

This effect has been used by Doutre (1998) who injected helium gas, with a conductivity nearly identical to hydrogen (approximately seven times the conductivity of air), into the air gap of a large Al alloy intake manifold cast in a permanent mould. The gas was introduced 30 s after pouring at a rate in the region of 10 mLs^{-1}. In a full-scale works trial, the production rate of the casting was increased by 25%. An even larger potential productivity increase, 45%, was found

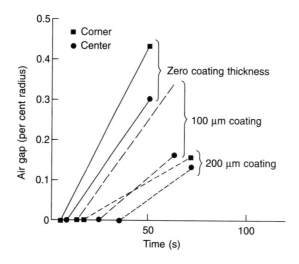

FIGURE 5.5

Results averaged from various dies (Isaac et al., 1985), illustrating the start of the air gap at the corners, and its spread to the centre of the mould face. Increased thickness of mould coating is seen to delay solidification and to reduce the growth of the gap.

by Grandfield's team (2007) when casting horizontal ingots in an open iron ingot mould. Gebelin and Griffiths (2007) found a reduced effect when attempting to cast resin-bonded sand moulds in an evacuated chamber back-filled with He. The explanation of the poor result in this latter case is almost certainly the result of the outgassing of sand moulds under vacuum. This outward wind of volatiles would prevent the ingress of He.

In passing, it seems worth commenting that He is expensive, and world supplies are limited. Naturally pure hydrogen introduced into the gap would perform similarly, but involve an unacceptable danger in the work place. Almost the same effect would be expected if steam, or better still, a water mist, were introduced into the air gap of a permanent mould. It would react with the metal to form hydrogen in situ, precisely where it is needed. Alternatively, many sand moulds, such as greensand or sodium silicate bonded sands, have sufficient water that the hydrogen atmosphere is provided automatically, and free of charge.

The heat transfer coefficient

The authors Ho and Pehlke (1984) from the University of Michigan have reviewed and researched this area thoroughly. We shall rely mainly on their analytical approach for an understanding of the heat transfer problem.

When the metal first enters the mould the macroscopic contact is good because of the conformance of the molten metal to the detailed shape of the mould. Gaps exist on a microscale between high spots as shown in Figure 5.6. At the high spots themselves, the high initial heat flux causes nucleation of the solid metal by local severe undercooling (Prates and Biloni, 1972). The nucleated solid then spreads to cover most of the surface of the casting because the thin layer of liquid adjacent to the cool mould surface will be expected to be undercooled. Conformance and overall contact between the surfaces is expected to remain good during all of this early period, even though both the casting and mould will now be starting to move rapidly because of distortion.

After the creation of a solidified layer with sufficient strength, further movements of both the casting and the mould are likely to cause the good fit to be broken, so that contact is maintained across only a few widely spaced random high spots (Figure 5.6(b)). The total transfer of heat across the interface h_t may now be written as the sum of three components:

$$h_t = h_s + h_c + h_r$$

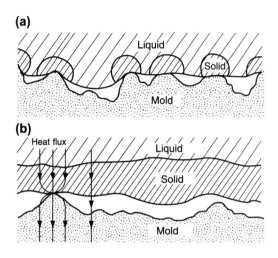

FIGURE 5.6

Metal-mould interface at an early stage when solid is nucleating at points of good thermal contact. Overall macroscopic contact is good at this stage (a). Later (b) the casting gains strength, and casting and mould both deform, reducing contact to isolated points at greater separations on non-conforming rigid surfaces.

where h_s is the conduction through the *solid* contacts, h_c is the *conduction* through the gas phase and h_r is that transferred by *radiation*. Ho and Pehlke produce analytical equations for each of these contributors to the total heat flux. We can summarise their findings as follows:

1. While the casting surface can conform, the contribution of solid–solid conduction is the most important. In fact, if the area of contact is enhanced by the application of pressure, then values of h_t up to 60,000 $Wm^{-2}K^{-1}$ are found for aluminium in squeeze casting. Such high values are quickly lost as the solid thickens and conformance is reduced, the values falling to more normal levels of 100–1000 $Wm^{-2}K^{-1}$ (Figure 5.7).
2. When the interface gap starts to open, the conduction via any remaining solid contacts becomes negligible. The point at which this happens is clear in Figure 5.7(b). (The actual surface temperature of the casting and the mould in this figure are reproduced from the results calculated by Ho and Pehlke.) The rapid fall of the casting surface

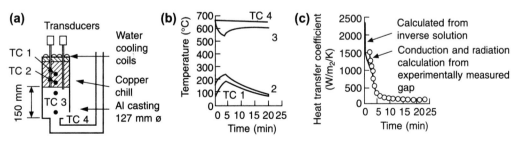

FIGURE 5.7

The experiment by Ho and Pehlke (1984) showing (a) their bottom gated mould; (b) the temperature history recorded across a casting-chill interface; and (c) the inferred heat transfer coefficient.

temperature is suddenly halted, and reheating of the surface starts to occur. An interesting mirror image behaviour can be noted in the surface temperature of the mould that, now out of contact with the casting, starts to cool. The estimates of heat transfer are seen to simultaneously reduce from over 1000 to around 100 $Wm^{-2}K^{-1}$ (Figure 5.7(c)).

3. After solid conduction diminishes, the important mechanism for heat transfer becomes the conduction of heat through the gas phase. This is calculated from:

$$h_c = k/d$$

where k is the thermal conductivity of the gas and d is the thickness of the gap. An additional correction is noted by Ho and Pehlke for the case where the gap is smaller than the mean free path of the gas molecules, which effectively reduces the conductivity. Thus heat transfer now becomes a strong function of gap thickness. As we have noted previously, it will also be a strong function of the composition of the gas. Even a small component of hydrogen will greatly increase the conductivity. Note also that it is assumed, almost certainly accurately, that the gas is stationary, providing heat flow by conduction only, not by convection.

For the case of light alloys, Ho and Pehlke find that the contribution to heat transfer from radiation is of the order of 1% of that from conduction by gas. Thus radiation can be safely neglected at these temperatures.

For higher-temperature metals, results by Jacobi (1976) from experiments on the casting of steels in different gases and in vacuum indicate that radiation becomes important to heat transfer at these higher temperatures.

Turning now to experimental work on the effect of die coatings on permanent moulds on heat transfer coefficients, a comprehensive review has been made by Nyamekye et al. (1994). Chiesa (1990) found that the conductance of a black coat was roughly twice that of a white coat of moderate thickness in the region of 120 μm. Also, the insulating effect of a white coat increased only marginally with thickness. Their findings that coats with high surface roughness were more effective insulators have been confirmed by the calculations of Hallam and Griffiths (2000) for the case of Al alloy castings. They demonstrate excellent predictions based on the assumption that the resistance of the die coating is mainly from the gas voids between the casting and the coating surface. Thus the structure of the coating surface was a highly influential factor in determining the heat transfer across the casting/mould interface.

The effect of gravity on the contact between the casting and mould has already been discussed previously. Woodbury (1998) finds that for a lightweight horizontal plate casting of only 6 mm thickness the heat transfer settles to a constant level of 70 $Wm^{-2}K^{-1}$. However, for a 25 mm thick plate the heat transfer from the top is approximately unchanged at 70, but the value from the underside of this heavier plate is now approximately doubled at 140 $Wm^{-2}K^{-1}$.

For sand moulds the use of pressure to enhance metal/mould contact is of course limited by penetration of the metal into the sand. The use of pressure has no such limitation in the case of metal moulds. Tadayon (1992) reports the freezing time of a squeeze casting to be 84 s at zero applied pressure, but reducing to 56 s at 5 MPa. A further increase of pressure to 10 MPa reduced the time only minimally further to 54 s.

Finally, it seems necessary to draw attention to the comprehensive review by Woolley and Woodbury (2007). These authors critically assess the vast literature on the determination of heat transfer coefficients, concluding that they are sceptical of the accuracy of all of the published data, and cite a number of key reasons for unreliability. In a later paper (2009), for instance, these authors find that the use of thermocouples to measure heat flow in these experiments alone introduces an error of 65%. It seems, we still have a long way to go to achieve reliable heat transfer coefficients.

Resistance 3: The mould

The rate of freezing of castings made in silica sand moulds is generally controlled by the rate at which heat can be absorbed by the mould. In fact, compared with many other casting processes, the sand mould acts as an excellent insulator, keeping the casting hot. However, of course, ceramic investment and plaster moulds are even more insulating, avoiding premature cooling of the metal, and aiding fluidity to give the excellent ability to fill thin sections for which these casting processes are renowned. It is a pity that the extremely slow cooling generally contributes to rather poorer mechanical properties, but this is to some extent a self-inflicted problem. If the metal were free from bifilms it is predicted that mechanical properties would be not noticeably affected by slower rates of freezing (see Chapter 9.4).

5.1 HEAT TRANSFER

Considering the simplest case of unidirectional conditions once again, and metal poured at its melting point T_m against an infinite mould originally at temperature T_0, but whose surface is suddenly heated to temperature T_m at $t = 0$, and that has thermal diffusivity α_m, we now have:

$$\frac{\partial T}{\partial t} = \alpha_m \frac{\partial^2 T}{\partial x^2} \qquad (5.9)$$

Following Flemings, the final solution is:

$$S = \frac{2}{\sqrt{\pi}} \underbrace{\left(\frac{T_m - T_0}{\rho_s H}\right)}_{\text{metal}} \underbrace{\sqrt{K_m \rho_m C_m}}_{\text{mould}} \sqrt{t} \qquad (5.10)$$

This relation is most accurate for the highly conducting non-ferrous metals, aluminium, magnesium and copper. It is less good for iron and steel, particularly those ferrous alloys that solidify to the austenitic (face centred cubic) structure that has especially poor conductivity. The relation quantifies a number of interesting outcomes as discussed next.

Note that at a high temperature heat is lost more quickly, so that a casting in steel should solidify faster than a similar casting in grey iron. This perhaps surprising conclusion is confirmed experimentally, as seen in Figure 5.8.

Low heat of fusion of the metal, H, similarly favours rapid freezing because less heat has to be removed. Magnesium castings therefore freeze faster than similar castings in aluminium despite their similar freezing points (Table 5.1).

The product $K_m \rho_m C_m$ is a useful parameter to assess the rate at which various moulding materials can absorb heat. The reader needs to beware that some authorities have called this parameter the heat diffusivity, and this definition was followed in CASTINGS (Campbell, 1991). However, originally the definition of heat diffusivity b was $(K_m \rho_m C_m)^{1/2}$ as described for instance by Ruddle (1950). In subsequent years, the square root seems to have been overlooked in error. Ruddle's definition is therefore accepted and followed here. However, of course, both b and b^2 are useful quantitative measures. What we call them is merely a matter of definition. (I am grateful to John Berry of Mississippi State University for pointing out this fact. As a further aside from Professor Berry, the units of b are even more curious than the units of toughness; see Table 5.2.)

For simple shapes, if we assume that we may replace S with V_s/A where V_s is the volume solidified at a time t, and A is the area of the metal-mould interface (i.e. the cooling area of the casting), then when $t = t_f$ where t_f is the total freezing time of a casting of volume V we have:

$$\frac{V}{A} = \frac{2}{\sqrt{\pi}} \left(\frac{T_m - T_0}{\rho_s H}\right) \sqrt{K_m \rho_m C_m} \sqrt{t_f} \qquad (5.11)$$

and so:

$$t_f = B(V/A)^2 \qquad (5.12)$$

where B is a constant for given metal and mould conditions. The ratio (V/A) is the useful parameter generally known as the *modulus m;* thus, Eqn (5.12) indicates that the parameter m^2 is the important factor that controls the solidification time of the casting. Approximate values of m for simple shaped castings as illustrated in Table 5.3 are usefully memorised.

Equation (5.12) is the famous Chvorinov rule. Convincing demonstrations of its accuracy have been made many times. Chvorinov himself showed in his paper published in 1940 that it applied to steel castings from 12 to 65,000 kg weight made in greensand moulds. This superb result is presented in Figure 5.9. Experimental results for other alloys are illustrated in Figure 5.8.

Chvorinov's rule is one of the most useful guides to the student. It provides a powerful general method of tackling the feeding of castings to ensure their soundness.

However, the previous derivation of Chvorinov's rule is open to criticism in that it uses one-dimensional theory but goes on to apply it to three-dimensional castings. In fact, it is quickly appreciated that the flow of heat into a concave

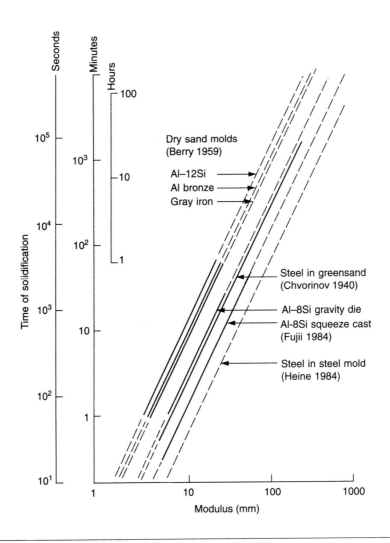

FIGURE 5.8

Freezing times of plate shaped castings in different alloys and moulds.

mould wall will be divergent, and so will be capable of carrying away heat more rapidly than in a one-dimensional case. We can describe this exactly (without the assumption of one-dimensional heat flow), following Flemings once again:

$$\frac{\partial T}{\partial t} = \alpha_m \left(\frac{\partial^2 T}{\partial r^2} + \frac{n \partial T}{r \partial r} \right) \tag{5.13}$$

where $n = 0$ for a plane, 1 for a cylinder, and 2 for a sphere. The casting radius is r. The solution to this equation is:

$$\frac{V}{A} = \left(\frac{T_m - T_0}{\rho_s H} \right) \left(\frac{2}{\sqrt{\pi}} \sqrt{K_m \rho_m C_m} \sqrt{t_f} + \frac{n K_m t_f}{2r} \right) \tag{5.14}$$

Table 5.1 Mould and Metal Constants

Material	Melting Point (m.p.) (°C)	Latent Heat Melting (J/g)	Liquid–Solid Contraction (%) a	Liquid–Solid Contraction (%) b	Specific Heat (J/kg K) Solid 20°C	Specific Heat (J/kg K) Solid m.p.	Specific Heat (J/kg K) Liquid m.p.	Density (kg/m³) 20°C	Density (kg/m³) Solid m.p.	Density (kg/m³) Liquid m.p.	Thermal Conductivity (J/m K s) 20°C	Thermal Conductivity (J/m K s) Solid m.p.	Thermal Conductivity (J/m K s) Liquid m.p.
Pb	327	23	3.22	3.20	130	(138)	152	11,680	11,020	10,678	39.4	(29.4)	15.4
Zn	420	111	4.08	4.08	394	(443)	481	7140	(6843)	6575	119	95	9.5
Mg	650	362	4.2	4.21	1038	(1300)	1360	1740	(1657)	1590	155	(90)?	78
Al	660	388	7.14	6.92	917	(1200)	1080	2700	(2550)	2385	238	–	94
Cu	1084	205	5.30	4.78	386	(480)	495	8960	8382	8000	397	(235)	166
Fe	1536	272	3.16	3.56	456	(1130)	795	7870	7265	7015	73	(14)?	–
Graphite	–	–	–	–	1515	–	–	2200	–	–	147	–	–
Silica sand	–	–	–	–	1130	–	–	1500	–	–	0.0061	–	–
Mullite	–	–	–	–	750	–	–	1600	–	–	0.0038	–	–
Plaster	–	–	–	–	840	–	–	1100	–	–	0.0035	–	–

References: Brandes (1991), Flemings (1974)
[a]Wray (1976)
[b]From densities presented here.

Table 5.2 Thermal Properties of Mould and Chill Materials at Approximately 20°C

Material	Heat Diffusivity $(K\rho C)^{1/2}$ $(Jm^{-2}K^{-1}s^{-1/2})$	Thermal Diffusivity $K/\rho C$ (m^2s^{-1})	Heat Capacity per Unit Volume ρC $(JK^{-1}m^{-3})$
Silica sand	3.21×10^3	3.60×10^{-9}	1.70×10^6
Investment	2.12×10^3	3.17×10^{-9}	1.20×10^6
Plaster	1.8×10^3	3.79×10^{-9}	0.92×10^6
Magnesium	16.7×10^3	85.8×10^{-6}	1.81×10^6
Aluminium	24.3×10^3	96.1×10^{-6}	2.48×10^6
Copper	37.0×10^3	114.8×10^{-6}	3.60×10^6
Iron (pure Fe)	16.2×10^3	20.3×10^{-6}	3.94×10^6
Graphite	22.1×10^3	44.1×10^{-6}	3.33×10^6

Table 5.3 Moduli of Some Common Shapes

Shape			Modulus 100% Cooled Area		Base Uncooled	
Sphere	D		D/6	0.167D	—	—
Cube	D		D/6	0.167D	D/5	0.200D
Cylinder	H, D	H/D 1.0	D/6	0.167D	D/5	0.200D
		1.5	3D/16	0.188D	3D/14	0.214D
		2.0	D/5	0.200D	2D/9	0.222D
Infinite cylinder		∞	D/4	0.250D	—	—
Infinite plate	D		D/2	0.500D	—	—

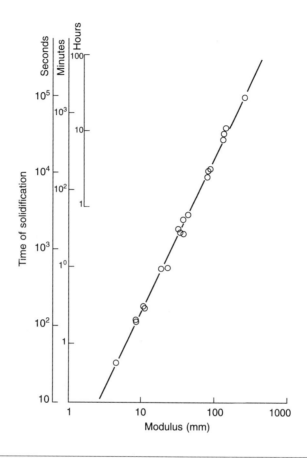

FIGURE 5.9

Freezing time of steel castings in greensand moulds as a function of modulus (Chvorinov, 1940).

The effect of the divergency of heat flow predicts that for a given value of the ratio V/A (i.e. a given modulus m) a sphere will freeze quickest, the cylinder next, and the plate last. Katerina Trbizan (2001) provides a useful study, confirming these relative freezing rates for these three shapes. For aluminium in sand moulds, Eqn (5.14) indicates these differences to be close to 20%. This is part of the reason for the safety factor 1.2 recommended when applying Chvorinov's feeding rule, because the feeding rules tacitly assume that all shapes with the same modulus freeze at the same time.

The simple Chvorinov link between modulus and freezing time is capable of great sophistication. One of the great exponents of this approach has been Wlodawer (1966), who produced a famous volume devoted to the study of the problem for steel castings. This has been a source book for the steel castings industry ever since.

The subject has been advanced further by the work of Tiryakioglu in 1997 (interestingly using his deceased father's excellent doctoral research at the University of Birmingham, UK, in 1964) that showed secondary, but important, effects of shape, volume and superheat on the freezing time of the casting.

A final aspect relating to the divergency of heat flow is important. For a planar freezing front, the rate of increase of thickness of the solidified metal is parabolic, gradually slowing with thickness, as described by Equations such as 5.3 and 5.4 relating to 1-dimensional heat flow. However, for more compact shapes such as cylinders, spheres, cubes, etc., the heat flow from the casting is 3-dimensional. Thus initially for such shapes, when the solidified layer is relatively thin, the solid thickens parabolically. However, at a much later stage of freezing, when little liquid remains in the centre of

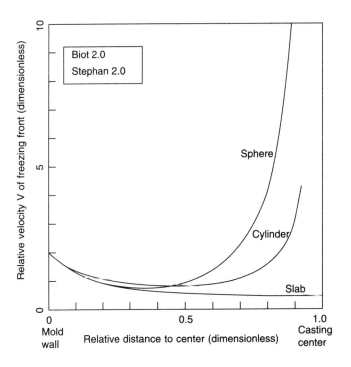

FIGURE 5.10

Acceleration of the freezing front in compact castings as a result of three-dimensional extraction of heat (sphere and cylinder curves calculated from Santos and Garcia, 1998).

the casting, the extraction of heat in all three directions greatly accelerates the rate of freezing. Santos and Garcia (1998) show that the effect, accurately predicted theoretically by Adams in 1956, is general. Whereas in a slab casting the velocity of the front slows progressively with distance according to the well-known parabolic law, for cylinders and spheres the growth rate is similar until the front has progressed to about 40% of the radius. From then onwards the front accelerates rapidly (Figure 5.10).

This increase of the rate of freezing in the interior of many castings explains the otherwise baffling observation of 'inverse chill' as seen in cast irons. Normal intuition would lead the caster to expect fast cooling near the surface of the casting, and this is true to a modest degree in all castings. From this point onwards the front slows progressively in uniform plate-like sections. But in bars and cylinders, as the residual liquid shrinks in size towards the centre of the casting, the front speeds up dramatically, causing grey iron to change to carbidic white iron. The accelerated rate has been demonstrated experimentally by Santos and Garcia on a Zn-4Al alloy by measurement of the increasing fineness of dendrite arm spacing towards the centre of a cylindrical casting.

It is interesting that the effect of accelerated freezing appears never to have been seen in Al alloys. This seems to be the result of the high thermal conductivity of these alloys, causing dendritic freezing over the whole cross-section of the casting, and thus smoothing and obscuring the acceleration of solidification towards the centre of castings.

5.1.2 INCREASED HEAT TRANSFER

In practice, the casting engineer can manipulate the rate of heat extraction from a casting using a number of tricks. These include the placement of blocks of metal in the mould, adjacent to the casting, to act as local chills. Similarly, fins or pins attached to the casting increase the local surface area from which heat can be conducted deep into the mould where it can be dissipated. These techniques will be described next.

The action of chills, fins and pins can provide localised cooling of the casting to assist directional solidification of the casting towards the feeder, thus assisting in the achievement of soundness. This is one of the important actions of these chilling devices. It is, however, not the only action, as discussed next.

The use of any of these chilling techniques acts to increase the ductility and strength of that locality of the casting. It seems most probable that this occurs because the faster solidification freezes in the bifilms in their compact form before they have chance to unfurl. (Recall that the bifilms are compacted by the extreme bulk turbulence during pouring and during their travel through the running system. However, they subsequently unravel if they have sufficient time or sufficient driving force, gradually opening up in the mould cavity when conditions in the melt become quiet once again, so as to gradually reduce properties.) However, there is a small contribution towards strength and ductility from the refinement of structure. Al-Si alloys and Mg-based alloys particularly benefit. The remaining Al alloys and other fcc structured metals such as Cu-based alloys and austenitic steels would not be expected to benefit usefully from finer grain or finer DAS.

The interesting corollary of this fact is that if chills are seen to increase ductility and strength of these alloys, it confirms that the cast material is defective, containing a high percentage of bifilms. Another interesting corollary is that if the alloy can be cast without bifilms, its properties will already be high, so that chilling should not increase its properties further. This rather surprising prediction is fascinating, and, if true, indicates the huge potential for the increase of the properties of cast alloys. All castings without bifilms are therefore predicted to have extraordinary ductility and strength. It also explains our lamentable current condition in which most of us constantly struggle in our foundries to achieve minimum mechanical properties for castings even when the casting is plastered in chills. The common experience when attempting to achieve minimum properties is that some days we win, other days we continue to struggle. The message is clear, we need to focus on technologies for the production of castings with reduced bifilm content, preferably zero bifilm content. The rewards are huge.

Another action of chills is to straighten bifilms. This unfortunate action occurs because the advancing dendrites cannot grow through the air layer between the double films, and so push the bifilms ahead. Those that are somehow attached to the wall will be partially pushed, straightened and unravelled by the gentle advance of grains. This effect is reported in Section 2.3 on furling and unfurling of bifilms. Thus although a large percentage of bifilms will be pushed ahead of the chilled region, concentrating (and probably reducing the properties) in the region immediately ahead, some bifilms will remain aligned in the dendrite growth direction, and so be largely perpendicular to the mould wall. The mechanism is presented in Figure 2.43(a) and (b), and an example is seen in the radiograph in Figure 2.43(c).

The overall effects on mechanical properties of the pushing action are therefore not easily predicted. The reduction in density of defects close to the chill will raise properties, but the presence of occasional bifilms aligned at right angles to the surface of the casting would be expected to be severely detrimental. These complicated effects require to be researched. However, we can speculate that they seem likely to be the cause of troublesome edge cracking in the rolling of cast materials of many types, leading to the expense of machining off the surface of many alloys before rolling can be attempted. The superb formability of electroslag remelted (ESR) compared to vacuum arc remelted (VAR) alloys is almost certainly explained in this way. The ESR process produces an extremely clean material because oxide films are dissolved during remelting under the layer of liquid slag, and will not re-form in the solidifying ingot. In contrast, the relatively poor vacuum of the VAR process ensures that the lapping of the melt over the liquid meniscus at the mould wall will create excellent double oxide films. If considerable depths of the VAR ingot surface are not first removed (a process sometimes called 'peeling') oxide lap defects will open as surface cracks when subjected to forging or rolling.

Although, as outlined previously, the chilling action of chills and fins is perhaps more complicated than we first thought, the chilling action itself on the rate of solidification is well documented and understood. It is this thermal aspect of their behaviour that is the subject of the remainder of this section.

External chills

In a sand mould, the placing of a block of metal on the pattern, and subsequently packing the sand around it to make the rest of the mould in the normal way, is a widely used method of applying localised cooling to that part of the casting.

180 CHAPTER 5 SOLIDIFICATION STRUCTURE

A similar procedure can be adopted in gravity and low-pressure die-casting by removing the die coat locally to enhance the local rate of cooling. In addition, in dies of all types, this effect can be enhanced by the insertion of metallic inserts into the die to provide local cooling, especially if the die insert is highly conductive (such as made from copper alloy) and/or artificially cooled, for instance by air, oil or water.

Such chills placed as part of the mould, and that act against the outside surface of the casting are strictly known as external chills, to distinguish them from internal chills that are cast in, and become integral with, the casting.

In general terms, the ability to chill is a combination of ability to absorb heat and to conduct it away. It is quantitatively assessed by

$$\text{heat diffusivity} = (K\rho C)^{1/2}$$

where K the thermal conductivity, ρ the density, and C is the specific heat of the mould. It has complex units $\text{Jm}^{-2}\text{K}^{-1}\text{s}^{1/2}$. Take care not to confuse with

$$\text{thermal diffusivity} = K/\rho C$$

normally quoted in units of m^2s^{-1}.

From the room temperature data in Table 5.1 (unfortunately, high temperature values are less easily obtained) we can obtain some comparative data on the chilling power of various mould and chill materials, shown in Table 5.2. It is clear that the various refractory mould materials—sand, investment and plaster—do not act effectively as chilling materials. The various chill materials are all in a league of their own, having chilling powers orders of magnitude higher than the refractory mould materials. They improve marginally, within a mere factor of 5, in the order steel, graphite and copper.

The heat diffusivity value indicates the action of the material to absorb heat when it is infinitely thick, being unconstrained in the amount of heat it can conduct away and store in itself i.e. as would be reasonably well approximated by constructing a thick-walled mould from such material.

This behaviour contrasts with that of a relatively small lump of cast iron or graphite used as an external chill in a sand mould. A small chill does not develop its full potential for chilling as promised by the heat diffusivity because it has limited capacity for heat. Thus although the initial rate of freezing of a metal may be in the order given by the previous list, for a chill of limited thickness its cooling effect quickly becomes limited because it becomes saturated with heat; after a time it can absorb no more. The amount of heat that it can absorb is defined as its heat capacity. We can formulate the useful concept of heat capacity per unit volume ρC in terms of its density ρ and its specific heat C, so that the heat capacity of a chill of volume V is simply

$$\text{volumetric heat capacity} = V\rho C$$

In the SI system, its units are JK^{-1}. Figure 5.11 illustrates the fact that if chills are limited by their relatively low heat capacity, there is little difference between copper, graphite and iron. However, for larger chills that are able to conduct heat away without saturating, copper is by far the best material. The next best, graphite, is only half as good, and iron is only a quarter as effective. For situation intermediate between small and large chills conditions partway between the extremes can be read off the nomogram.

Aluminium chills are interesting, in that if rather small, they are relatively poor compared with steel, graphite and copper, whereas larger blocks capable of carrying heat away without saturation are better than steel or graphite, but still only half as good as copper.

The results by Rao and Panchanathan (1973) on the casting of 50 mm thick plates in Al-5Si-3Cu reveals that the casting is insensitive to whether it is cooled by copper, graphite or steel chills, provided that the volumetric heat capacity of the chill is taken into account. We can conclude that their chills were rather limited in size, and so limited by their heat capacity.

These authors show that for a steel chill 25 mm thick its heat capacity is 900 JK^{-1}. A chill with identical capacity in copper they claimed to be 32 mm thick, and in graphite 36 mm. These values originally led the author to conclude that copper may therefore not always the best chill material (CASTINGS 1991). However, using somewhat more accurate data (Figure 5.11) copper is found, after all, to be the most effective whether limited by heat diffusivity or heat capacity.

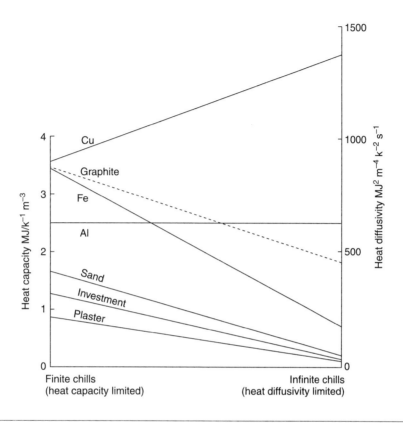

FIGURE 5.11

Relative diffusivities (ability to diffuse heat away if a large chill) and heat capacities (ability to absorb heat if relatively small) of chill materials.

Lerner and Kouznetsov (2004) confirm for copper and iron chills on a 20 mm thick Al alloy casting the rather surprising fact that the volume of external chills is far more important than their surface area of contact with the casting.

Figure 5.12 illustrates that the chills are effective over a considerable distance, the largest chills greatly influencing the solidification time of the casting even up to 200 mm (four times the section thickness of the casting) distant. This large distance is perhaps typical of such a thick-section casting in an alloy of high thermal conductivity, providing excellent heat transfer along the casting. A steel casting would respond less at this distance. This work by Rao and Panchanathan (1973) reveals the widespread sloppiness of much present practice on the chilling of castings. General experience of the chills generally used in foundrywork nowadays shows that chill size and weight are rarely specified, and that chills are in general too small to be fully effective in any particular job. It clearly matters what size of chill is added.

Computational studies by Lewis and colleagues (2002) have shown that the number, size and location of chills can be optimised by computer. These studies are among the welcome first steps towards the intelligent use of computers in casting technology.

Finally, in detail, the action of the chill is not easy to understand nor to predict. The surface of the casting against the chill will often contract, distorting away and thus opening up an air gap. The chilled casting surface may then reheat to such an extent that the surface of the casting remelts. The exudation of eutectic is often seen between the casting and the

182 CHAPTER 5 SOLIDIFICATION STRUCTURE

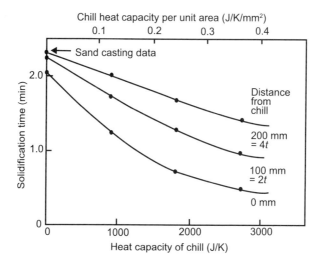

FIGURE 5.12

Freezing times of a plate 225 × 150 × 50 mm in Al-5Si-3Cu alloy at various distances from the chill is seen to shrink steadily as the chill is approached and as the chill size is increased (Rao and Panchanathan, 1973).

chill (Figure 5.13). The new contact between the eutectic and chill probably then starts a new burst of heat transfer and thus a new rapid phase of solidification of the casting. Thus the history of cooling in the neighbourhood of a chill may be a succession of stop/start, or slow/fast events.

Internal chills

The placing of chills inside the mould cavity with the intention of casting them in place is an effective way of localised cooling. Such chills are usually carefully chosen to be of the same chemical composition as the melt so that they will be essentially invisible when finally frozen into the casting. The technique is an excellent solution to the challenge of making sound a heavy boss or thick section in the centre of a complex casting that cannot be accessed by conventional feeding, especially if the centre of the boss, together with the internal chill, is to be subsequently machined out.

Internal chills take a great variety of shapes as illustrated in Figure 5.14.

When estimating what size of chill might be required (i.e. its weight), the simple method of mixtures approach (Campbell and Caton, 1977) indicates that to cool superheated pure liquid iron to its freezing point, and freezing a proportion of it, will require various levels of addition of cold, solid low carbon steel depending on the extent that the addition itself actually melts. These estimations take no account of other heat losses from the casting. Thus for normal castings the predictions are likely to be incorrect by up to a factor of 2. This is broadly confirmed by Miles (1956), who top-poured steel into dry sand moulds 75 mm square and 300 mm tall. In the centre of the moulds was positioned a variety of steel bars ranging from 12.5 mm round to 25 mm square, covering a range of chilling from 2% to 11% solid addition. His findings reveal that the 2% solid addition nearly melted, compared with the predicted value for complete melting of 3.5% solid. The 11% solid addition caused extensive (possibly total) freezing of the casting judging by the appearance of the radial grain structure in his macrosections. He found 5% addition to be near optimum; it had a reasonable chilling effectiveness and caused relatively few defects.

These additions are plotted in graphical form in Figure 5.15. They illustrate that additions up to about 3.5% completely melt and disappear (although clearly cool the melt by this process). Further additions linearly increase the amount of liquid that is solidified, finally becoming completely frozen at about 10% addition.

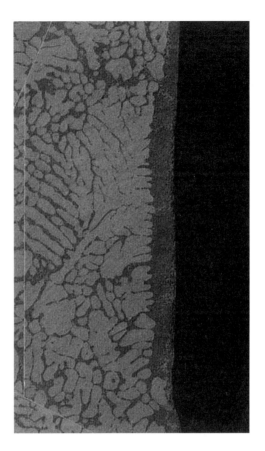

FIGURE 5.13

Al-Si eutectic liquid segregation by exudation at a chilled interface of an Al-7Si alloy.

In the case of additions higher than 10%, where the heat input is not sufficient to melt the chill, the fusing of the chill surface into the casting has to be the result of a kind of diffusion bonding process. This would emphasise the need for cleanness of the surface, requiring the minimum presence of oxide films or other debris against the chill during the filling of the mould. If Miles had used a better bottom gated filling technique he may have reduced the observed filling defects further, and found that higher percentages were practical.

The work by Miles does illustrate the problems generally experienced with internal chills. If the chills remain for any length of time in the mould, particularly after it is closed, and more particularly if closed overnight, then condensation is likely to occur on the chill, and blow defects will be caused in the casting. Blows are also common from rust spots or other impurities on the chill such as oil or grease. The matching of the chemical composition of the chill and the casting is also important; mild steel chills will, for instance, usually be unacceptable in an alloy steel casting.

Internal chills in aluminium alloy castings have not generally been used, almost certainly as a consequence of the difficulty introduced by the presence of the oxide film on the chill. This appears to be confirmed by the work of Biswas et al. (1985), who found that at 3.5% by volume of chill and at superheats of only 35°C the chill was only partially melted and retained part of its original shape. It seems that over this area it was poorly bonded. At superheats of more than 75°C, or at only 1.5% by volume, the chill was more extensively melted, and was useful in reducing internal porosity and in

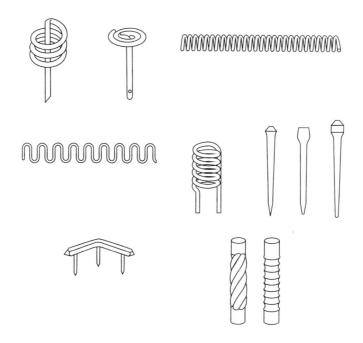

FIGURE 5.14

A variety of internal chills.

From Heine and Rosenthal (1955).

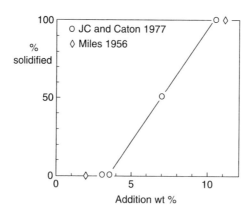

FIGURE 5.15

The addition of increasing percentage of cold mild steel particles to liquid mild steel cools the melt by melting of the additions up to about 4% addition, then starts to promote freezing, finally causing complete freezing of the melt at about 10% addition.

raising mechanical properties. The lingering presence of the oxide film from the chill (now having floated off to lurk elsewhere in the casting) remains a concern however.

The development of a good bond between the internal chill and the casting is a familiar problem with the use of chaplets – the metal devices used to support cores against sagging because of weight, or floating because of buoyancy. A one-page review of chaplets is given by Bex (1991). To facilitate the bond for a steel chaplet in an iron or steel casting the chaplet is often plated with tin. The tin serves to prevent the formation of rust, and its low melting point (232°C) and solubility in iron assists the bonding process.

There has been much work carried out on the casting of steel inserts into cast irons. Xu (2007) studied thin steel chill plates in a heavy grey iron block casting, measuring the diffusion of alloying elements between them. He notes that carbon content is smoothed across the interface, but the silicon concentration forms a sharp step at the interface. This is behaviour to be expected in terms of the rates of diffusion given in Figure 1.4(c).

Noguchi (1993) used a thermally sprayed Ni-based self-fluxing alloy on mild steel inserts in both flake and ductile iron. Although he reports that the bonding of inserts is sensitive to the volume ratio of the poured metal and insert (as we have seen in Figure 5.15) and the pouring temperature, the thermal spray greatly expands the regime of successful bonding. In later work, Noguchi (2005) cast steel sleeves and end rings into cast irons, avoiding the use of sprayed interlayers. In this work he raised the temperature of the insert higher than could be achieved by simple immersion. He arranged for an extended period of flow of hot metal around the insert, in some cases running the excess metal into flow-offs. He found that if he could raise the interface temperature to 1150°C, the melting point of the iron, he could achieve what appeared to be a perfect bond over large areas of the insert.

The bond between steel and titanium inserts in Al alloy castings has been investigated in Japan (Noguchi et al., 2001) who found only a 10 μm silver coating was effective to achieve a good bond, although even this took up to 5 min to develop at the Al-Ag eutectic temperature 566°C. Attempts to achieve a bond with gold plating and Al-Si sprayed alloy were largely unsuccessful.

Biswas and co-workers (1994) have researched Al alloy chills in Al alloys as a function of relative volume and casting superheat. However, these authors overlook the problem of the oxide films, not appreciating that it represents an ever-present danger. It will persist as a double film (having acquired its second layer during the immersion of the chill by the liquid) and so pose the risk of leakage or crack formation. Such risks are only acceptable for low duty products.

Brown and Rastall (1986) take advantage of the oxide on the surface of heavy aluminium inserts in aluminium castings, using this to avoid any bonding between the casting and the insert. They use a cast aluminium alloy core inside an aluminium alloy casting to form re-entrant details that could not easily be provided in a pressure die cast product. Also, of course, because the freezing time is shortened, productivity is enhanced. The internal core is subsequently removed by disassembly or part machining, or by mechanical deformation, peeling apart the oxide bifilm separating the core and the casting.

Fins

If we add an appendage such as a fin to a casting, we inevitably form a T-junction between the two. Depending on the thickness of the appendage, the junction may be either a hot spot or a cold spot. Therefore, before we look specifically at fins on castings, it is worth spending some time to consider the concepts involved in junctions of all types.

Kotschi and Loper (1974) were among the first to evaluate junctions. Their results are summarised in Figures 5.16 and 5.17 and further interpreted in Figure 5.18 to show the complete range of junctions and their effect on the residual liquid in the main cast plate. Considering the range in Figure 5.18, starting at the thinnest appendage:

1. When the wall forming the upright of the T is thin, it acts as a cooling fin, chilling the junction and the adjacent wall (the top cross of the T) of casting. We shall return to a more detailed consideration of fins shortly.
2. When the upright of the T-section has increased to a thickness of half the casting section thickness, then the junction is close to thermal balance, the cooling effect of the fin balancing the hot-spot effect of the concentration of metal in the junction.
3. By the time that the upright of the T has become equal to the casting section, the junction is a hot spot. This is common in castings. Foundry engineers are generally aware the 1:1 T-junction is a problem. It is curious therefore

186 CHAPTER 5 SOLIDIFICATION STRUCTURE

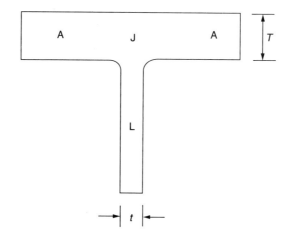

FIGURE 5.16

Geometry of a T-shaped junction.

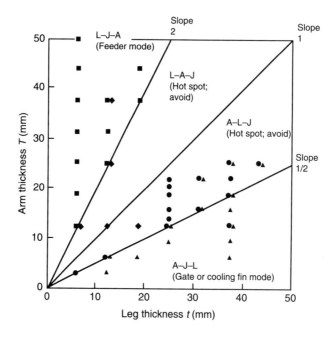

FIGURE 5.17

Solidification sequence for T-shaped castings (A = arm, J = junction, L = leg).

Experimental data from Hodjat and Mobley (1984).

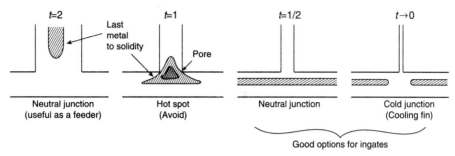

FIGURE 5.18

Array of T-junctions showing the thermal effects at the junction with different relative thickness of casting.

that castings with uniform wall thickness are said to be preferred, and that designers are encouraged to design them. Such products necessarily contain 1:1 junctions that will be hot spots. (Techniques for dealing with these are dealt with later when considering the feeding of castings; Section 6.)

4. When finally the section thickness of the upright of the T is twice the casting section, then the junction is balanced once again, with the casting now acting as the mild chill to counter the effect of the hot spot at the junction. We have considered these junctions merely in the form of the intersections of plates. However, we can extend the concept to more general shapes, introducing the use of the geometric modulus $m =$ (volume)/(cooling area). It subsequently follows that an additional requirement when a feeder forms a T-junction on a casting is that the feeder must have a modulus two times the modulus of the casting. The hot spot is then moved out of the junction and into the feeder, with the result that the casting is sound. This is the basis behind Feeding Rule 4 discussed later (Chapter 10.6.3).

Pellini (1953) was one of the first experimenters to show that the siting of a thin 'parasitic' plate on the end of a larger plate could improve the temperature gradient in the larger plate. However, the parasitic plate that he used was rather thick, and his experiments were carried out only on steel, whose conductivity is poor, reducing useful benefits.

Figure 5.19 shows the results from Kim et al. (1985) of pour-out tests carried out on 99.9% pure aluminium cast into sand moulds. The faster advance of the freezing front adjacent to the junction with the fin is clearly shown. (As an aside, this simple result is a good test of some computer simulation packages. The simulation of a brick-shaped casting with a cast-on fin should show the cooling effect by the fin. Some relatively poor computer algorithms do not take into account the conduction of heat in the casting, thus predicting, erroneously, the appearance of the junction as a hot spot, clearly revealed by the contours near the junction curving the wrong way.)

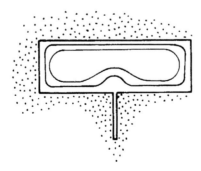

FIGURE 5.19

T-junction casting in 99.9Al by Kim et al. (1985) showing successive positions of the solidification front.

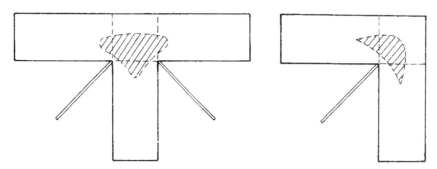

FIGURE 5.20

T- and L-junctions in pure aluminium cast in oil-bonded greensand. The shape of porosity in these junctions is shown, and the region of the junction used to calculate the percentage porosity is shown by the broken lines. The position of fins added to eliminate the porosity is shown. Results are presented in Figure 5.21.

Creese and Sarfaraz (1987) demonstrate the use of a fin to chill a hot spot in pure Al castings that was difficult to access in other ways. They cast on fins to T- and L-junctions as shown in Figure 5.20. The reduction in porosity achieved by this technique is shown in Figure 5.21. For these casting sections of 50 mm there was no apparent difference between fins of 2.5 and 3.3 mm thickness so these results are treated together in this figure. These fins at 5 and 6% of the casting section happen to be close to optimum as is confirmed later below. The reason that they conduct away perhaps less effectively than might be expected is because of their unfavourable location at 45° between two hot components of the junction.

Returning to the case where the upright of the T is sufficiently thin to act as a cooling fin, one further case that is not presented in Figure 5.18 is the case where the fin is so thin that it does not exist. This, you will say, is a trivial case. But think what it tells us. It proves that the fin can be too thin to be effective, because it will have insufficient area to carry away enough heat. Thus we arrive at the important conclusion that there is an optimum thickness of fin for a given casting section.

Similarly, an identical argument can be made about the fin length. A fin of zero length will have zero effect. As length increases, effectiveness will increase, but beyond a certain length, additional length will be of reducing value. Thus the length of fins will also have an optimum.

These questions have been addressed in a preliminary study by Wright and the author (1997) on a horizontal plate with a symmetrical fin (Figure 5.22). Symmetry was chosen so that thermocouple measurements could be taken along the centre line (otherwise the precise thermal centre was not known so that the true extension in freezing time may not have been measured accurately). In addition the horizontal orientation of the plate was selected to suppress any complicating effects of convection so far as possible. The thickness of the fin was t.H and the length L.H where t and L are the dimensionless numbers to quantify the fin in terms of H, the thickness of the plate. From this study it was discovered that there was an optimum thickness of a fin, and this was less than one tenth of H. Figure 5.22(a) interpolates an optimum in the region of 5% of the casting section thickness. The optimum length was 2H, and longer lengths were not significantly more effective (Figure 5.22(b)). For the optimum conditions the freezing time of the casting was increased by approximately 10 times. Thus the effect is useful. However, the effect is also rather localised, so that it needs to be used with caution. Eventually, non-symmetrical results for a chill on one side of the plate would be welcome.

Even so, the practical benefits to the use of a fin as opposed to a chill are interesting, and possibly even compelling. They are.

1. The fin is always provided on the casting because it is an integral part of the tooling. Thus, unlike a chill, the placing of it cannot be forgotten.

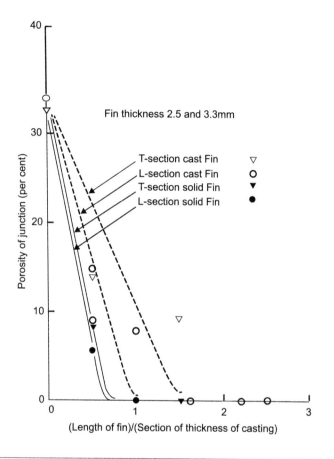

FIGURE 5.21

Results from Creese and Sarfaraz (1987, 1988) showing the reduction in porosity as a result of increasing length of fins.

2. It is always exactly in the correct place. It cannot be wrongly sited before the making of the mould. (The incorrect positioning of a chill is easily appreciated, because although the location of the chill is normally carefully painted on the pattern, the application of the first coat of mould release agent usually does an effective job in eliminating all traces of this. Furthermore of course, the chill can easily move during the filling of the moulding box with sand.)
3. It cannot be displaced or lifted during the making of the mould. If a chill lifts slightly during the filling of the tooling with sand, the resulting sand penetration under the edges of the chill, and the casting of additional metal into the roughly shaped gap, makes an unsightly local mess of the casting surface. Displacement or complete falling out of the chill from the mould is a common danger, sometimes requiring studs to support the chill if awkwardly angled or on a vertical face. Displacement commonly results in sand inclusion defects around the chill or can add to defects elsewhere. All this is costly to dress off.
4. An increase in productivity has been reported as a result of not having to find, place and carefully tuck in a block chill into a sand mould (Dimmick, 2001).

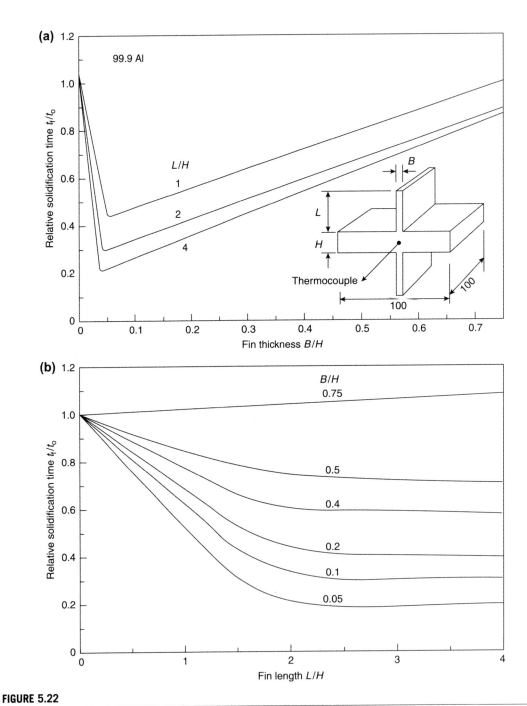

FIGURE 5.22
The effect of a symmetrical fin on the freezing time at the centre of a cast plate of 99.9Al as a function (a) the thickness, and (b) the length of the fin.

Averaged results of simulation and experiment from Wright and Campbell (1997).

5. The fin is easily cut off. In contrast, the witness from a chill also usually requires substantial dressing, especially if the chill was equipped with v-grooves, or if it became misplaced during moulding, as mentioned previously.
6. The fin does not cause scrap castings because of condensation of moisture and other volatiles, with consequential blow defects, as is a real danger from chills.
7. The fin does not require to be retrieved from the sand system, cleaned by shot blasting, stored in special bins, on special shelves, re-located, counted, losses made up by re-ordering new chills, casting new chills (particularly if the chill is shaped) and finally ensuring that the correct number in good condition, re-coated, and dried, is delivered to the moulder on the required date.
8. The fin does not wear out. Old chills become rounded to the point that they are effectively worn out. In addition, in iron and steel foundries, grey iron chills are said to 'lose their nature' after some use. This seems to be the result of the oxidation of the graphite flakes in the iron, thus impairing the thermal conductivity of the chill. This is understandable in terms of Figure 5.11 because graphite has significantly better conductivity than pure iron or steel which would materially affect sufficiently large chills.
9. Sometimes it is possible to solve a localised feeding problem (the typical example is the isolated boss in the centre of the plate) by chilling with a fin instead of providing a local supply of feed metal. In this case the fin is enormously cheaper than the feeder, and sometimes the feeder would be located where it cannot be subsequently removed, and its continued presence would be objectionable to the customer. The customer might accept a fin remaining in an obscure part of the casting, or the fin might be more easily removed.

This lengthy list represents considerable costs attached to the use of chills that are not easily accounted for, so that the real cost of chills is often underestimated.

Even so, the chill may be the correct choice for technical reasons. Fins perform poorly for metals of low thermal conductivity such as zinc, Al-bronze, iron and steel. The computer simulation result in Figure 5.23 illustrates for the rather low thermal conductivity material, Al-bronze, that there are extensive conditions in which the chill is far more effective. Lerner and Kouznetsov (2004) compare the action of chills and fins, concluding that the two are sometimes interchangeable but the action of the fin is necessarily rather localised, whereas the chill give a better distribution of cooling effect.

The kind of result shown in Figure 5.23 would be valuable if available for a variety of casting alloys varying from high to low thermal conductivity, so that an informed choice could be made whether a chill or fin were best in any particular case. Such data have yet to be worked out and published.

Fins are most easily provided on a joint line of the mould or around core prints. Sometimes, however, there is no alternative but to mould them at right angles to the joint. From a practical point of view, these upstanding fins on patternwork are of course vulnerable to damage. Dimmick (2001) records that fins made from flexible and tough vinyl plastic solved the damage problem in their foundry. They would carry out an initial trial with fins glued onto the pattern. If successful, the fins would then be permanently inserted into the pattern. In addition, only a few standard fins were found to be satisfactory for a wide range of patterns; a fairly wide deviation from the optimum ratios did not seem to be a problem in practice.

Safaraz and Creese (1989) investigated an interesting variant of the cast-on fin. They applied loose metal fins to the pattern, and rammed them up in the sand as though applying a normal external chill, in the manner shown in Figure 5.20. The results of these 'solid' or 'cold' fins (so-called to distinguish them from the empty cavity that would, after filling with liquid metal, effectively constituting a 'cast' or 'hot' fin) are also presented in Figure 5.21. It is seen that the cold fins are more effective than the cast fins in reducing the porosity in the junction castings. This is the consequence of the heat capacity of the fin being used in addition to its conducting role. This effect clearly over-rides any disadvantage of heat transfer resistance across the casting/chill interface.

The cold fin is, of course, really a chill of rather slim shape. It raises the interesting question, that as the geometry of the fin and the chill is varied, which can be the most effective. This question has been tackled in the author's laboratory (Wen and colleagues, 1997) by computer simulation. The results are summarised in Figure 5.23. Clearly, if the cast fin is sufficiently thin, it is more effective than a thin chill. However, for normal chills that occupy a large area of the casting (effectively approaching an 'infinite' chill as shown in the figure), as opposed to a slim contact line, the chill is massively more effective in speeding the freezing of the casting.

192 CHAPTER 5 SOLIDIFICATION STRUCTURE

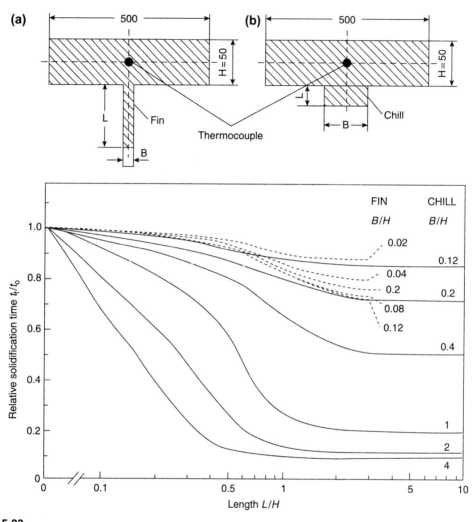

FIGURE 5.23

Comparison of the action of (a) cooling fins, and (b) chills on the rate of freezing of the aluminium bronze alloy AB1 (Wen, Jolly and Campbell 1997).

Other interesting lessons to be learned from Figure 5.23 are that a chill has to be at least equal to the section thickness of the casting to be really effective. A chill of thickness up to twice the casting section is progressively more valuable. However, beyond twice the thickness, increasingly thick chills show progressively reducing benefit.

It is to be expected that in alloys of higher thermal conductivity than aluminium bronze, a figure such as Figure 5.23 would show a greater regime of importance for fins compared to chills. The exploration of these effects for a variety of materials would be instructive and remains as a task for the future.

The business of getting the heat away from the casting as quickly as possible is taken to a logical extreme by Czech workers (Kunes et al., 1990) who show that a heat pipe can be extremely effective for a steel casting. Canadian workers (Zhang et al., 2003) explore the benefits of heat pipes for aluminium alloys. The conditions for successful application of the principle are not easy, however, so I find myself reluctant at this stage to recommend the heat pipe as a general purpose technique in competition to fins or chills. In special circumstances, however, it could be ideal.

Pins

Pins are analogous to fins and can be surprisingly effective. They take the form of simple pencil-like projections from a casting. Their narrow form allows them to be accommodated in spaces where chills or fins have insufficient room. Furthermore, compared to fins, they are fundamentally more effective as a result of the powerful local effect of their doubled rate of local heat extraction; they lose heat radially (in x and y dimensions, if the fin axis is the z dimension) as opposed to the fin which can lose its heat only unidirectionally (only in the single x dimension).

The geometry of pins, their diameter and length in terms of the thickness of section to be cooled, has never been investigated. This is a loss to the foundry engineer because pins have all the advantages listed for fins, plus the additional special features that can make them uniquely useful. Cooling pins deserve wide application in castings, and deserve the immediate attention of researchers to research and publish optimised use for different cast alloys. Even steel castings which do not respond to fins might benefit from pins. This would be a welcome benefit.

5.1.3 CONVECTION

Convection is the bulk movement of the liquid under the driving force of density differences in the liquid. In Section 5.3.4 we shall consider the problems raised by convection driven by solutes; heavy solutes cause the liquid to sink, and the lighter solutes cause flotation. In this section, we shall confine our discussion simply to the effects of temperature: hot liquid will expand, becoming less dense, and will rise; cool liquid will contract, becoming denser, and so will sink.

The existence of convection has been cited as important because it affects the columnar to equiaxed transition (Smith et al., 1990). There may be some truth in this. However, in most castings, grain structure is of no importance. Hardly any customer specifies the grain structure of their castings, the usual only critical features being soundness and leak tightness. Only in very few castings is grain size specified or is in any way noticeably significant in terms of affecting properties. Turbine blades are the exceptional example.

It is of course understood that soundness is of vital importance to nearly every casting, but it is not well known that in certain circumstances, convection can give severe unsoundness problems. This is especially true in counter-gravity systems, and sometimes in investment castings. This important phenomenon is dealt with at length in Chapter 10, Rule 7 for the manufacture of good castings. It is recommended reading.

5.1.4 REMELTING

When considering the solidification of castings it is easy to think simply of the freezing front as advancing. However, there are many times when the front goes into reverse! Melting is common during the filling and solidification of castings and needs to be considered at many stages.

On a microscale, melting is known to occur at different points on the dendrite arms. In a temperature gradient along the main growth direction of the dendrite the secondary arms can migrate down the temperature gradient by the remelting of the hot side of the arms and the freezing of the cold side. Allen and Hunt (1979) show how the arms can move several arm spacings. Similar microscopic remelting occurs as the small arms shrink and the larger arms grow to reduce the overall energy of the system. This is the mechanism of dendrite arm coarsening, leading to the dendrite arm spacing (DAS) being a useful indicator of the solidification time of a casting.

Slightly more serious thermal perturbations can cause the secondary dendrite arms to become detached when their roots are remelted (Jackson et al., 1966). The separated secondaries are then free to float away into the melt to become nuclei for the growth of equiaxed grains. If, however, there is too much heat available, then the growth front stays in reverse, with the result that the nuclei vanish, having completely remelted!

On a larger scale in the casting, the remelting of large sections of the solidification front can occur. This can happen as heat flows are changed as a result of changes in heat transfer at the interface, as the casting flexes and moves in the mould,

194   CHAPTER 5 SOLIDIFICATION STRUCTURE

changing its contact points and pressures at different locations and at different times. It is likely that this can happen as parts of the mould, such as an undersized chill, become saturated with heat, while cooling continues elsewhere. Thus the early rapid solidification in that locality is temporarily reversed.

Local remelting of the solid is seen to occur as a result of the influx of fresh quantities of heat from forced convection during filling. The so-called 'flow lines' seen on the radiographs of magnesium alloy castings are clearly a result of the local washing away of the solidification front, as a curving river erodes its outer bank. These features appear quite distinct from other linear defects such as hot tears or cracks as a result of their smooth outlines and gradual shading of radiographic density (Skelly and Sunnucks 1954; Lagowski 1967). Interestingly, no damaging loss of properties has ever been reported from such linear features of castings, nor would any loss of properties be expected.

The existence of continuous fluidity is a widely seen effect resulting directly from the remelting of the metal which has solidified in the filling system, keeping the metal flowing despite an unfavourable modulus. Without the benefits of this phenomenon it would be difficult to make castings at all!

Other convective flows produced by solute density gradients in the freezing zone take time to get established. Thus channels are formed by the remelting action of low-melting-point liquid flowing at a late stage of the freezing process. The A and V segregated channels in steel ingots, and freckle defects in nickel- and cobalt-based alloys, are good examples of this kind of defect.

5.1.5 FLOW CHANNEL STRUCTURE

Consider the direct-gated vertical plate shown in Figure 5.24(a). If this casting is filled quickly, the twin rotating vortices ensure that the plate fills with liquid at a uniform temperature. This may now be a problem to feed, and might exhibit the usual problems of distributed microporosity or surface sinks etc.

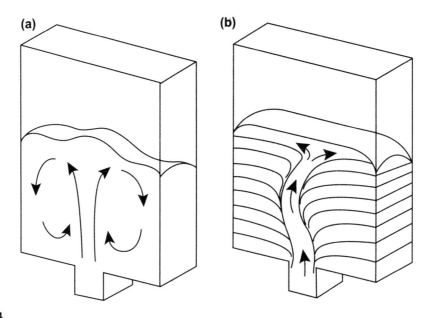

FIGURE 5.24

(a) Fast filling showing mixing by vortices, giving a uniform temperature in the filled plate. (b) Slow filling, showing layered freezing during filling and the development of a central flow channel.

This scenario contrasts with that of the slow filling of the casting, perhaps being filled slowly to reduce the potential for surface turbulence. In this case slow filling means *filling in a time commensurate with the solidification time of the section being filled*. If the filling rate were reduced to the point that the metal just reached the top of the mould by the time the metal had just cooled to its freezing point, then it might be expected that the top of the casting would be at its coldest, and freezing would then progress steadily down the plate, from the top to the gate. Nothing could be further from the truth.

In reality, the slow filling of the plate causes metal to flow sideways from the gate into the sides of the plate (Figure 5.24(b)), cooling as it goes, and freezing near the walls. Thus while more distant parts of the casting are freezing, layers of fresh hot metal continue to arrive through the gate. The successive positions of the freezing front are shown in Figure 5.24(b). The final effect is a flow path kept open by the hot metal through a casting that by now has mainly solidified. The well-fed panels either side of the flow channel are usually extremely sound and fine-grained, and contrast with the flow channel. The final freezing of the flow channel is slow because of the preheated mould around the path, and so its structure is coarse and porous. The porosity will be encouraged by the enhanced gas precipitation under the conditions of slow cooling, and shrinkage may contribute if local feeding is poor because either the flow path is long or it happens to be distant from a source of feed metal. Rabinovich (1969) describes these patterns of flow in thin vertical plates, calling them *jet streams*. *Flow channel* is suggested as a good name, if somewhat less dramatic.

Figure 5.25 shows an extreme example. This figure is a radiograph of an Al-7Si-0.4Mg alloy plate cast in an acid catalysed furan bonded silica sand mould. The flow channel is outlined by minute bubbles (most probably of hydrogen) that appear to have formed just under the surface of the casting as a result of reaction with the mould, but have floated

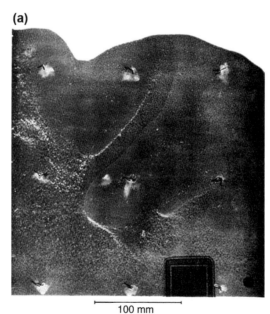

FIGURE 5.25

(a) Radiograph of an Al-7Si-0.4Mg alloy vertical plate filled via a side riser and slot gate shown in Figure 3.1(a), showing unexpected flow channel filling behaviour when cast particularly cool. Remains of thermocouples can be seen (Runyoro, 1992). (b) Computer simulation of two flow channels developing in the walls of a grey iron casting despite the provision of five ingates in an effort to spread heat as evenly as possible.

Courtesy Puhakka (2009).

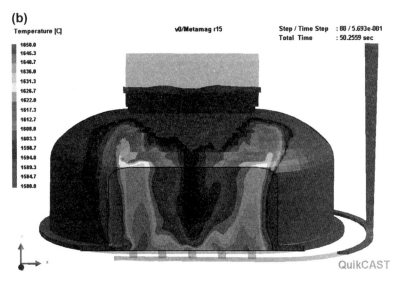

FIGURE 5.25 Cont'd

upwards to rest against and decorate the upper surface of the channel. They have probably grown sufficiently large to become significantly buoyant because of the extra time and temperature for reaction in the flow channel region. Those parts of the casting that have solidified quickly do not show bubbles on the radiograph. The flow channel in this case arises because the casting was made at a particularly low pouring temperature, so that the metal was flowing into the mould as a partly solidified slurry. The flow therefore adopted the form of a magma vent in the earth's crust, the pasty material resembling viscous lava, forming a volcano-like structure at the top surface.

Although Figure 5.25 illustrates a clear example because of the metal/mould reaction, in general the defect is not easily recognisable. It can occasionally be seen as a region of coarse grain and fine porosity in radiographs of large plate-like parts of castings. The structure contrasts with the extensive areas of clear, defect-free regions of the plate on either side. It is possible that many so-called shrinkage problems (for which more or less fruitless attempts are made to provide a solution by extra feeders or other means) are actually residual flow channels that might be cured by changing ingate position or size, or raising fill rate. No research appears to have been carried out to guide us out of this difficulty.

The flow channel structure is a standard feature of castings that are filled slowly from their base. This serious limitation to structure control seems to have been largely overlooked, and is one of the phenomena that limit the choice of a very long filling time.

Nevertheless, in general, the problem is reduced by filling faster, if that is possible without introducing other problems. More precisely, the velocity into the mould is initially controlled below approximately 0.5 m/s to avoid jetting through the ingate. After the base of the mould is covered to a sufficient depth the velocity can be ramped up. This is easily accomplished with a counter-gravity system, or even gravity system using a surge control design of filling system. A sufficiently high velocity will drive large circulating eddies (Figure 5.24) and finer scale circulating eddies (bulk turbulence) beneath the relatively tranquil surface. In this way the temperature of a large area, if not the whole casting, will be relatively uniform, completely free of a flow channel.

However, even fast filling does not cure the other major problem of bottom gating, which is the adverse temperature gradient, with the coldest metal being at the top and the hottest at the bottom of the casting, particularly concentrated in the ingate. Where feeders are placed at the top of the casting this thermal regime is clearly unfavourable for efficient feeding. However, of course, feeding can be made to be effective by oversized, if inefficient, feeders.

The problems of directly gating into the casting arise because the hot metal has to travel through the casting to reach all parts. One solution is *not* to gate into the casting but create a separate flow path called an *up-runner, outside* the casting. The melt is transferred by sideways flow off the up-runner via a slot gate into the mould cavity. The technique is described in Section 12.4.16. Another technique to counter adverse temperature gradient as a result of filling is the inversion of the casting after filling. The 'roll over' technique, such as used in the Cosworth casting process, is highly effective and especially recommended for volume production.

It is well to remember that on occasions remelting to create flow channels can occur without the transfer of heat. The channels seen commonly in Ni-base superalloys, known as 'freckles' because of their appearance when etched revealing the random orientations of fragments of grains, and the 'A' and 'V' segregates in heavy steel castings and ingots, have been formed by residual liquid, strongly segregated in C, S and P, that has a lower melting point than the matrix (Section 5.3.4). The matrix therefore melts when in contact with this liquid as a result of the diffusion of alloying elements into the matrix, lowering its melting point. Melting therefore occurs in this case by the transfer of solutes rather than the transfer of heat.

5.2 DEVELOPMENT OF MATRIX STRUCTURE
5.2.1 GENERAL

The liquid phase can be regarded as a randomly close-packed heap of atoms, in ceaseless random thermal motion, with atoms vibrating, shuffling, and jostling a meandering route, shoulder to shoulder, among and between their neighbours. The haphazard motion and random overall direction of travel of individual atoms has been termed 'the drunkard's walk'.

In contrast, the solid phase is an orderly array, or lattice, of atoms arranged in more or less close-packed rows and layers called a lattice. Atoms arranged in lattices constitute the solid bodies we call crystals. Iron in its alpha phase takes on the body-centred-cubic lattice known as ferrite (Figure 5.26(a)). This is a rather less close packed lattice than the face-centred-cubic lattice known as the gamma-phase, or as austenite. Figure 5.26(b) shows only a single '*unit cell*' of the lattice. The concept of the lattice is that it repeats such *unit cells*, replicating the symmetry into space millions of times in all directions. Macroscopic lattices can sometimes be seen in castings as crystals having sizes from 1 μm to 100 mm, representing arrays 10^3–10^8 atoms across.

The transition from liquid to solid, the process of solidification, is not always easy, however. For instance, in the case of glass, the liquid continues to cool, gradually losing the thermal motion of its atoms, to the point at which it becomes incapable of undergoing sufficient atomic rearrangements for it to convert to a lattice. It has therefore become a supercooled liquid, capable of remaining in this state for ever.

Metals, too, are sometimes seen to experience this reluctance to convert to a solid, despite on occasions being cooled hundreds of degrees Celsius below their equilibrium freezing temperature. This is easily demonstrated for clean metals in a clean container, for instance liquid iron in an alumina crucible.

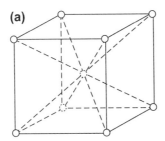
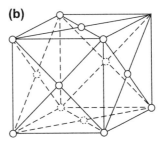

FIGURE 5.26

Body-centred cubic form of α-iron (a) exists up to 910°C. Above this temperature, iron changes to (b) the face-centred cubic form. At 1390°C the structure changes to β-iron, which is bcc once again (a).

If and when the conversion from liquid to solid occurs, it is by a process first of nucleation, and then of growth.

Nucleation is the process of the aggregation of clusters of atoms that represent the first appearance of the new solid phase. *Growth* is, self-evidently, getting bigger. However this process is subject to factors that encourage or discourage it as we shall discuss.

In fact, the complexities of the real world dictate not only that the main solid phase appears during solidification, but also that alloys and impurities concentrate in ways to trigger the nucleation and growth of other phases. These include solid and liquid phases that we call second phases or inclusions, and gas or vapour phases which we call gas pores or shrinkage pores. It is convenient to treat the solid and liquid phases together as condensed (i.e. practically incompressible) matter that we shall consider in this Section 5. The gas and vapour phases, constituting the non-condensed (and very compressible) matter such as gas and shrinkage porosity respectively, will be treated separately in Section 7.

For those readers who are enthusiastic about nucleation theory, there are many good formal accounts, some highly mathematical. A readable introduction relating to the solidification of metals is presented in Flemings (1974). We shall consider only a few basic aspects here, enough to enable us to understand how the structures of castings originate.

5.2.2 NUCLEATION OF THE SOLID

At first, as the temperature of a liquid is reduced below its freezing point, nothing happens. This is because in clean liquids the conversion to the solid phase involves a nucleation problem. We can gain an insight into the nature of the problem as follows.

As the temperature continues to fall, the thermal agitation of the atoms of the liquid reduces, allowing small random aggregations of atoms into crystalline regions. For a small cubic cluster of size d the net energy to form this new phase is reduced in proportion to its volume d^3 because of the lower free energy per unit volume ΔG_v of the solid. At the same time however the creation of new surface area $6d^2$ involves extra energy because of the interfacial energy γ per unit area of surface. The net energy to form our little cube of solid is therefore:

$$\Delta G = 6d^2\gamma - d^3 \Delta G_v \tag{5.15}$$

Figure 5.27 shows plots of the factors in this equation, showing that the net energy to grow the embryo increases at first, reaching a maximum. Embryos that do not reach the maximum require more energy to grow, they are unstable, so normally they will shrink and redissolve in the liquid.

Only when the temperature is sufficiently low to allow a chance chain of random additions to grow an embryo to the critical size will further growth be encouraged by a reduction in energy; thus growth will enter a 'runaway' condition. The temperature at which this event can occur is called the *homogeneous nucleation temperature*. For metals like iron and nickel, it is hundreds of degrees Celsius below the equilibrium freezing point as has been demonstrated many years ago by Walker (1961) and by many more recent researchers such as Valdez and colleagues (2007).

Such low temperatures may in fact be attained when making castings, because the liquid metal sits inside its own surface oxide film, never making contact with its container. At microscopic points of contact of the metal with the container, for moulds with high conductivity, the cooling through the surface film may be so intense that homogeneous nucleation in the liquid may occur. Nucleation is not likely to be a heterogeneous event in this situation because the liquid contained in its oxide skin is not in actual atomic contact with the surface of the mould, and its oxide is not a favourable nucleating substrate. (We shall see later that although oxides, such as some crucible materials, are *not* good heterogeneous nuclei to initiate nucleation of the solid, they are *excellent* nuclei for many second phases, and are efficient, indirectly as bifilms, for the initiation of porosity and cracks. These two entirely different behaviours of oxides are a major factor in the development of cast structure which will be returned to repeatedly.)

Dispinar and Campbell (2007) have reported evidence of massive undercoolings in comparatively large aluminium alloy castings. The observations that drew our attention were studies of solidified filters from the filling systems of castings, in which adjacent regions of melt, separated by an apparent grain boundary, had vastly different DAS on either side of the boundary (Figure 5.28). This seemingly indicated the impossible situation of two different rates of freezing in adjacent regions of metal. The fine microstructure was finally interpreted as undercooled regions isolated from the bulk

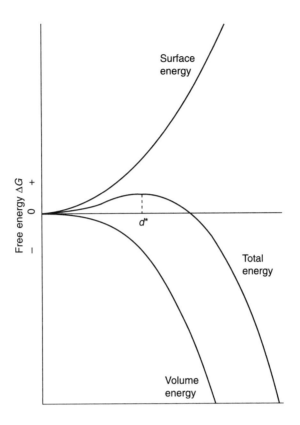

FIGURE 5.27

Surface and volume energies of an embryo of solid growing in a liquid give the total energy as shown. Below the critical size d^* any embryos will tend to shrink and disappear. Above d^* increasing size reduces the total energy, so growth will be increasingly favoured, becoming a runaway process.

melt by oxide bifilms. Thus some of these separated regions had a high probability of containing no suitable inclusion on which to nucleate. The surrounding matrix, having solidified some time previously and so much cooler, acted as an efficient chill when freezing finally occurred. The enclosed droplet therefore exhibits its quenched-in, rapidly frozen fine dendritic structure. This phenomenon, first seen in regions of melt trapped among oxide film inclusions, has now been identified in many castings poured turbulently (including, for instance, in Ni-base superalloys cast in vacuum), where droplets from splashes cannot be re-assimilated by the melt because of their surface (double) oxide, and consequently have their own independent freezing behaviour.

It is more common for the liquid to contain other solid particles in suspension on which new embryo crystal can form. In this case the interfacial energy component of Eqn (5.15) can be reduced or even eliminated. Thus the presence of foreign nuclei in a melt can give a range of heterogeneous nucleation temperatures; the more effective nuclei requiring less undercooling. It is even conceivable that some solid might exist on an extremely favourable substrate at temperature above its freezing point. This seems to be the case, for instance, with silicon particles nucleating on AlP particles on oxide bifilms in Al-Si alloys. Graphite flakes appear to form similarly above the general Fe-graphite eutectic in cast irons. These examples will be discussed further in Chapter 6.

200 CHAPTER 5 SOLIDIFICATION STRUCTURE

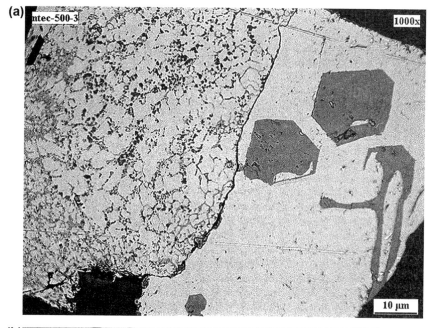

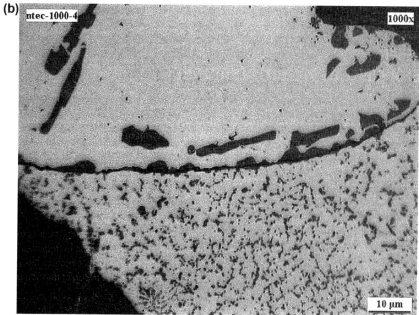

FIGURE 5.28

The separation and isolation of regions of Al-7Si alloy by oxide bifilms, resulting in (a) widely different solidification conditions, and (b) formation of Si particles on the bifilms (Dispinar, 2007).

5.2 DEVELOPMENT OF MATRIX STRUCTURE

Nevertheless, it is important to keep in mind that not all foreign particles in liquids are favourable nuclei for the formation of the solid phase. In fact it is likely that the liquid is indifferent to the presence of much of this debris. Only rarely will particles be present that reduce the interfacial energy term in Eqn (5.15). Thus, as far as most metals are concerned, oxides are not good nuclei. It is worth noting that it makes no difference whether the lattice structure and spacing of the oxide and metal are closely matched. The oxides are not wetted. This indicates that their electronic contribution to the interfacial energy with the metal is not favourable for nucleation. The covalently bonded oxides sit reluctantly against the metallic bonds of the metals, taking no part in the liquid to solid transformation of the metal.

Grain refinement

Materials with more metallic properties are good nuclei for the initiation of the solid phase. For the nucleation of steels, these include some borides, nitrides and carbides. For Al alloys, an intermetallic compound, $TiAl_3$, is the key inoculant, together with TiB_2. These details are discussed in the section on casting alloys in Chapter 6.

It seems, however, that the action of an effective grain refining addition to a melt is not only that of nucleating new grains. An important secondary role is that of inhibiting the rate of growth of grains, thereby allowing more grains the opportunity to nucleate.

Even so, as a cautionary note, the addition of TiB_2 to Fe-3Si alloys (Campbell and Bannister, 1975) exhibited a profound grain refinement action, almost certainly enhanced by the thick layer of borides that surrounded each grain. The mechanical properties were expected to be seriously impaired by this brittle grain boundary phase, illustrating the probability that not all grain refinement of cast alloys is beneficial.

For more complex systems, where many solutes are present, the rate of growth of grains was assumed by Greer and colleagues (2001) to be controlled by the rate at which solute can diffuse through the segregated region ahead to the advancing front. They carried out a detailed exercise for aluminium alloys, using a *growth restriction parameter*. This concept was refined by Australian workers (StJohn et al., 2007) as reported later in Section 6.3.3.

Although the grain refinement of Al alloys now appears to be well understood, the same cannot be said for the hexagonal close packed (hcp) alloys of Zn and Mg. Liu et al. (2013) admit that current theories cannot explain the grain refinement of Zn, and Birol (2012) suggests the same conclusion for Mg, even though StJohn and colleagues (2013) suggest that Fe and Mn impair the grain refinement of Mg, explaining contradictory previous outcomes. It does not seem clear whether we really understand the grain refinement of the common hcp metals. The presence of bifilms has so far been overlooked in these researches even though it is not evident that bifilms could play any part in the nucleation of the primary solid phase. The extent of our ignorance leaves wide scope for further research.

5.2.3 GROWTH OF THE SOLID

Once nucleated, the initial grow of the solid will release latent heat, the ΔGv term in Eqn (5.15), causing the temperature of the melt to rise rapidly to near its equilibrium freezing point. This is called 'recalescence' meaning 'reheating'. The primary solid will spread relatively quickly through the undercooled liquid but will be slowed by recalescence. After this point, further growth will be controlled by the much slower loss of heat from the casting.

Commonly in castings it is only the metal in contact with the face of the mould which experiences any cooling below its equilibrium freezing temperature. Thus when nucleation occurs in this thin layer, a solid skin is quickly grown which envelops the casting. The interesting question now is 'How does it then continue its progress into the melt'?

Progress will only occur if heat is extracted through the solidified layer, cooling the freezing front below the equilibrium freezing point. The actual amount of undercooling experienced at the freezing front is usually only a few degrees Celsius. If the rate of heat extraction is increased, the temperature of the solidification front will fall further, and the velocity of advance, V_s, of the solid will increase correspondingly.

For pure metals (assuming relative freedom from over-enthusiastic grain refining additions), as the driving force for solidification increases, so the front is seen to go through a series of transitions. Initially, it is planar; at higher rates of advance it develops deep intrusions, spaced rather regularly over the front. These are parts of the front that have been left far behind. At higher velocities still, this type of growth transforms into cigar-like projections called cells, which finally

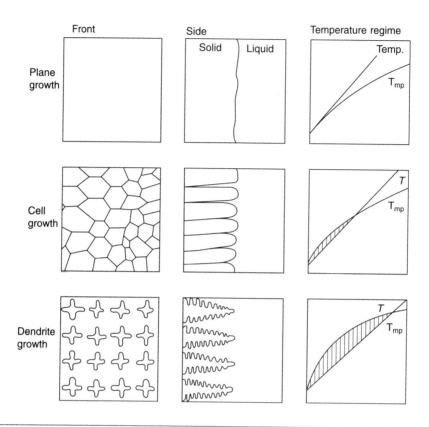

FIGURE 5.29

Transition of growth morphology from planar, to cellular, to dendritic, as compositionally induced undercooling increases (equivalent to G/V being reduced).

develop complex geometry involving side branches (Figure 5.29). These tree-like forms have given them the name dendrites (after the Greek word for tree, *Dendros*).

For the more important case of alloys, however, the three growth forms (planar, cellular and dendritic) are similarly present (Figure 5.29). However, the driving force for promoting the instability of planar growth, encouraging cellular or dendritic growth, is a kind of effective undercooling that arises because of the segregation of alloying elements ahead of the front. The presence of this extra concentration of alloying elements reduces the melting point of the liquid. If this reduction is sufficient to reduce the melting point to below the actual local temperature, then the liquid is said to be locally *constitutionally undercooled* (that is, effectively undercooled because of a change in the constitution of the liquid).

Figure 5.30 shows how detailed consideration of the phase diagram can explain the relatively complicated effects of segregation during freezing. It is worth examining the logic carefully.

The original melt of composition C_0 starts to freeze at the liquidus temperature T_L. The first solid to appear has composition kC_0 where k is known as the partition coefficient. This coefficient k usually has a value less than 1 (although the reader needs to be aware of the existence of the less common but important cases where k is greater than 1). For instance, for $k = 0.1$ the first solid has only 10% of the concentration of alloy compared with the original melt; the first solid to appear is therefore usually rather pure.

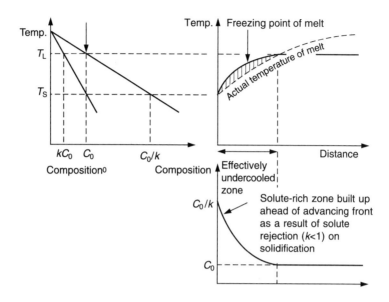

FIGURE 5.30
Link between the constitutional phase diagram for a binary alloy, and constitutional undercooling on freezing.

In general, k defines how the solute alloy partitions between the solid and liquid phases. Thus:

$$k = C_S/C_L \tag{5.16}$$

For those equilibrium phase diagrams for which the solidus and liquidus lines are straight, k is accurately constant for all compositions. However, even where they are curved, the relative matching of the curvatures often means that k is still reasonably constant over wide ranges of composition. When k is close to 1, the close spacing of the liquidus and solidus lines indicates little tendency towards segregation. When k is small, then the wide horizontal separation of the liquidus and solidus lines warns of a strongly partitioning alloying element.

On forming the solid that contains only kC_0 amount of alloy, the alloy remaining in the liquid has to be rejected ahead of the advancing front. Thus although the liquid was initially of uniform composition C_0, after an advance of about a millimetre or so the composition of the liquid ahead of the front builds up to a peak value of C_0/k. The build-up effect is like that of snow ahead of a snow plough. This is the steady-state condition shown in Figure 5.30.

In common with all other diffusion-controlled spreading problems, we can estimate the spread of the solute layer ahead of the front by the order-of-magnitude relation for the thickness d of the layer. If the front moves forward by d in time t, this is equivalent to a rate V_s. We then have:

$$d = (Dt)^{1/2} \quad \text{where} \quad V_s = d/t$$

so:

$$d = D/V_s \tag{5.17}$$

where D is the coefficient of diffusion of the solute in the liquid. It follows that constitutional undercooling will occur when the temperature gradient, G, in the liquid at the front is:

$$G \leq -\frac{T_L - T_S}{D/V_S} \tag{5.18}$$

or

$$G/V_S \leq -(T_L - T_S)/D \tag{5.19}$$

from Figure 5.30, assuming linear gradients. Again, from elementary geometry which the reader can quickly confirm, assuming straight lines on the equilibrium diagram, we may eliminate T_L and T_S and substitute C_0, k and m, where m is the slope of the liquidus line, to obtain the equivalent statement:

$$\frac{G}{V_S} = \frac{-mC_0(1-k)}{kD} \tag{5.20}$$

which is the classical solution derived from more rigorous diffusion theory by Chalmers in 1953, nicely summarised by Flemings (1974). This famous result marked the breakthrough in the history of the understanding of solidification by the application of physics. It marked the historical revolution from *qualitative description* to *quantitative prediction*. Computers have encouraged an acceleration of this radical transformation.

Figure 5.29 illustrates how the progressive increase in constitutional undercooling causes progressive instability in the advancing front, so that the initial planar form changes first to form cells, and with further instability ahead of the front will be finally provoked to advance as dendrites.

Notice that the growth of dendrites is in response to an *instability condition* in the environment ahead of the growing solid, not the result of some influence of the underlying crystal lattice (although, of course, the crystal structure will subsequently influence the details of the shape of the dendrite). In the same way, stalactites will grow as dendrites from the roof of a cave as a result of the destabilising effect of gravity on the distribution of moisture on the roof. Icicles are a similar example; their forms being, course, independent of the crystallographic structure of ice. Droplets running down window panes are a similar unstable-advance phenomenon that can owe nothing to crystallography. There are numerous other natural examples of dendritic advance of fronts that are not associated with any long-range crystalline internal structure. It is interesting to look out for such examples. Remember also the converse situation that the planar growth condition also effectively suppresses any influence that the crystal lattice might have. It is clear therefore that the constitutional undercooling, assessed by the ratio G/R, is the factor that measures the degree of stability of the growth conditions, and so controls the type of growth front (*not*, primarily, the crystal structure).

Figure 5.31 shows a transition from planar, through cellular, to dendritic solidification in a low-alloy steel that had been directionally solidified in a vertical direction. The speeding up of the solidification front has caused increasing instability. Figures 5.32 and 5.33 show different types of dendritic growth. Both types are widely seen in metallic alloy systems. In fact, dendritic solidification is the usual form of solidification in castings.

A columnar dendrite nucleated on the mould wall of a casting will grow both forwards and sideways, its secondary arms generating more primaries, until an extensive 'raft' has formed (Figure 5.34). All these arms will be parallel, reflecting the internal alignment of their atomic planes. Thus on solidification the arms will 'knit' together with almost atomic perfection, forming a single-crystal lattice known as a grain. A grain may consist of thousands of dendrites in a raft. Alternatively a grain may consist merely of a single primary arm, or, in the extreme, merely an isolated secondary arm.

The boundaries formed between rafts of different orientation, originating from different nucleation events, are known as grain boundaries. Usually these are high-angle grain boundaries, so-called to distinguish them from the low-angle boundaries within grains. Low angle boundaries result from small imperfections in the way the separate arms of the raft may grow, or suffer slight mechanical damage, so that their lattices join slightly imperfectly, at small but finite angles.

Given a fairly pure melt, and extremely quiescent conditions, it is not difficult to grow an extensive dendrite raft sufficient to fill a mould having dimensions of 100 mm or so, producing a single crystal. Nordland (1967) describes an unusual and fascinating experiment in which he solidifies bismuth at high undercooling and high rates, but preserves the fragile dendrite in one piece. He achieves this by adding weights to the furnace that contained his

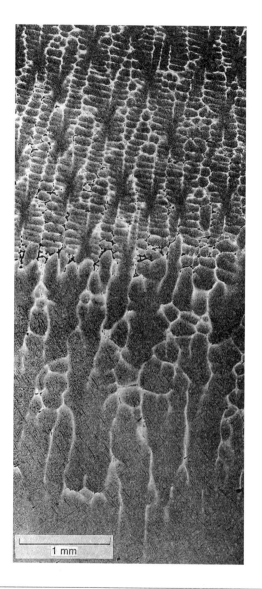

FIGURE 5.31

Structure of a low-alloy steel subjected to accelerating freezing from bottom to top, changing from planar, through cellular, to dendritic growth.

sample of solidifying metal, and suspends the whole assembly in **mid-air**, using long lengths of polypropylene tubing from the walls and ceiling of the room. In this way he was able to absorb and dampen any outside vibrations.

In a review of the effects of vibration on solidifying metals, the author (1981) confirms that Nordland's results fall into a regime of frequency and amplitude where the vibrational energy is too low for damage to occur to the dendrites (Figure 5.35).

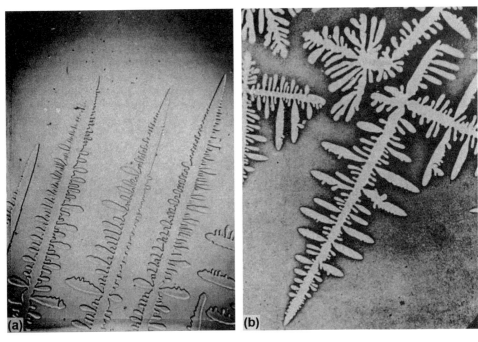

FIGURE 5.32

Transparent organic alloy showing dendritic solidification. Columnar growth (a) and equiaxed growth (b) with a modification to the alloy by the addition of a strongly partitioning solute, with k « 1 which can be seen to be segregated ahead of the growing front.

Courtesy J. D. Hunt; see Jackson et al. (1966).

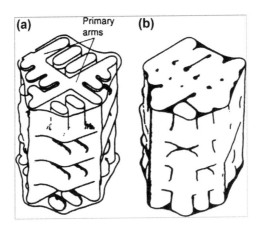

FIGURE 5.33

Rather irregular dendrites common in aluminium alloys at (a) 50% and (b) 90% solidified. The secondary arms spread laterally, joining to form continuous plates.

After Singh et al. (1970).

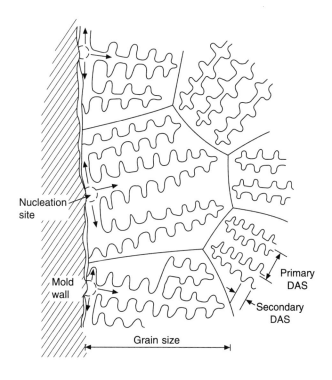

FIGURE 5.34

Schematic illustration of the formation of a raft of dendrites to make grains. The dendrite stems within any one raft or grain are all crystallographically related to a common nucleus.

Dendrite arm spacing

In the metallurgy of wrought materials, it is the *grain size* of the alloy that is usually the important structural feature. Most metallurgical textbooks therefore emphasise the importance of grain size.

In castings, however, grain size is sometimes important (as will be discussed later), but more often it is the *secondary dendrite arm spacing* (often shortened to *dendrite arm spacing or DAS*) that appears to be the most important structural length parameter.

The mechanical properties of most cast alloys are usually seen to be strongly dependent on secondary arm spacing. As DAS increases, so ultimate strength, ductility and elongation fall. Also, because homogenisation heat treatments are dependent on the time required to diffuse a solute over a given average distance d, if the coefficient of diffusion in the solid is D, then from the order-of-magnitude relation, Eqn 1.5, we predict, quantitatively, the finer DAS to give us specifically shorter homogenisation times, or better homogenisation in similar times; the cast material is more responsive to heat treatment, giving better properties or faster treatments.

It is now known that the secondary DAS is controlled by a coarsening process, in which the dendrite arms first grow at very small spacing near the tip of the dendrite. As time goes on, the dendrite attempts to reduce its surface energy by reducing its surface area. Thus small arms preferentially go into solution whilst larger arms grow at their expense, increasing the average spacing between arms. The rate of this process appears to be limited by the rate of diffusion of solute in the liquid as the solute transfers between dissolving and growing arms. From a relation such as Eqn 1.5, and assuming the alloy solidifies in a time t_f, we would expect that DAS would be proportional to $t_f^{1/2}$, because t_f is the time available for coarsening. In practice, it has been observed that DAS is actually proportional to t^n where n usually lies between 0.3 and 0.4 (Young and Kirkwood 1975). Figure 5.36 shows the magnificent research result, illustrating the

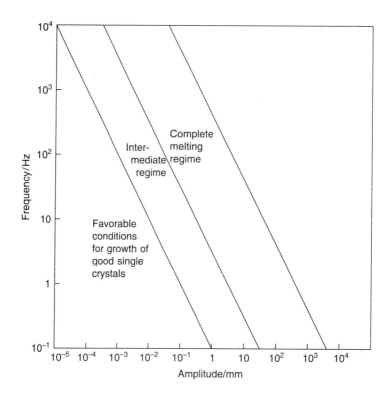

FIGURE 5.35

Grain refinement threshold as a function of amplitude and frequency of vibration (Campbell, 1981).

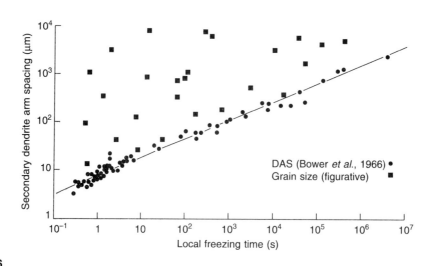

FIGURE 5.36

Relation between dendrite arm spacing (DAS), grain size, and local solidification time for Al-4.5Cu alloy.

relation between DAS and t_f continuing to hold for Al-4.5Cu alloy over eight orders of magnitude (Bower et al., 1966). Interestingly, however, a plot of grain size on the same figure shows that grain size is completely scattered above the DAS line. Clearly, grain size is completely independent of solidification time. In addition, of course, a grain cannot be smaller than a single dendrite arm, but can grow to unlimited size in some situations.

The primary DAS is of course significantly coarser than the secondary DAS, and appears controlled by rather different parameters. Young and Kirkwood (1975) proposed it is proportional to the parameter $(GV)^{-1/2}$ where G is the temperature gradient and V is the rate of advance of the solidification front. A little later, Hunt and Thomas (1977) present more accurate analytical solutions as a function of G, V and k, which were fitted to the results generated by a sophisticated computer program. Readers are recommended to the original papers for details of this piece of exemplary work.

In summary, primary DAS appears linked to solidification parameters and secondary DAS is controlled by solidification time. Grain size, on the other hand, is controlled by a number of quite separate processes, some of which are discussed further in the following section.

5.2.4 DISINTEGRATION OF THE SOLID (GRAIN MULTIPLICATION)

As the growers of single-crystal turbine blades know only too well, a single knock or other slight disturbance during freezing can damage the growing dendrite, breaking off secondary arms that then constitute nuclei for new separate grains. The growing crystal is especially vulnerable when strongly partitioning alloys are present that favour the growth of dendrites with secondary arms with weak roots. Thus some single crystals are fragile, and much more difficult to grow than others.

Figure 5.35 indicates that for any disturbance that can be characterised by a vibration of frequency f and/or amplitude a, a critical threshold exists at which the grains are fragmented. In a review of the mechanism of fragmentation by vibration (Campbell, 1981) it was not clear whether the dendrite roots melted, or whether they were mechanically sheared, because these two processes could not be distinguished by the experimental results. Whatever mechanism was operating, the experimental results on a wide variety of metals that solidify in a dendritic mode from Al alloys to steels, could be summarised to a close approximation by

$$f \cdot a = 0.10 \text{ mms}^{-1}$$

This relation describes the product of frequency (Hz) and amplitude (mm) that represents a critical velocity threshold for grain fragmentation. It seems to be valid over the complete range of experimental conditions ever tested, from subsonic to ultrasonic frequencies, and from amplitudes of micrometres to centimetres. Thus at ultrasonic frequency 25 kHz the amplitude for refinement is only 4 nm, whereas at the frequency of mains electricity 50 Hz the amplitude needs to be 2 µm.

In single-crystal growth, it seems that the damage to a dendrite arm may not be confined to breaking off the arm. Simply bending an arm will cause that part of the crystal to be misaligned with respect to its neighbours. Its subsequent growth might be in a direction favourable for its continued growth, causing it to grow to the size of a significant defect. Vogel and colleagues (1977) propose that given a sufficiently large angle of bend, the plastically deformed material will recrystallise rapidly. The newly formed high angle grain boundary, having a high energy, will be preferentially wetted by the melt, so that liquid will therefore propagate along the boundary and detach the arm. The arm now becomes a free-floating grain.

In the pouring of conventional castings, Ohno (1987) has drawn attention to the way in which the grains of some metals grow from the nucleation site on the mould wall. The grains grow from narrow stems that are vulnerable to plastic deformation and detachment. Thus as metal washes over the mould surface, thousands of crystals are swept into the melt. The nucleation sites with their remnant of dendrite root continue to be attached to the mould wall, quickly re-growing to 'seed' strings of replacement crystals, one after another. There is an element of runaway catastrophe in this process; as one dendrite is felled, it will lean on its neighbours and encourage their fall.

The fragments of crystals that are detached in this way may dissolve once again as they are carried off into the interior of the melt if the casting temperature is too high. The interior of the casting may therefore become free of so-called equiaxed grains. If so, the structure of the casting will consist only of columnar grains that grow inwards from the mould wall.

However, if the casting temperature is not too high, then the detached crystals will survive, forming the seeds of grains that subsequently grow freely in the melt. The lack of directionality and the *equal* length of the *axes* of these crystals has given them the name '*equiaxed*' grains. At very low casting temperatures, perhaps together with sufficient bulk turbulence, the whole of the casting may solidify with an equiaxed structure.

In mixed situations where modest quantities of equiaxed grains exist, they may be caught up as isolated grains in the growing forest of columnar dendrites. The directional heat flow that they will then experience will grow them unidirectionally, converting them to columnar grains. However, a sufficient deluge of equiaxed grains will swamp the progress of the columnar zone, converting the structure to equiaxed.

The *columnar-to-equiaxed transition* has been the subject of much solidification research. In summary. it seems that the transition is controlled by the numbers of equiaxed grains that are available. The recent model by Bisuola and Martorano (2008) indicates that the advance of the columnar zone is blocked when only 20% of the liquid ahead has transformed to equiaxed grains. The modelling work by Spittle and Brown (1989) illuminates the concepts admirably (Figure 5.37).

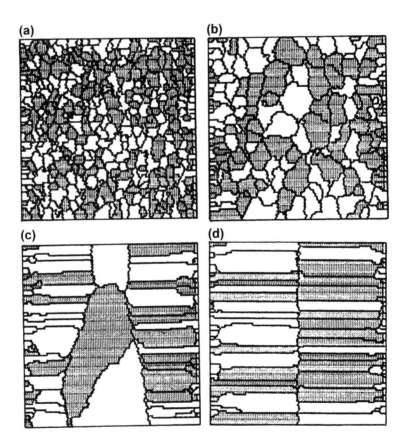

FIGURE 5.37

Computer simulated macrostructure of growth inwards from the sides of an ingot for progressively increasing casting temperature (a) to (d).

Reprinted with permission from J. Materials Science, Chapman & Hall, London.

5.2 DEVELOPMENT OF MATRIX STRUCTURE

In large steel ingots the columnar grains can reach lengths of 200 mm or more. These long cantilevered projections from the mould wall are under considerable stress as a result of their weight, and the additional weight of equiaxed grains, en route to the bottom of the ingot, that happen to settle, clustering on their tips. Under this weight, the grains will therefore bend by creep, possibly recrystallising at the same time, allowing the grains that grow beyond a certain length to sag downwards at various angles. This mechanism seems consistent with the structure of the columnar zone in large castings, and explains the so-called *branched columnar zone*. The straight portions of the columnar crystals near the base have probably resisted bending by the support provided by secondary arms, linked to form transverse web-like walls, providing the excellent rigidity of a box-girder supporting structure.

The bending of dendrites under their own weight when growing horizontally is not confined to steel ingots. Newell and colleagues (2009) found that dendrites of Ni-base superalloys growing across the horizontal platforms of turbine blade single crystal castings sagged under their own weight, but remaining straight as if creating a 'hinge' at their base, causing misalignments of up to $10°$.

Growing dendrites can be damaged or fragmented in other ways to create the seeds of new grains.

Mould coatings that contain materials that release gas on solidification, and so disturb the growing crystals, are found to be effective grain refiners (Cupini et al., 1980). Although these authors do not find any apparent increase of gas take-up by the casting, it seems prudent to view the 10% increase in strength as hardly justifying such a risk until the use of volatile mould dressings is assessed more rigorously.

The application of vibration to solidifying alloys is also successful in refining grain size. The author admits that it was hard work reviewing the vast amount of work in this area (Campbell, 1981). It seems that all kinds of vibration, whether subsonic, sonic, or ultrasonic, are effective in refining the grain size of most dendritically freezing materials providing the energy input is sufficient (Figure 5.35). The product of frequency and amplitude has to exceed 0.01 mms^{-1} for 10% refinement, 0.02 mms^{-1} for 50% refinement and 0.1 mms^{-1} for 90% refinement (Campbell, 1980). It is possible that at the free liquid surface of the metal the energy required to fragment dendrites is much less than this, as Ohno (1987) points out. This is sometimes known as *shower nucleation*, as proposed by the Australian researcher, Southin (1967), although it is almost certainly not a nucleation process at all. Most probably it is a dendrite fragmentation and multiplication process, resulting from the damage to dendrites growing across the cool surface liquid. These are possibly actually attached to the floating oxide film, or growing from the side walls, but are disturbed by the washing effect of the surface waves.

If we return to Figure 5.36 grain sizes are dotted randomly all over the upper half of the diagram, above the DAS size line. Occasionally some grains will be as small as one dendrite arm, and so will lie on the DAS line. No grain size can be lower than the line. This is because, if we could imagine a population of grains smaller than the DAS, and which would therefore find itself below the line, then in the time available for freezing, the population would have coarsened, reducing its surface energy to grow its average grain size up to the predicted size corresponding to that available time. Thus although grains cannot be smaller than the dendrite arm size, the grain size is otherwise independent of solidification time. Clearly, totally independent factors control the grain size.

It is clear, then, that the size of grains in castings results not only from nucleation events, such as homogeneous events on the side walls, or from chance foreign nuclei, or intentionally added grain refiners. Grain nuclei are also subject to further chance events such as re-solution. Further complications, mostly in larger-grained materials, result from chance events of damage or fragmentation from a variety of causes.

A further effect should be mentioned. The grains formed during solidification may not continue to exist down to room temperature. Many steels, for instance, as discussed in Section 6.6, undergo phase changes during cooling. Even in those materials that are single phase from the freezing point down to room temperature can experience grain boundary migration, grain growth or even wholesale recrystallisation. Figure 5.38 shows an example of grain boundary migration in an aluminium alloy.

It bears emphasising once again that dendrite arm spacing is controlled principally by freezing time, whereas grain size is influenced by many independent factors.

Before leaving the subject of the as-cast structure, it is worth giving a warning of a few confusions concerning nomenclature in the technical literature.

First, there is a widespread confusion between the concept of a grain and the concept of a dendrite. It is necessary to be on guard against this.

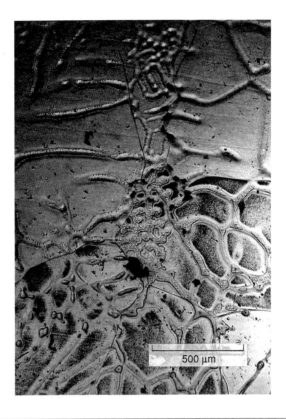

FIGURE 5.38

Micrograph of Al-0.2Cu alloy showing porosity and interdendritic segregation. Some grain boundary migration during cooling is clear. (Electropolished in perchloric and acetic acid solution and etched in ferric chloride. Dark areas are etch pitted.)

Second, the word 'cell' has a number of distinct technical meanings that need to be noted:

1. A cell can be a general growth form of the solidification front, as used in this book.
2. Cell is the term used to denote graphite 'rosettes' in grey cast irons. Strictly, these are graphite grains; crystals of graphite which have grown from a single nucleation event. They grow within, and appear crystallographically unrelated to, the austenite grains that form the large dendritic rafts of the grey iron structure.
3. The term 'cell count' or 'cell size' is sometimes used as a measure of the fineness of the microstructure, particularly in aluminium alloys. In Al alloys the distinction between primary arm, secondary arm, and grain is genuinely difficult to make in randomly oriented grains, where primary and secondary arms are not clearly differentiated (see Figure 5.33). To avoid the problem of having to make any distinction, a count is made of the number of rounded, bright features (that could be primary or secondary arms or grains) in a measured length. This is arbitrarily called 'the cell count', giving a measure of something called a 'cell size'.
In these difficult circumstances it is perhaps the only practical quantity that can be measured, whatever it really is!

5.3 SEGREGATION

Segregation may be defined as any departure from uniform distribution of the chemical elements in the alloy. Because of the way in which the solutes in alloys partition between the solid and the liquid during freezing, it follows that all alloy castings are segregated to some extent.

Some variation in composition occurs on a microscopic scale between dendrite arms, known as microsegregation. It can usually be significantly reduced by a homogenising heat treatment lasting only minutes or hours because the distance over which diffusion has to take place to redistribute the alloying elements is sufficiently small, usually in the range 10–100 μm. Equation 1.5 can assist to estimate times required for homogenisation for different spacings.

Macrosegregation cannot be removed. It occurs over distances ranging from 1 cm to 1 m, and so cannot be removed by diffusion without geological time scales being available! In general, therefore, whatever macrosegregation occurs has to be lived with.

In this section, we shall consider explicitly only the case for which the distribution coefficient k (the ratio of the solute content of the solid compared to the solute content of the liquid in equilibrium) is less than 1. This means that solute is rejected on solidification and therefore builds up ahead of the advancing front. The analogy, used repeatedly before, is the build-up of snow ahead of the advancing snow plough.

(It is worth keeping in mind that all the discussion can, in fact, apply in reverse, where k is greater than 1. In this case, extra solute is taken into solution in the advancing front, and a depleted layer exists in the liquid ahead. The analogy now is that of a domestic vacuum cleaner advancing on a dusty floor, and sucking up dust that lies ahead.)

For a rigorous treatment of the theory of segregation the interested reader should consult the standard text by Flemings (1974), which summarises the pioneering work in this field by the team of researchers at the Massachusetts Institute of Technology. We shall keep our treatment here to a minimum, just enough to gain some insight into the important effects in castings.

5.3.1 PLANAR FRONT SEGREGATION

There are two main types of normal segregation that occur when the solid is freezing on a planar front: one that results from the freezing of quiescent liquid and the other of stirred liquid. Both are important in solidification, and give rise to quite different patterns of segregation.

Figure 5.39 shows the way in which the solute builds up ahead of the front if the liquid is still (quiescent liquid, case 1). The initial build-up to the steady-state situation is called the initial transient. This is shown rather spread out for clarity. Flemings (1974) shows that for small k the initial transient length is approximately $D/V_s k$ where D is the coefficient of diffusion of the solute in the liquid and V_s is the velocity of the solidification front. In most cases, the transient length is only of the order of 0.1–1 mm.

After the initial build-up of solute ahead of the front, the subsequent freezing to solid of composition C_0 takes place in a steady, continuous fashion until the final transient is reached, at which the liquid and solid phases both increase in segregate. The length of the final transient is even smaller than that of the initial transient because it results simply from the impingement of the solute boundary layer on the end wall of the container. Thus its length is of the same order as the thickness of the solute boundary layer D/R. For many solutes, this is therefore between 5 and 50 times thinner than the initial transient.

For the case in which the liquid is stirred, moving past the front at such a rate to sweep away any build-up of solute, Figure 5.39 (stirred liquid case 2) shows that the solid continues to freeze at its original low composition kC_0. The slow rise in concentration of solute in the solid is, of course, only the result of the bulk liquid becoming progressively more concentrated.

The important example of the effect of normal segregation, building up as an initial transient, is that of subsurface porosity in castings. The phenomenon of porosity being concentrated in a layer approximately 1 mm beneath the surface of the casting is a clear example of the build-up of solutes. In this case the solute is a dissolved gas, which only after 1 mm

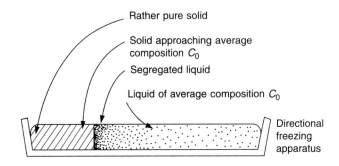
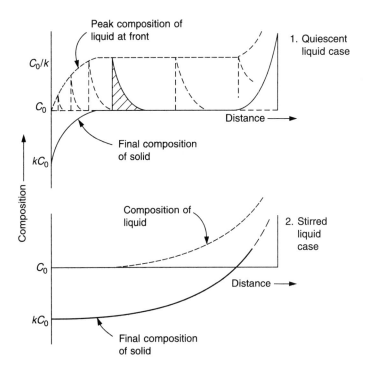

FIGURE 5.39

Directional solidification on a planar front giving rise to two different patterns of segregation depending on whether solute is allowed to build up at the advancing front or is swept away by stirring.

or so advance of solidification its concentration finally exceeds the threshold to nucleate pores. The nucleation and growth of gas pores is discussed in Section 7.

Moving on now to consider an example of segregation where the liquid is rapidly stirred, the classic case was that of the rimming steel ingot seen in Figure 5.40. Although rimming steels are now a phenomenon of the past, their behaviour is instructive. During the early stages of freezing, the high temperature gradient against the mould walls favoured a planar front. The rejection of carbon and oxygen resulted in bubbles of carbon monoxide. These detached from the planar front and rose to the surface, driving a fast upward flow of metal, effectively scouring the interface clean of any solute that

5.3 SEGREGATION 215

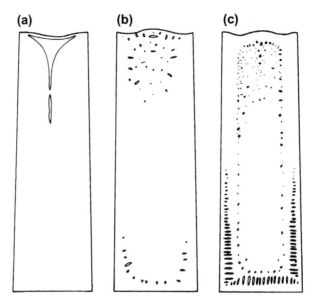

FIGURE 5.40

Ingot structures: (a) a killed steel; (b) a balanced steel; and (c) a rimming steel.

attempted to build up. Thus the solid continued to freeze with its original low impurity content, forming the pure iron 'rim', giving the steel its name. At lower levels in the ingot there was a lower density of bubbles to scour the front, so some bubbles succeeded in remaining attached, explaining the array of wormhole-like cavities in the lower part of the ingot. During this period, the incandescent spray from the tops of the ingots as the bubbles emerged and burst at the surface was one of the great spectacles of the old steelworks, almost ranking in impressiveness with the blowing of the Bessemer converters. A good spray was said to indicate a good rimming action. I have fond memories of these times from my early days casting ingots in the steel industry. Continuous casting of steels is, I regret, comparatively boring (although it is necessary to admit its impressive productivity and quality of product).

As the rim thickened, the temperature gradient fell so that the front started to become dendritic, retaining both bubbles and solute. Thus the composition then adjusted sharply to the average value characteristic of the remaining liquid, which was then concentrated in carbon, sulphur and phosphorus.

Rimming steel was widely used for rolling into strips, and for such purposes as deep drawing, where the softness and ductility of the rim assisted the production of products with high surface finish.

The oxygen levels in rimming steels were in excess of 0.02%, and were strongly dependent on the carbon and manganese contents. Typically these were 0.05–0.20%C and 0.1–0.6%Mn, giving a useful range of hardness, ductility and strength.

With the development of continuously cast steel, the casting of steel into ingots has now become part of steelmaking history. The challenge to the deep-drawing qualities was to regain so far as possible the benefits of the rimmed steel from a continuous casting process, where the rimming action could no longer be used. This has been achieved by attention to the surface quality and inclusion content of the new steels.

As an interesting diversion it is worth including at this point the two other major classes of steel that were produced as ingots, because these still have lessons for us as producers of shaped castings. The two other types of ingots were produced as (a) killed steels and (b) balanced steels (Figure 5.40).

The balanced, or semi-killed steel, was one that, after partially deoxidising, contained 0.01–0.02% oxygen. This was just enough to cause some evolution of carbon monoxide towards the end of freezing, to counter the effect of solidification shrinkage, thus yielding a substantially level top. (The deep shrinkage pipe typical of fully killed steels required to be cropped off and remelted). The whole ingot could be utilised. The great advantage of this quality of steel was the high yield on rolling, because the dispersed cavities in the ingot tended to weld up. For this reason bulk constructional steel could be produced economically. However, it was a difficult balancing act to maintain such precise control of the chemistry of the metal. It was only because balanced steels were so economical that such feats were routinely attempted. Some steelmakers were declared foolhardy for attempting such tasks!

In contrast, killed steels were easy to manufacture. They included the high-carbon steels and most alloy steels. They contain low levels of free oxygen, normally less than 0.003% because of late additions of deoxidisers such as Si, Mn and Al. The Al addition was often made together with other components such as Ca, and sometimes Ti etc. Consequently, there was no evolution of carbon monoxide on freezing, and a considerable shrinkage cavity was formed as seen in Figure 5.40. If allowed to form in this way, the cavity opened up on rolling as a fish-tail, and had to be cropped and discarded. Alternatively the top of the ingot was maintained hot during solidification by special hot-topping techniques. Either way, the shrinkage problem involved expense above that required for balanced or rimming steels. Fully killed steels were generally therefore reserved for higher priced, low and medium alloy applications.

5.3.2 MICROSEGREGATION

As the dendrite grows into the melt, and as secondary arms spread from the main dendrite stem, solute is rejected. The solute is effectively pushed aside and concentrates in the tiny regions enclosed by the secondary dendrite arms. Because these liquid regions are smaller than the diffusion distance, we may consider them more or less uniform in composition. The situation, therefore, is closely modelled by Figure 5.39, case 2. Remember, the uniformity of the liquid phase in this case results from diffusion within its small size, rather than any bulk motion of the liquid.

The interior of the dendrite therefore has an initial composition close to kC_0, whereas, towards the end of freezing, the centre of the residual interdendritic liquid has a composition corresponding to the peak of the final transient. This gradation of composition from the inside to the outside of the dendrite earned its common description as 'coring' because, on etching a polished section of such dendrites, the progressive change in composition is revealed, appearing as onion-like layers around a central core. The concentration of chromium and nickel in the interdendritic regions of the low-alloy steel shown in Figure 5.31 has caused these regions to be relatively 'stainless', resisting the etch treatment, and so causing them to be revealed in the micrograph.

Some diffusion of solute in the dendrite will tend to smooth the initial as-cast coring. This is often called back-diffusion. Additional smoothing of the original segregation can occur as a consequence of other processes such as the remelting of secondary arms as the spacing of the arms coarsens.

The partial homogenisation resulting from back-diffusion and other factors means that, for rapidly diffusing elements such as carbon in steel, homogenisation is rather effective. For plain carbon steels, therefore, the final composition in the dendrite and in the interdendritic liquid is not far from that predicted from the equilibrium phase diagram. The maximum freezing range from the phase diagram is clearly at about 2.0% carbon and would be expected to apply.

Even so, in steels where the carbon is in association with more slowly diffusing carbide-forming elements, the carbon is not free to homogenise: the resulting residual liquid concentrates in carbon to the point at which the eutectic is formed at carbon contents well below those expected from the phase diagram. In steels that contained between 1.3 and 2.0% manganese the author found that the eutectic phase first appeared between 0.8 and 1.3% carbon (Campbell, 1969) as shown in Figure 5.41. Similarly, in 1.5Cr-1C steels, Flemings et al. (1970) found that the eutectic phase first appeared at about 1.4% carbon (Figure 5.42). This point was also associated with a peak in the segregation ratio, S, the ratio of the maximum to the minimum composition; this is found between the interdendritic liquid and the centre of the dendrite arm. (Note: The interpretation of these diagrams as two separate curves intersecting in a cusp rather than a smoothly curved maximum is based on the fact that the two parts of the curve are expected to follow different laws. The first part represents the solidification of a solid solution; the second part represents the solidification of a solid solution plus some eutectic. As we have seen before, this is much more common in freezing problems than appears to have been generally recognised.)

5.3 SEGREGATION

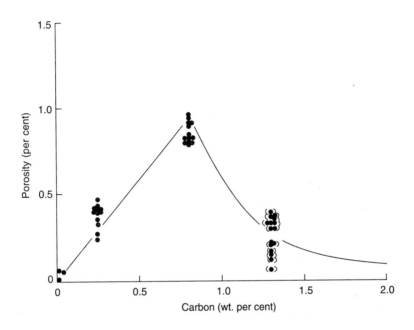

FIGURE 5.41

Porosity in Fe-C-Mn alloys, showing the reduction associated with the presence of non-equilibrium eutectic liquid (data points in brackets) (Campbell, 1969).

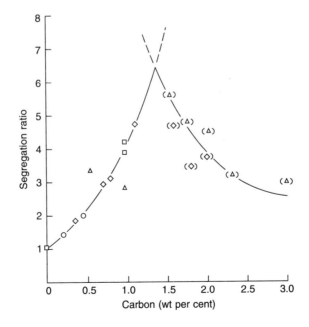

FIGURE 5.42

Severity of microsegregation in C-Cr–bearing steels, illustrating the separate regime for structures containing eutectic (points in brackets).

Data from Flemings et al. (1970).

218 CHAPTER 5 SOLIDIFICATION STRUCTURE

The segregation ratio S is a useful parameter when assessing the effects of treatments to reduce microsegregation. Thus the progress of homogenising heat treatments can be followed quantitatively.

It is important to realise that S is only marginally affected by changes to the rate of solidification in terms of the rates that can be applied in conventional castings. This is because although the dendrite arm spacing will be reduced at higher freezing rates, the rate of back diffusion is similarly reduced. Both are fundamentally controlled by diffusion, so that the effects largely cancel. (During any subsequent homogenising heat treatment, however, the shorter diffusion distances of the material frozen at a rapid rate will be a useful benefit in reducing the time for treatment.)

Where microsegregation results in the appearance of a new liquid interdendritic phase, there are a number of consequences that may be important:

1. The appearance of a eutectic phase reduces the problem for fluid flow through the dendrite mesh because the most difficult part of the mesh, the dendrite root region, is eliminated by the planar front of the advancing eutectic. Shrinkage porosity is thereby reduced after the arrival of the eutectic, as seen in Figure 5.41. This important effect on porosity is discussed in greater detail in Chapter 7.
2. The alloy may now be susceptible to hot tearing, especially if there is only a very few percent of the liquid phase. This effect is discussed further in Chapter 8.
3. A low-melting-point phase may limit the temperature at which the material can be heat treated.
4. A low-melting-point phase may limit the temperature at which an alloy can be worked because it may be weakened, disintegrating during working because of the presence of liquid in its structure.

5.3.3 DENDRITIC SEGREGATION

Figure 5.43 shows how microsegregation, the sideways displacement of solute as the dendrite advances, can lead to a form of macrosegregation. As freezing occurs in the dendrites, the general flow of liquid that is necessary to feed solidification shrinkage in the depths of the pasty zone carries the progressively concentrating segregate towards the roots of the dendrites.

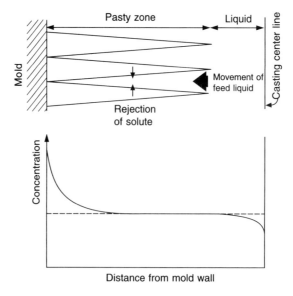

FIGURE 5.43

Normal dendritic segregation (usually misleadingly called inverse segregation) arising as a result of the combined actions of solute rejection and shrinkage during solidification in a temperature gradient.

In the case of a freely floating dendrite in the centre of the ingot that may eventually form an equiaxed grain, there will be some flow of concentrated liquid towards the centre of the dendrite if in fact any solidification is occurring at all. This may be happening if the liquid is somewhat undercooled. However, the effect will be small, and will be separate for each equiaxed grain. Thus the build-up of long-range segregation in this situation will be negligible.

For the case of dendritic growth against the wall of the mould, however, the temperature gradient will ensure that all the flow is in the direction towards the wall, concentrating the segregation here. Thus the presence of a temperature gradient is necessary for a significant build-up of segregation.

It will by now be clear that this type of segregation is in fact the usual type of segregation to be expected in dendritic solidification. The phenomenon has in the past suffered the injustice of being misleadingly named '*inverse segregation*' on account of it appearing anomalous in comparison to planar front segregation and the normal pattern of positive segregation seen in the centres of large ingots. In this book we shall refer to it simply as '*dendritic segregation*'. It is perfectly normal and to be expected in the normal conditions of dendritic freezing.

Dendritic segregation is observable but is not normally severe in sand castings because the relatively low temperature gradients allow freezing to occur rather evenly over the cross-section of the casting; little directional freezing exists to concentrate segregates in the direction of heat flow.

In castings that have been made in metal moulds, however, the effect is clear and makes the chill casting of specimens for chemical analysis a seriously questionable procedure. Chemists should beware! The effect of positively segregating solutes such as carbon, sulphur and phosphorus in steel is clearly seen in Figure 5.44 as the high concentration around the edges and the base of the ingot; all those surfaces in contact with the mould.

In some alloys with very long freezing ranges, such as tin bronze (liquidus temperature below 1000°C and solidus close to 800°C), the contraction of the casting in the solid state and/or the pressure in the liquid from the metallostatic head, plus perhaps the evolution of dissolved gases in the interior of the casting causes eutectic liquid to be squeezed out on to the surface of the casting. This exudation is known as *tin sweat*. It was described by Biringuccio in the year 1540 as a feature of the manufacture of bronze cannon. Similar effects can be seen in many other materials; for instance when making sand castings in the commonly-used Al-7Si-0.3Mg alloy, eutectic (Al-11Si) is often seen to exude against the surface of external chills (Figure 5.13). The surface exudation of eutectic liquid gives problems during the manufacture of Ni-base superalloy single crystal turbine blades as is discussed in the section on Ni-based alloys.

5.3.4 GRAVITY SEGREGATION

In the early years of attempting to understand solidification, the presence of a large concentration of positive segregation in the head of a steel ingot was assumed to be merely the result of normal segregation. It was simply assumed to be the same mechanism as illustrated in Figure 5.39, case 1, where the solute is concentrated ahead of a planar front.

This assumption overlooked two key factors: (1) the amount of solute that can be segregated in this way is negligible compared to the huge quantities of segregate found in the head of a conventional steel ingot and (2) this type of positive segregation applies only to planar front freezing. In fact, having now realised this, if we look at the segregation that should apply in the case of dendritic freezing then an opposite pattern (previously called inverse segregation) applies such as that shown in Figure 5.43! Clearly, there was a serious mismatch between theory and fact. That this situation had been overlooked for so long illustrates how easy it is for us to be unaware of the most glaring anomalies. It is a lesson for us all in the benefits of humility!

This problem was brilliantly solved by McDonald and Hunt (1969). In work with a transparent model, they observed that the segregated liquid in the dendrite mesh moved under the influence of gravity. It had a density that was in general different from that of the bulk liquid. Thus the lighter liquid floated, and the heavier sank.

In the case of steel, they surmised that as the residual liquid travels towards the roots of the dendrites to feed the solidification contraction, the density will tend to rise as a result of falling temperature. Simultaneously, of course, its density will tend to decrease as a result of becoming concentrated in light elements such as carbon, sulphur and phosphorus. The compositional effects outweigh the temperature effects in this case, so that the residual liquid will tend to rise. Because of its low melting point, the liquid will tend to dissolve dendrites in its path as solutes from the stream diffuse into and reduce the melting point of the dendrites. Thus as the stream progresses it reinforces its channel, as a flooding river carves obstructions from its path. This slicing action causes the side of the channel that contains the flow to be straighter, and its

220 CHAPTER 5 SOLIDIFICATION STRUCTURE

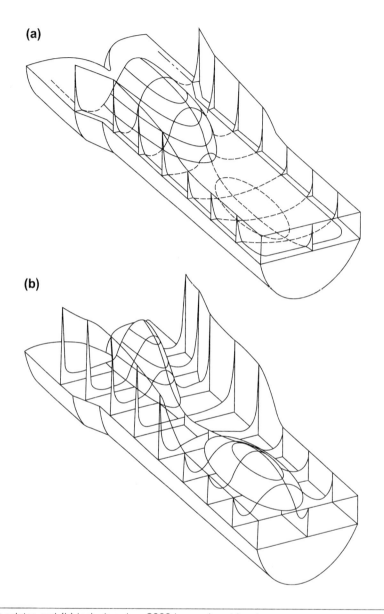

FIGURE 5.44

Segregation of (a) solutes and (b) inclusions in a 3000 kg sand cast ingot.

Information mainly from Nakagawa and Momose (1967).

opposite side to be somewhat ragged. It was noted by Northcott (1941) when studying steel ingots that the edge nearest to the wall (i.e. the upper edge) was straighter. This confirms the upward flow of liquid in these segregates.

The 'A' segregates in a steel ingot are formed in this way (Figure 5.45). They constitute an array of channels at roughly mid-radius positions and are the rivers that empty segregated, low density liquid into the sea of segregated liquid floating at the top of the ingot.

5.3 SEGREGATION

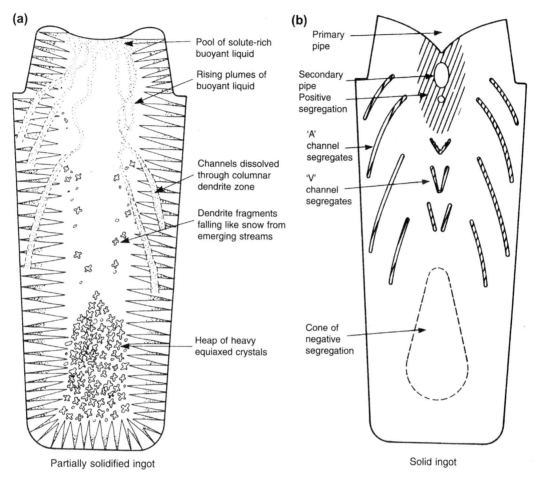

FIGURE 5.45

Development of segregation in a killed steel ingot (a) during solidification and (b) in the final ingot.

At the same time these channels are responsible for emptying the debris from partially melted dendrites into the bulk liquid in the centre of the ingot. These fragments tumble along the channels, finally emerging in the sea of segregation in the head of the ingot, from where they subsequently fall down the centre of the ingot at a rate somewhere between that of a stone and a snowflake. They are likely to grow as they fall if they travel through the undercooled liquid just ahead of the growing columnar front, possibly by rolling or tumbling down this front. The heap of such fragments at the base of the ingot has a characteristic cone shape. In some ingots, as a result of their width, there are heaps on either side, forming a double cone. Because such cones are composed of dendritic fragments their average composition is that of rather pure iron, having less solute than the average for the ingot. The region is therefore said to exhibit *negative segregation*. It is clearly seen in Figure 5.44(a). The equiaxed cone at the base of ingots is a variety of gravity segregation arising as a result of the sedimentation of the solid, in contrast with most other forms of gravity segregation that arise because of the gravitational response of the liquid.

A further contributing factor to the purity of the equiaxed cone region probably arises as a result of the divergence of the flow of residual liquid through this zone at a late stage in solidification, as suggested by Flemings (1974).

The 'V' segregates are found in the centre of the ingot. They are characterised by a sharply delineated edge on the opposite side to that shown by the A segregates. This clue confirms the pioneering theoretical work by Flemings and co-workers that indicated that these channels were formed by liquid flowing downwards. It seems that they form at a late stage in the freezing of the ingot, much later than the formation of the 'A' segregates, when the segregated pool of liquid floating at the top of the ingot is drawn downwards to feed the solidification shrinkage in the centre and lower parts of the ingot.

On sectioning the ingot transversely, and etching to reveal the pattern of segregation, the A and V segregates appear as a fairly even distribution of clearly defined spots, having a diameter in the range 2–10 mm. Probably depending on the size and shape of the ingot, they may be concentrated at mid-radial to central positions in zones, or evenly spread. The central region of positive segregation is seen as a diffuse area of several hundred millimetres in diameter. In both areas, the density of inclusions is high. These channel segregates, seen as spots on the cross-section, survive extensive processing of the ingot, and may be still be seen even after the ingot has been rolled and finally drawn down to wire!

It is interesting to note that in alloys such as tool steels that contain high percentages of tungsten and molybdenum, the segregated liquid is higher in density than the bulk liquid, and so sinks, creating channel segregates that flow in the opposite direction to those in conventional carbon steels. The heavy concentrated liquid then collects at the base of the ingot, giving a reversed pattern to that shown in Figure 5.44.

In nickel-based and high-alloy steel castings the presence of partially melted and collapsed crystals in the channels has the effect of a localised grain refinement, so that on etching longitudinal sections of the castings, the channels seem to sparkle with numerous grains at different angles. In this industry, channel segregates are therefore widely known as 'freckle defects'. The production rate of nickel-based ingots weighing many tonnes produced by secondary remelting processes, such as electroslag and vacuum arc processes, is limited by the unwanted appearance of freckles. It is their locally enhanced concentration of inclusions such as sulphides, oxides and carbides etc. that make freckles particularly undesirable.

Channel segregates are also observed in Al-Cu alloys. In fact workers at Sheffield University (Bridge et al., 1982) have carried out real-time radiography on solidifying Al-21Cu alloy. They were able to see that channels always started to form from defects in the columnar dendrite mesh. These defects were regions of liquid partially entrapped by either the sideways growth of a dendrite arm, or the agglomeration of equiaxed crystals at the tips of the columnar grains. Channels developed both downstream and upstream of these starting points.

The author has even observed channel defects on radiographs of castings in Al-7Si-0.3Mg alloy. Despite the small density differences in this system, the conditions for the formation of these defects seem to be met in sand castings of an approximately 50 mm cross-section.

Although few ingots are cast in modern steelworks, large steel castings continue to be made in steel foundries. Such castings are characterised by the presence of channel segregates, in turn causing extensive and troublesome macrosegregation.

From the point of view of understanding channel formation, Valdes (2010) proposes a condition to predict freckle formation based on the assumptions (1) the Rayleigh number; (2) Darcy flow through the pasty zone; and (3) the condition by Flemings and Nereo (1967) that the liquid flow from colder to hotter regions is faster than the rate of crystal growth. This comprehensive approach to this complex problem appears to yield impressive results. Over succeeding years, there have been a number of excellent studies of the formation of channel defects following Valdes, leading to what appears to be a mature understanding of these defects. Torabi Rad, Kotas and Beckermann (2013) use similar fundamental theory to achieve good predictions calibrated against 27 different steel compositions.

In summary, in terms of parameters which may be within the foundry engineer's control, channel segregates may be controlled by:

1. Decreasing the time available for their formation by increasing the rate of solidification. This action also reduces the spacing of the primary dendrites creating conditions of higher drag forces on the movement of liquid, thus suppressing the rate of formation and growth of channels.
2. Adjusting the chemical composition of the alloy to give a solute-rich liquid that has a more nearly neutral buoyancy at the temperature within the freezing zone.

In practice, both these approaches have been used successfully.

CHAPTER 6

CASTING ALLOYS

The metallurgy of the casting alloys is presented here, interpreted according to bifilm theory. There will be those educated as physicists and metallurgists, like myself, who may be uncomfortable with this new approach. It is with regret that I view the new interpretation as revolutionary. I request the readers' patience to study and understand the approach.

It has to be recognised that at the time of writing the bifilm interpretation is more speculative than I would wish. However, it is a theory that fits the facts with uncanny accuracy. It has to be kept in mind that theories are useful only if they can explain and if they can predict. As a well-known instance, the flat Earth theory was useful, being accurate to predict both distances and directions while humans could only walk limited distances on the Earth's surface. It is a theory that served mankind for thousands of years until the arrival of global travel. The bifilm approach is similarly presented as a useful theory until the time that it is proven inadequate. But at this time, it seems more than adequate to provide many useful insights into the behaviour of liquid metals and castings.

The reader is invited to form his or her own opinion as the metals are described in turn throughout the chapter. For traditional reviews of the metals listed in this section, but not including bifilm theory, nor the benefits of improved methoding techniques such as naturally pressurised filling systems, the reader is recommended the interesting and detailed accounts in the ASM Handbook 2008, Volume 15 'Casting'.

The very low melting point metals such as Pb, Sn and Bi etc. are not dealt with here. We shall concentrate on the structural engineering metals. Even so, in passing, we should note that the casting of lead and its alloys for battery grids is currently a major application that should not be overlooked. The new area of lead-free solders is also of interest, noting that a Sn-based solder containing Ag and Cu exhibits planar voids at the base of ductile dimples on a fracture surface, suggestive that the bifilms in this alloy had been straightened by their formation on the planar intermetallic Ag_3Sn at the base of each dimple. It seems probable that for all metals, not only the high melting point engineering metals, the theory of the control of microstructure and properties of castings by bifilms remains applicable.

6.1 ZINC ALLOYS

For a general introduction to zinc casting alloys, particularly covering the practical details of routine melting and casting, the reader is strongly recommended to the readable account by Arthur C. Street in his famous work *The Diecasting Book* (1986) because approximately 80% of zinc alloys are cast by high pressure die casting (HPDC).

Relatively little is continuously cast, or gravity cast into permanent or sand moulds. The alloy Kirksite is used for prototypes for cast tooling for sheet metal working and plastic injection moulding. The higher strength zinc–aluminium (ZA) alloys are similarly sometimes used for low-volume products cast in permanent or sand moulds. Graphite moulds are sometimes machined from solid blocks for ease and relatively low cost for low volume moulds.

Zinc alloys were first cast by pressure die casting in about 1914. However, these early days were plagued by quality problems which undermined public confidence. The problems were eventually claimed to be brought under some kind of control from research carried out by Brauer and Peirce and published in 1923. As a result, modern zinc-based casting alloys are made from 99.99% pure Zn in an attempt to guard against the problems of contamination from the heavy elements including Pb, Bi, Tl, Cd, In and Sn that were found migrate to grain boundaries and seemed to be associated with the loss of mechanical properties, particularly under conditions of heat or moisture. We shall return to this issue later in the chapter.

A feature of most Zn-based alloys is that practically all contain Al. The two major hot chamber casting alloys were developed in the 1920s as the ZAMAK 3 and ZAMAK 5 alloys (Z = zinc; A = aluminium; MA = magnesium; K = kopper, the German for copper). These are denoted alloys A and B, respectively, in UK specifications. They all contain 4% Al. Alloy A is most common, but alloy B contains a little Cu and is significantly harder. The nominal compositions of the casting alloys are given in Table 6.1. In the US alloy number system, nos. 2, 3, 5 and 7 coincide approximately at nos. 3 and 5 with ZAMAK 3 and 5 and alloys A and B.

The addition of Al to the hot chamber casting alloys appears to have originally been designed to reduce the attack of the mild steel crucible and swan neck components of the pressure die casting machine because of the development of the thin, protective oxide formed by the Al. Also, the 4%Al addition produced an alloy close to the Zn/Al eutectic at 5%Al, and thus making a useful reduction in the melting temperature from 419°C to 382°C, and at the same time giving a valuable increase in strength.

The International Zinc Association have recently publicised a recent development in an effort to achieve even thinner-walled castings, to aid its competition with the light pressure die cast alloys based on Al and Mg. Whereas the traditional 4%Al alloys are claimed to be limited to 0.75 mm thickness sections, the new composition claims 0.3 mm section thickness potential (Goodwin, 2009). Clearly, some benefit will have arisen from the new composition at 4.5%Al being nearer to the eutectic at 5%Al as is clear from Figure 6.1. Furthermore, the limit imposed on the copper content at only 0.07% seems also to be intended to limit the development of a freezing range which would have inhibited fluidity as explained in Chapter 3. This alloy has yet to prove itself commercially at this time, but it has a good chance, because the logic of its design seems sound.

The ZA hyper-eutectic series of alloys containing 8%, 12% and 27% Al were a significant advance achieved by research carried out in the late 1960s. Their high strength, toughness and bearing properties increasing with increasing Al content, make these alloys among the strongest and toughest of commonly available low-cost casting alloys. ZA27 can achieve 440 MPa in the as-cast condition, which is higher than practically all Al and Mg alloys even after expensive heat treatments.

The microsegregation and macrosegregation accompanying the solidification of the higher ZA alloys is impressive and can be troublesome. ZA27 starts as a homogenous liquid of 27%Al, but the first metal to solidify

Table 6.1 Zn Casting Alloys

Alloy	Nominal Composition (wt%)				Comments
Alloy 2	4Al	0.03Mg	3Cu		'Kirksite' if gravity cast
Zamak 3 (alloy A)	4Al	0.03Mg			Common hot chamber alloy
Zamak 5 (alloy B)	4Al	0.05Mg	1Cu		Common hot chamber alloy
Alloy 7	4Al	0.01Mg	0.01Ni		
Internat. Zinc Assoc	4.5Al	0.01Mg	0.07Cu		Eutectic fluid alloy
ZA8	8Al	0.02Mg	1Cu		Hot chamber or gravity cast
ZA12	11Al	0.02Mg	1Cu		Cold chamber or gravity in permanent mould or sand
ZA27	27Al	0.015Mg	2Cu		
AlCuZinc 5	4Al	<0.05Mg	5Cu		General Motors creep-resistant alloys
AlCuZinc 10	4Al	<0.05Mg	10Cu		
Superplastic zinc	22Al	0.01Mg	0.5Cu		(Wrought alloy)
ZM 11	22Al	1.5Si	0.5Cu	0.3Mn	Mitsubishi, high strength, high damping, lightweight alloys
ZM 3	40Al	3.0Si	1.0Cu	0.3Mn	
Super cosmal	60Al	6.0Si	1.0Cu	0.3Mn	

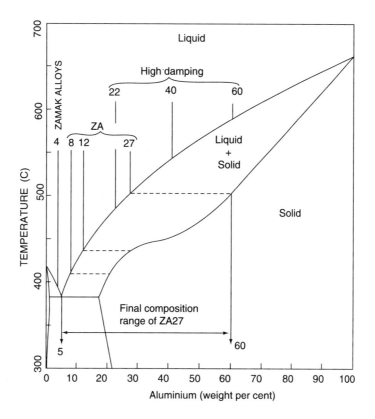

FIGURE 6.1

The Zn-Al phase diagram with some common zinc-based casting alloys.

contains 60%Al, whereas the final liquid to solidify contains 5%Al (Figure 6.1). These wide differences lead to a highly cored dendritic structure that a homogenising heat treatment is sometimes used to encourage some limited redistribution of Al.

Although ZA8 can be cast by the hot chamber pressure die casting process, the higher temperatures required for ZA12 and ZA27 would result in unacceptable rates of attack of the casting machine. The liquidus temperature for ZA27 is 500°C.

The superplastic zinc should not really be included in the casting alloys because it needs plastic working and heat treatment to reduce its grain size before it can develop its full superplastic properties. It is included here simply to illustrate its interesting similarity with Zn casting alloys. The spectacular 1000% elongation that this alloy can develop is limited by microscopic cavitation, almost certainly the result of oxide bifilms entrained during melting and casting. Better melting and casting might deliver even better properties. Its composition is nevertheless close to other casting alloys, indicating that perhaps heat treatment might deliver some part of the superplastic properties, giving a huge benefit to toughness. This does not seem to be generally recognised.

One of the standard problems with Zn-based alloys is their poor creep resistance; the relaxation of steel bolts in holes threaded in Zn-based castings, especially at slightly raised temperatures in the region of 100–150°C has been a central issue for automotive application in the engine compartment of vehicles. Research by General Motors (Rashid and Hanna, 1989) has resulted in the two high copper alloys in Table 6.1 that exhibit improved creep performance.

A related problem is the continued ageing of zinc alloys at room temperature over weeks or months. This change of structure is accompanied by changes in properties, and by minute changes in the size of the casting. The size issues are discussed in Chapter 18.

In a separate quest, this time to optimise noise reduction in vehicles by improving the damping properties of zinc alloys, Mitsubishi has proposed three new alloys at very high levels of Al, 22%, 40% and 60% (Mae and Sakonooka, 1987). These are, of course, really Al-based alloys, their volume additions corresponding approximately to 40%, 60% and 80%, respectively, explaining their improved strength and increased lightness.

Unfortunately, however, from the point of view of the casting quality, the formation of the alumina film on the surface of the liquid alloys gives the standard problems, involving the entrainment of air bubbles and films during the extreme surface turbulence of filling. Although the melting and casting temperatures of zinc pressure die casting alloys are low, therefore restricting the rate of thickening of films, making them even less visible than in Al alloys, this effectively merely sweeps the problem out of sight. The films have to be double, necessarily have an unbonded inner interface, and therefore still perform efficiently as cracks, even though less detectable than those in Al alloys.

The high level of Al in the three Mitsubishi alloys probably means the alloys contain high levels of aluminium oxide in the form of bifilms, so it may be the bifilms that are the reason for the high damping properties; during the shearing across the bifilm as the matrix is strained (see Section 9) the friction involved will dissipate energy in the form of heat. The elastic stiffness (Young's Modulus) of the alloy is similarly significantly reduced by bifilms (as demonstrated by Hall and Shippen (1994) for a bronze alloy). It seems reasonable to conclude therefore that although the damping properties may be found to disappear if the alloys are cleaned from oxides, the alloys will benefit in other ways: their tensile and corrosion properties are likely to improve. They might then enjoy a wider commercial acceptance.

Bifilms seem almost certainly present in Figure 2.31, as witnessed by the sharp changes in microstructure across the boundaries formed by oxide flow tubes around the jets of metal that filled the die. In many situations, the continued flow over previous jets that have solidified can re-melt the solid, allowing it to lose its grip on its oxide tube which is stripped away in the flow. The prior solidified jet then continues to re-melt back somewhat, blurring the original sharp divisions between jets. The longitudinal layering of structures in HPDC products is common, being a characteristic feature of HPDC zinc pressure die cast alloys (Goodwin, 2008).

Romankiewicz (1976) found that zinc alloys can contain up to 1% or 2% of oxides. These reduced mechanical properties. Furthermore, in damp conditions, they became a source of selective corrosion. He offered no explanation of this behaviour. Clearly, in terms of bifilms, the answer is that the films, when entrained by turbulence, are actually double and act as cracks and will usually be pushed by dendrite growth into the interdendritic or intergranular regions. In damp conditions, corrodants such as rainwater can penetrate into the matrix along the unbonded interiors of the bifilms. The corrosion will be enhanced by the precipitation of other elements on the back of the films, explaining the effects of heavy elements such as Pb in promoting the often catastrophic degradation of properties over time. The degradation is often called intercrystalline or intergranular corrosion. It is easily demonstrated after only 10 days of testing in steam (Colwell, 1973) or perhaps as long as a year in tropical locations.

The deterioration of properties with time eventually cause the casting to fail is an experience most of us have with such items as zip fasteners on travel goods, handles on brief cases and window fasteners in the home. Thus although the research of Brauer and Pierce made huge improvements to Zn alloys, the problems are clearly not totally solved.

Other Canadian workers (Dionne et al., 1984) found that toughness of ZA27 was limited by brittle $CuZn_4$ following high temperature homogenisation treatment. This is typical behaviour of a compound precipitating on bifilms (otherwise the failure by cracking is difficult to explain).

Blisters commonly form on the surface of pressure die cast alloys immediately after opening the die. Kaiser and Groenveld (1975) described how when the die is opened, the wall of the hot casting may be insufficiently strong to withstand the pressure of gases entrapped in the casting during die filling. Although the authors report that the defect can be controlled by reducing metal and die temperatures to increase the strength of the casting wall, they do not recommend this practice. Temperatures may rise in service to allow the alloy to creep, slowly forming a blister.

Blisters are known to form on Zn alloys at ambient temperature in the open air after a period varying from 1 to 6 months. The blisters almost certainly grow by the pressure of gases trapped in defects just under the surface of the casting. The defects are most likely bifilms and/or porosity (there really is practically no difference between the two!). The origin of the high pressure of gas inside the defect is less certain. Possibilities include:

1. high pressures of air or other gases entrained during the turbulence of die filling.
2. high levels of hydrogen from plating processes are common in many plated zinc castings
3. hydrogen generated from a surface corrosion processes can diffuse into the metal. The presence of moisture creates a gently corroding environment, releasing hydrogen that will diffuse into the metal, precipitating in the centres of bifilms and/or bubbles.

The blister may grow by the gradual precipitation and increasing pressure in the defect, leading to the gradual creep of the alloy. A flow of hydrogen into the defect seems necessary, because after the initial expansion of the blister, the pressure will be expected to fall rapidly with increasing volume, bringing the rate of growth to a stop. In practice, the blisters appear to continue to grow over a period of months or years.

That Birch (2000) found that the properties fall as dendrite arm spacing increases is further evidence that Zn die casting alloys, particularly alloy 3, contain bifilms that have chance to straighten, reducing properties, as the time for solidification lengthens. This effect is discussed in more detail in Section 9.4.3.

A quite different problem arises in heavy section castings. As the Al content of the ZA series of alloys increases some casting problems emerge which are unique, and which is found in all these alloys, but particularly ZA27. During solidification, the Al-rich dendrites form first and float to the top of the casting. Thus the Zn-rich liquid, being much denser than the Al, is displaced to the bottom of the casting. Because of its much lower melting point, the Zn-rich liquid freezes much later and therefore forms a shrinkage cavity on the base of the casting (completely opposite to a normal casting where the shrinkage would be concentrated in a top feeder). It is known as *underside shrinkage*. Naturally, this problem is more severe in heavier sections and where freezing times are long because of the use of a sand mould as opposed to a metal die.

Canadian workers have explored a variety of alloy additions that appear capable of eliminating underside shrinkage, of which the most practical and cost-effective is Sr (Sahoo et al., 1984, 1985). This worked well—provided an adequate feeder was sited on the top of the casting. These authors were not able to explain why Sr was effective, but surmised that a skin of metal formed on the base of the casting which was sufficiently strong to withstand collapse (even though an internal cavity was now likely to occur in place of the external sink). The pressurisation of the casting interior by a generous feeder will assist to prevent collapse during the early stages of solidification. Furthermore, the pressurisation of the casting skin against the drag surface of the mould will increase the rate of heat transfer, helping to thicken and strengthen the skin. The detailed action of Sr remains a mystery, even though it is tempting to draw an analogy with the modification of Al-Si alloys in which Sr acts to suppress nucleation of the solid ahead of the freezing front, thereby straightening the front, and so forming a stronger solidified skin.

The hot chamber pressure die cast Zn-based alloys yield relatively small components of high accuracy, good finish requiring no machining and thin walls. It also has unequalled productivity and fulfils the requirements of a wide market. Larger castings at somewhat slower rates are produced by the cold chamber pressure die technique. Both types of HPDC parts have practically unique features that make them popular with designers, such as integrally cast rivets or cast threads to facilitate joining, or integrally cast steel inserts to eliminate assembly.

However, even at this date, it is clear that there remains great scope for the improvement of the properties and behaviour of Zn alloys. In my opinion, this centres on the problem to reduce oxide bifilms by improved melt handling and casting. Here once again there is plenty of work to do.

6.2 MAGNESIUM ALLOYS

Magnesium is the lightest structural metal with a density of only 1738 kgm^{-3} at room temperature. Its light weight drives exploration of the use of its alloys for automotive and aerospace in competition with aluminium-based alloys. The competition is finely balanced, with Mg having the highest specific stiffness, and relatively high specific strength, with good damping characteristics (i.e. quiet castings), but remains limited in some applications because its relatively low strength means that full advantage cannot usually be taken of its attractive low density. Numerous other factors weigh against Mg compared with Al.

Historically, the relative volatility of the cost of Mg has been a serious disadvantage in comparison with Al for volume applications. Even so, the aircraft industry in particular finds Mg alloys essential to its designs of numerous castings, and the electronics and communications industries (e.g. mobile telephones) opt for Mg for weight saving of chassis and frames.

Another major factor hampering the wider use of Mg at this time is the difficulty of recycling Mg alloys. Rapid oxidation and the numerous alloys, all in relatively small quantities, make recycling problematic. This contrasts with Al alloys for which an efficient worldwide recycling industry is already in place, making a big contribution towards the reduction in costs of Al castings using these so-called 'secondary alloys', meaning recycled alloys.

The relatively poor corrosion resistance of some Mg alloys means that expensive surface protection is required for some components. However, the more recent introduction of a purer variety of AZ91 alloy (Mg-9Al-1Zn) has made a major contribution towards the growth potential of a volume Mg market. It seems likely that if Mg could be cast without bifilm defects corrosion resistance would be enhanced further at practically zero cost. This prediction remains to be put into practice.

In the meantime, for readers who want a wealth of practical advice on the melting and casting of Mg alloys, two works stand out: *Principles of Magnesium Technology* by Emley (1966) and *Product Design and Development for Magnesium Die Castings* by the Dow Chemical Company (1985). For a more detailed and more up-to-date account of the metallurgy of Mg alloys, the excellent text by Polmear (2006) is recommended. Even so, it is necessary to bear in mind that these renowned texts neither include the concept of bifilms nor such concepts as naturally pressurised filling systems for castings. Thus the reader needs caution.

In fact, when starting to consider the metallurgy of the light casting alloys, Al- and Mg-based, the role of oxide films appears so fundamental that it seems sensible to make a start with first understanding the oxides to be expected on such melts.

6.2.1 FILMS ON LIQUID Mg ALLOYS + PROTECTIVE ATMOSPHERES

In their comprehensive review of the oxidation of liquid metals made in 1969, Drouzy and Mascre mentioned some of the benefits of oxidation, but caution 'oxidation is something of a hindrance in the handling of liquid metals, particularly in casting'. This masterpiece of understatement 'something of a hindrance' seemed worth quoting. Such disarming and admirable restraint summarises the awesome, towering importance of perhaps the central problem in metallurgy.

The oxidation of magnesium metal in air is not easily studied, because the metal often ignites before it reaches its melting point. This emphasises that for foundries, the metal has to be handled at all times in an atmosphere and in a manner that helps to suppress ignition.

The oxide, MgO, has a density greater than that of solid Mg, with the result that the prediction based on the Pilling-Bedworth ratio is less than 1, with the consequence that the oxide does not cover the metal from which it forms, and therefore does not protect it. Clearly, these considerations only strictly apply to solid Mg and its alloys, but require to be kept in mind during the heating of the Mg charge in the furnace before melting.

Once molten, it has been traditional to protect the metal by fluxes based on chlorides and fluorides. However, this practice is less popular today as a result of the widespread problems of the entrainment of flux inclusions in the castings, and as a result of the contamination of the environment with fluorides and other spent fluxes.

There has therefore been a major move to so-called 'dry melting', using protection from gases, plus possibly a little sprinkled sulphur. One of the gases used to suppress the burning of Mg has been sulphur hexafluoride (SF_6). At a dilution of only 0.1–1.0% in air or other gas such as CO_2, the mixture was extremely effective. Xiong and Liu (2007) found that the film of MgO on the melt surface was supplemented by islands of MgF_2 under the oxide. The islands gradually spread to about 50% of the total area. Emami and co-workers (2014) confirmed the formation of a stable MgF_2 film in a mixture of air and 0.5% SF_6, but find that this takes up to 7 min to achieve good protection.

Although SF_6 is now rightly outlawed because of its huge effect on the Earth's ozone layer, other chlorine- and/or fluorine-containing gases are being actively evaluated that are less harmful to the environment, and their mode of action may be similar (International Magnesium Association, 2006). Polmear (2006) describes the use of the refrigerant gas HFC 134a, one of the several forms of tetrafluoroethane. Another promising new protective atmosphere contains boron trifluoride; the toxicity problems of this gas are addressed by using the gas in extreme dilutions generated in line from a solid fluoroborate (Revankar, 2000). Other developments are under way and appear to be having some success to refine further the effectiveness of the traditional dilute SO_2 mixtures in CO_2. Perhaps the old remedies will be proven best after all.

The action of CO_2 alone is perhaps more active than might be expected because it is reduced by Mg to form MgO in two stages: (1) CO_2 is reduced to CO and (2) CO is reduced to carbon (Wightman and Fray, 1983). It is possible that the deposition of carbon may help to suppress the rate of oxidation although the formation of magnesium carbide would also be expected to complicate the situation further. Cochran and colleagues (1977) confirmed that CO_2 strongly inhibited the onset of breakaway oxidation although the mechanism appeared mysterious; they looked for but failed to find any carbon-containing phase in the surface of the metal after solidification. For low-Mg alloys, they found even a small amount of CO_2 was valuable to reduce oxidation, although the amount required was sensitive to the Mg content and the temperature of the melt.

An important detail is that magnesium alloys are known to give off magnesium vapour at normal casting temperatures, the oxide film growing by oxidation of the vapour, effectively growing 'from the top downwards'. This mechanism seems to apply not only for magnesium-based alloys (Sakamoto, 1999) but also for Al alloys containing as little as 0.4 wt% Mg (Mizuno, 1996).

The microscopic structure of MgO films on AZ91 alloy has been studied by Mirak and colleagues (2007) in the limited oxidation conditions offered by the interior of bubbles trapped in the melt. The films are thick and are characterised by much fine-scale wrinkling, clearly capable of entrapping much air. Their general appearance is indistinguishable from that shown for the Al-5Mg alloy in Figure 2.10(b).

In some admirably comprehensive experiments to compare protective atmospheres for melting Mg alloys, Frueling and Hanawalt (1969) noted that a CO_2 atmosphere was better than N_2 or Ar in preventing the 'smoking' of Mg. The smoke is almost certainly the condensation of Mg vapour to Mg metal and its oxidation to MgO (or possibly the direct oxidation of the vapour to oxide) in the atmosphere over the melt. In fact, the attempted protection of molten Mg by Ar is potentially hazardous. I recall a tragic fatal accident in which an experimental low-pressure furnace containing Mg alloy was provided with an atmosphere of Ar. Unknown to anyone, including the operator, the furnace had filled with Mg vapour. On opening the door to top up the furnace with additional Mg alloy ingots, air entered and mixed with the vapour, triggering a massive explosion. Such accidents do not happen when the furnace atmosphere comprises some oxidising fraction, such as air, CO_2, SO_2 and/or gases such as freon containing chlorine and/or fluorine. The clear lesson is (no apologies for spelling this out) the treatment of Mg requires knowledge, expertise, possibly and preferably humility, but certainly caution.

A rather different approach to the protection of Mg melts from oxygen in the air has been proposed by Rossmann (1982) who used Ar in a way that appears to be free from the dangers mentioned previously. He pours liquid argon directly on to the surface of the melt contained in a crucible in the open air. One litre of liquid argon will generate 836 L of argon gas at 1 atm and 15°C, so that very small volumes of liquid argon are required to entirely eliminate oxygen from the surface of the melt. The heaviness of argon is a help to keep it on the surface of the melt, but may not be totally proof against air. At -233°C, the density of Ar gas is 1.66, at 20°C it is 1.28 and at 700°C it is 1.12 kgm^{-3}. Thus it becomes a little less dense than air at 20°C, approximately 1.20 kgm^{-3}. Even so, the laminar spreading of the liquid Ar over the melt, and its action as a wind to keep air away as it boils, may continue to be effective in keeping air away from the melt surface.

Other approaches to developing resistance to burning include the addition of Ca to the Mg alloy at levels in the region of 1% (Sakamoto, 1996). The oxide film on the melt consisted of two layers: a mix of MgO and CaO on the lower layer and a CaO-rich layer on the top. The development of the modified oxide increased the ignition temperature of the alloy by approximately 250 K. It seems curious that two metals that both fail the Pilling-Bedworth criterion when mixed, appear to satisfy it nicely when present on liquid Mg. It is not clear, and perhaps is not likely, that the benefits of such alloying additions can be extended to the more useful engineering alloys such as AZ91. Although such benefits might be welcome, there are, as we have seen, other approaches to the control of Mg ignition that are already available.

For sand moulds, the addition of inhibitors to prevent the ignition of Mg is universal. In particular, for silica-based greensands containing perhaps 4% clay and 3–4% water, additions of 2% boric acid plus 2% sulphur are typical. For a self-setting resin-bonded silica sand, the level of inhibition is similar. However, of course, heavier castings might require more and lighter casting less (Mandal, 2000). Furthermore, the skill of the designer of the filling system is also likely to influence the required amount of inhibition, as described later.

One final aspect of the ignition of Mg alloys is critically important. It is predictable that if a mould was filled with a liquid Mg alloy, but the filling was carried out without surface turbulence, so that the melt surface was at all times undisturbed, the alloy would probably not ignite. This proposal is based on the fact that the huge heat of formation of the oxide is easily conducted away into the bulk of the melt, limiting the rise of temperature of the metal surface to some trivial few degrees. However, if the melt surface is subjected to jumping and splashing, as occurs often when poured under gravity, each splash and droplet has a mass of only a few grams, and is only a few millimetres in thickness or diameter. Thus a sliver of liquid metal when travelling through the air would experience oxidation from both sides, the huge heat of oxidation now saturating the small mass and limited diffusion distance, with the result that the sliver, jet or droplet now reaches a huge temperature, well above the ignition temperature. Thus surface turbulence initiates burning.

If the melt starts to burn, its reactivity will cause it to take oxygen from the silica sand of the mould, and thus will consume its way through mould, possibly continuing through the concrete floor, taking oxygen from the concrete, until the Mg is entirely oxidised. Only then will the burning stop. The next clear lesson for the handling of Mg alloys is that it is absolutely necessary to ensure that burning of the metal does not start.

We shall turn now to the behaviour of Al-Mg alloys in the range of approximately 2–20% Mg. These Al-Mg alloys are so dominated by the oxidation of Mg that they act practically as though they were pure Mg. We shall therefore include them in this section on Mg.

Provided burning can be successfully suppressed, as would be normal for the competent handling of Al-Mg alloys, the surface oxide initially develops as a thin layer of amorphous MgO (Figure 6.2(a)). After an incubation time that seems to depend on chance events to nucleate a change, the oxide crystallises to MgO, and some spinel, the name for the mixed MgO and Al_2O_3 oxide, magnesium aluminate, which can be written $MgAl_2O_4$ (Cochran, 1977). This stage of oxidation is fast (Figure 6.3), particularly if the atmosphere is moist air, creating so much oxide that the film has to corrugate as in a concertina (Figure 6.2(b)). The development of the oxidation process proposed in Figure 6.2 is based on an interpretation of the observations by Rault (1996) and Haginoya (1976). These authors found that the hydrogen released into the melt by the surface oxidation causes some of the sub-surface bifilms to inflate and float, coming into contact with the underside of the surface film. Here they can break open the surface leading to irregular masses of spinel formation. However, Haginoya noticed that those films that were more deeply immersed in the liquid, and for some reason unable to float, perhaps being attached to a submerged surface of the casting, remained as films. This is perhaps to be expected because they will probably require at least a part of the bifilm to be within the diffusion distance to the surface to be successful to gain any hydrogen.

6.2.2 STRENGTHENING Mg ALLOYS

Magnesium metal, like most pure metals, is naturally rather weak, and requires strengthening for most engineering applications. The Mg-based alloys can be complicated (Unsworth, 1988), and only a brief summary can be included here. In general, in common with the other family of low melting point alloys based on zinc, the magnesium based alloys are not only mainly wanting in strength, but tend to creep rapidly as temperatures are raised.

Alloying with Al and Zn has produced a series of useful alloys, known as the AZ series including AZ31, AZ63 and AZ91. The latter alloy (its letters standing for 9 wt% aluminium and 1 wt% zinc) is widely popular. Many of these alloys contain up to 0.4% Mn, as also is the case for the Al and Mn containing alloys AM50 (Mg-5Al-0.4Mn) and AM60 (Mg-6Mg-0.4Mn). Other Al- and/or Zn-containing alloys include Mg-Al-Si, Mg-Al-Rare Earth and Mg-Zn-Cu. The introduction of high purity variants of these alloys, with lower levels of Fe, Cu and Ni has significantly improved corrosion resistance. The common sand-casting alloy AZ91C has now been largely replaced by AZ91E, its high-purity equivalent which has about 100 times better corrosion resistance. The alloy ZC63 (Mg-6Zn-3Cu-0.5Mn) was developed as an easy-to-process material for engine castings such as cylinder blocks and oil pans.

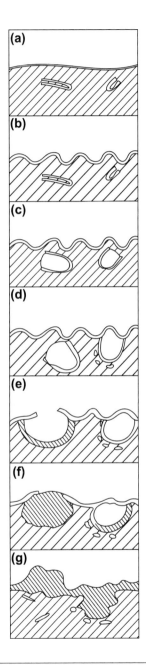

FIGURE 6.2

The stages in oxidation of Al-5Mg alloy. (a) MgO film on surface and folded into the melt; (b) growth of surface film leads to corrugations; (c) in moist air, OH^- ions diffuse through film, giving H^+ ions in the melt, swelling bifilms with H_2; (d) bubbles rise, disrupt surface film, starting break-away oxidation; (e) spinel $MgAl_2O_4$ (dark shaded) forms; and (f, g) complete conversion to thick, irregular spinel film.

Experimental data from Rault et al. (1996).

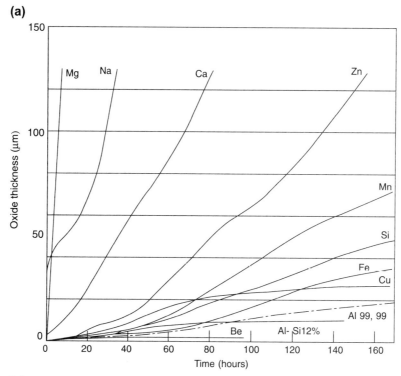
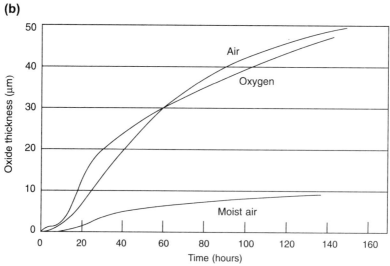

FIGURE 6.3

(a) Growth of oxide on Al and its alloys each containing 1 atomic percent alloying element at 800 °C. (b) Growth of oxide on 99.9Al at 800 °C in a flow of oxygen, and dry and moist air.

Data from Theile (1962).

So far as possible, it is important to grain refine the structure of Mg alloy castings. A typical grain refining action might reduce the grain size from 250 μm to 50 μm. The reason for the importance of grain refinement is that the yield strength σ_y in Mg alloys is strongly related to the grain size by the Hall–Petch relationship:

$$\sigma_y = \sigma_o + K_y d^{-1/2} \tag{6.1}$$

The relationship is strong for Mg alloys because of their hexagonal close packed crystal structure. The only plane allowing relatively easy slip is the basal 0001 plane. All other directions in the crystal represent significant difficulty to initiate slip. Thus, because it will be unlikely for two adjacent grains to have their basal planes aligned, propagating slip from one grain to the next will normally be difficult. Thus the K_y factor in the Hall–Petch equation for Mg alloys is unusually large. (This contrasts with αAl alloys, in which the effect of grain size is relatively weak because K_y is small. This follows because propagating slip from one grain to another is easy in the aluminium face centred cubic [fcc] lattice as a result of its high symmetry; an fcc crystal has many slip possibilities in all directions.)

Unfortunately, the grain refinement of Mg alloys is an unsatisfactory list of recipes of traditional practices, with no convincing science to provide explanations. For instance, the mysterious nature of the grain refinement of Mg-Al alloys by superheating to around 850°C is widely practised, but not understood. The role of carbon dissolved from the steel crucibles has been suspected. A rather wider range of alloys can be grain refined by carbon additions, most popularly by hexachloroethane plunged into the melt to effect some simultaneous de-gassing action. Other black art to grain refine these alloys is listed by Emley (1966). Zhang et al. (2010) described a new grain refining technique for AZ91 consisting of bubbling a mixture of Ar + CO_2 through a melt contained in a rotary de-gassing unit. The treatment takes 30 min, after which the grain size reduction seems better than competitive treatments. Although the authors suggest aluminium carbide, Al_4C_3, may be the effective nucleating agent, they neglect that the melt is almost certainly full of oxide films. Despite this, they report increased tensile strengths increased by up to 20% and elongation by up to 100%.

The Mg-Zr alloys

A completely different class of Mg alloys was opened up by the discovery of the Mg-Zr alloys by Sauerwald in 1947. This has been an outstanding development for Mg, giving alloys with enhanced properties at both room and elevated temperatures. The alloys derive their properties from the remarkable grain refining action of Zr.

Zirconium is only slightly soluble in Mg having a limit of only 0.60%Zr, and it is necessary to saturate the melt with Zr because only the Zr in solution at the time of casting is effective in refining the grain size. On solidification, undergoing a peritectic transformation close to 650°C (the available phase diagrams are not clear in this region), it appears that Zr-rich particles precipitate in the melt. The Zr-rich particles form an extremely effective nucleus for Mg as a result of their similar hexagonal lattice structure, and their nearly identical lattice spacings. The particles are often seen at the centres of Mg grains, confirming their role as nuclei for grain refinement. (The action of Zr is only useful in those alloys that do not contain Al because Al and Zr form a stable compound that effectively eliminates all free Zr from solution. Similar reactions occur with Mn, Si, Fe, Ni, Co, Sn and Sb.)

The main effect of Zr on the strength of Mg alloys (in those alloys free from the previous list of elements) is probably not the result of the usual mechanisms of precipitation hardening or solute hardening etc., but its impressive action on the refinement of grains according to Eqn (6.1). Grain size is typically reduced from 2 to 0.05 mm. This dramatic grain refining action appears to be the result not only of the presence of effective nuclei, but of a significant contribution to the subsequent restriction of the growth of the grains by other solutes, as made clear by StJohn (2005).

Historically, the addition of Zr to a Mg melt has proved to be difficult because of the low solubility of Zr and its loss by reaction with air and reaction with Fe from the steel crucible, particularly as the temperature increases from 730° to 780°C (Cao, Qian et al., 2004). Traditionally, one of the most successful techniques for the addition of Zr has been through the use of a master alloy containing Mg and 30%Zr, plus residual heavy flux, 'weighted down' often with a dense chloride such as $BaCl_2$. The alloy and its content of entrained chlorides and fluorides has to be 'puddled' (i.e. mechanically pummelled or stirred, and then left to settle again. It sits in excess at the base of the crucible to act as a reservoir of Zr, maintaining saturation as nearly as possible, replacing that being continuously lost by oxidation and by reaction with the iron crucible. The melt is poured leaving 10–20% of the melt with its residual flux undisturbed at the bottom of the crucible. (This material can be recycled in other Mg-Zr alloys.)

During studies of the separation of flux from Mg alloys, Reding (1968) confirmed that the flux settles to the bottom of the melt, but remains as separate droplets which refuse to coalesce into a single drop. This seems consistent with the drop being enclosed within a tenacious oxide skin. The SEM EDX studies revealing high Fe and Zr content at the drop/matrix interface corroborating the presence of an oxide skin on which Fe-rich and possibly Zr-rich intermetallics precipitate as on a favoured substrate (see the analogous behaviour in Al alloys, Section 6.3). The report by Reding makes sobering reading and is recommended to readers as a wakeup call in the problems of making Mg-Zr alloys. It is a concern that in the Mg-Zn-Zr alloy that he studies the flux droplets appear not to settle but remain dispersed uniformly through the melt. This suggests the presence of a network of large films, probably of MgO, that are resisting the settling of the flux. Whether this behaviour is exclusive to this alloy seems unlikely, and may reflect poor handling or melting practice. We have much to learn about Mg alloys.

In view of these uncertainties it is perhaps surprising that the flux technique for the addition of Zr has been more or less successful for the production of Mg-Zr alloys for many years (Emley, 1966). A more recent development by (Qian, 2003) showing Zr dissolution working well at only 730°C, stirring for only 2 min, and using melts protected from contamination from their iron crucibles by a simple wash of boron nitride promises future hope of a more controlled manufacture of these useful alloys.

6.2.3 MICROSTRUCTURE

The microstructural development of the very many Mg-based alloys is a complex and vast subject that is beyond our scope. However, some general points can be made about the simplest alloys, particularly the Mg-Al alloys.

The solid grains forming in a liquid Mg alloy melt have a flat (i.e. two-dimensional), hexagonal form, closely resembling the forms of snowflakes. These contrast with the dendritic forms of the cubic structured metals, whose dendrites are three-dimensional stars, with three sets of axes at right angles, their positive and negative directions creating six primary arms.

Slow cooling of the Mg-Al alloys causes a coarse precipitate of $Mg_{17}Al_{12}$ to precipitate at grain and dendrite boundaries (Figure 6.4). This curiously complex phase is a common feature of many Mg alloys. At even lower temperatures, the particles become surrounded by a fine lamellar eutectoid precipitate of Mg and $Mg_{17}Al_{12}$.

Whereas with Al alloys that have enjoyed massive attention from researchers, so that the central role of oxide bifilms in the control of microstructure and properties has now been established beyond doubt, Mg alloys have hardly started on

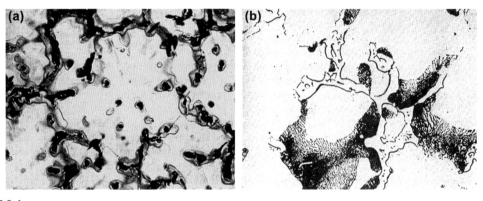

FIGURE 6.4

(a) Microstructure of chill cast Mg-8Al-0.5Zn-0.25Mn. The common intermetallic compound $Mg_{17}Al_{12}$ (dark) and interdendritic coring (grey) is shown by the electrolytic polish. (b) Slowly cooled alloy showing eutectoid precipitate from decomposition of Al-rich solid solution (nital etch) (Emley, 1966).

this path. Nevertheless, it is interesting to take an overview of an authoritative text such as Emley's *Principles of Magnesium Technology* (1966) to see that most of the defects appear to be films. At that time, of course, it was not known that all such films were in fact bifilms. It is tantalising therefore to consider that Mg alloys might similarly be controlled by the presence of bifilms because the bifilms are to be expected in Mg alloys. A few early hints that this may be so are listed later.

At this time, there is little evidence that $Mg_{17}Al_{12}$ nucleates on the outer surfaces of bifilms, even though closely analogous behaviour in Al-based and other alloy systems would lead us to expect this. Also, the presence of central linear features, including voids, and the well known brittleness of this compound are not easily explained in the absence of bifilms.

Srinivasan and colleagues (2006) investigated the structure and properties of AZ91 alloy to which had been added 0.5Si in an effort to improve high temperature strength and creep resistance by the precipitation of Mg_2Si particles. They found that the Mg_2Si formed as coarse 'Chinese script' and actually slightly reduced properties, but an 0.1Sr addition made a radical change to the microstructure, changing the Mg_2Si from Chinese script to a refined form, evenly distributed along the grain boundaries.

It is tempting to draw a comparison with Al alloys, where it is fairly certain that Sr deactivates oxide bifilms, preventing the nucleation and growth of other phases on the bifilms, and thereby forcing the formation of these phases at lower temperatures in eutectic growth modes. In this case, it seems probable therefore that Mg_2Si is precipitating on bifilms and taking a Chinese script form. After Sr addition, Mg_2Si no longer precipitates on bifilms and thus seeks the next most energetically favourable mode of precipitation, which appears to be a eutectic form associated with the grain boundaries. Other work on the addition of Sr at levels of only 0.01–0.02 wt% to AZ91 was reported by Aliravci (1992) reporting a number of changes such as finer grain size, reduced porosity and improved mechanical properties. These issues are unexplained and clearly require a significant further research effort.

Over the past few years, work has started to reveal the true role of bifilms in Mg metallurgy. As is now well known in Al metallurgy, in Mg metallurgy bifilms are also in control. Griffiths and Lai (2007) find large butterfly mirror images of bifilms on every tensile fracture surface and facets aligned by dendrites, suggesting bifilms cover practically the whole of their fracture surfaces. Using Weibull statistics to quantify the scatter of strength properties, they find the modulus of strength of top-filled castings only four compared with a modulus 12 for bottom-gated castings, confirming a threefold reduction in scatter of properties for carefully poured castings. Other recent work by the author on large Mg alloy castings for the frames of jet turbines for aero engines has clearly demonstrated the benefit of naturally pressurised filling systems which reduce the entrainment of bifilms and bubbles.

As in all cast metals at this time, the attainment of zero bifilms for Mg alloys remains an as-yet unrealised, but probably realisable, target. The first steps are being taken to reduce bifilm damage. It seems likely therefore that when founders start to cast Mg alloys entirely without bifilms, we should expect surprises.

6.2.4 INCLUSIONS

Flux inclusions can be troublesome in Mg alloys, reacting with moisture in the air and growing to create unsightly pustules of corrosion products on both the cast and machined surfaces. It seems likely that during surface turbulence, with a layer of flux on the surface, sitting on top of the MgO film, the flux inclusion always takes the form of a MgO bifilm which contain entrapped flux; it is probably a 'flux sandwich'.

Sand inclusions in Mg alloy sand castings are expected to be common if the filling system is sufficiently turbulent. However, because of an energetic mould–metal reaction Mg alloys can contain a unique defect: a reacted sand inclusion. Lagowski (1979) carried out a detailed study of these defects and their effects. It seems that the sand grains are detached from the mould wall and swept into the interior of the melt. The majority sink and sit on the drag surface of the casting. Here the 0.1 to 0.2 mm diameter grains react with the magnesium, swelling and dissolving until finally only a nebulous halo of the order of 1 mm diameter remains. On a radiograph the residue of a multitude of such inclusions can look like a snow storm. However, despite the alarming appearance, the mechanical properties appear to be hardly affected, provided the radiographic rating does not deteriorate beyond 4 on the ASTM radiographic standards. (The ASTM standard radiographs represent increasing severity of defects ranging from the least severe, rated 1, equivalent to a single inclusion of approximately 1 mm diameter in the representative 50 mm × 50 mm area, to the most severe, rated 8, depicting a blizzard.) Entrained sand inclusions are, of course, a clear signal that the filling system is turbulent; thus the problem is easily avoided by substituting a non-turbulent mode of filling.

6.3 ALUMINIUM

A special feature of the Al alloy market is that the metal is divided into primary and secondary markets in which the primary alloys are those made from metal freshly won from the ore, whereas secondary alloys are based on recycled material, recycled alloy economises on the huge energy required to smelt Al from its ore because melting it for the second time requires only about 5% of that energy. This economy is of course offset by the greater oxide and impurity contents, particularly iron. The Al alloy casting alloys for most applications are based on secondary metal. Only relatively few applications for aerospace and safety-critical items might specify primary alloys. Otherwise, most primary metal goes for wrought applications such as sheet and foil.

The Al alloys have enjoyed a huge expansion over the past few decades as a result of their accessible casting temperatures and light weight. Although the alloys have been presumed easily processable, probably as a result of their low melting temperature compared with cast iron, the industry has paid a high price for this presumption. Processing the alloys in the way that cast iron has been traditionally handled has resulted in major problems for Al castings and for the fledgling Al casting industry; the surface oxide has become entrained, leading to numerous and serious bifilm problems, and the hydrogen gas in solution in the liquid metal is partitioned uniquely strongly during solidification, inflating bifilms to become porosity. Thus there have been important and unsuspected problems with attempts to make castings in Al and its alloys. The metallurgical aspects of these issues will be dealt with in this section, and the practical engineering aspects of metal handling and casting later in Chapter 10 onwards.

Finally, these issues should not take our focus away from the central problem with the casting of all Al alloys; during the turbulence of pouring, the incorporation of the surface oxide into the bulk of the alloy, where it constitutes a bifilm crack that impairs properties. Al alloys have therefore gained a reputation for unreliability. It is necessary to understand this problem to ensure that it is effectively avoided. This is the reason for making an early start in this chapter with the oxidation of the melt immediately following the short introduction of the main alloy types.

The strengthening of the surprisingly weak and soft pure aluminium metal has led to the development of three families of casting alloys.

1. The Al-Si alloys form the mainstream of the industrially important engineering cast alloys. The presence of hard, metallic Si particles in the alloys gives the alloys the character of an in situ metal/matrix composite (MMC). Additions of Mg promote the precipitation of Mg_2Si in precipitation hardening heat treatments, creating one of the most widely used structural engineering alloys: Al-7Si-0.4Mg. Cu addition to Al-Si alloys similarly will precipitate and harden by the precipitation of $CuAl_2$ although the Cu addition impairs the corrosion resistance of the alloy. The Al-Si alloys, with perhaps Mg and/or Cu additions will feature significantly in this section.

In contrast to the two-phase Al-Si alloys, two alloys very different to the Al-Si types, essentially single-phase solid solution alloys, are also important:

2. The Al-Mg series, in which the Mg content may rise to around 10 wt% are reasonably strong ductile alloys with significant corrosion resistance, and capable of taking a bright white anodising coat much loved by the food industry. The choice of Mg levels is typically 5–6 wt%. The strongest Al-10 Mg alloy has been generally avoided over recent years as a result of people fearing rumours of it becoming brittle with age, particularly in warm conditions such as in the tropics or in engine compartments of cars. This is a travesty of the facts, and it is a pity that this excellent alloy has suffered such unjust criticism; it is true that it loses a little of its huge ductility after some years, possibly as a result of the precipitation of Mg_5Al_8 (or perhaps Mg_2Al_3) in grain boundaries (probably on bifilms—suggesting a remedy by eliminating bifilms). Even so, its generous reserves of ductility ensure that it always remains ductile, in fact, usually more ductile than most of the Al-Si alloys even when at their best. Much of its ductility arises as a result of its uniform single-phase solid solution structure (contrasting

sharply with the MMC structure of the Al-Si alloys). When liquid, the Al-Mg alloys react strongly with their environment, and usually require to have inhibitors in the mould to reduce metal/mould reactions if cast into sand.

3. The second important series of single phase solid solution alloys is that based on Al-Cu. Copper contents up to about 5.0 wt% are possible, with many alloys with a nominal composition around 4.5%. These alloys, when subjected to solution treatment and ageing can achieve high strength and ductility. A famous alloy, A201, contains approximately 0.7%Ag and is one of the strongest and toughest of all the cast Al alloys. Unfortunately, having silver lying about in the foundry introduces a security headache, adding to the daily problems the founder would rather avoid.

6.3.1 OXIDE FILMS ON Al ALLOYS

Considering first the reaction of liquid aluminium with oxygen, the equilibrium solubility of oxygen in aluminium is extremely small; less than one atom in about 10^{35} or 10^{40} atoms. This corresponds to less than one atom of oxygen in the whole world supply of the metal since Al was discovered and extraction began. Even allowing for the dynamic effects described in Chapter 1 in which the metal can have much higher levels of oxygen in solution, we can safely approximate its solubility to zero. Yet everyone knows that aluminium and its alloys are full of oxides. How is this possible? The oxides certainly cannot have been precipitated by reaction with any oxygen in solution. Oxygen can only reach and react with the metal surface. Furthermore, the surface can only access the interior of the metal if it is entrained or folded in. This is a mechanical, not a chemical process. The presence of oxygen in aluminium thereby has to be understood not in terms of chemistry but in terms of mechanical entrainment.

It makes sense therefore to start with an understanding of the reactions at the surface of the metal. For the process and mechanisms of oxidation, the reader is referred to the few original sources concerned with the oxidation of liquid aluminium alloys (Theile, 1962; Drouzy and Mascre, 1969). This short review of the oxidation of Al-based alloys is based on these works.

For the case of pure liquid aluminium, the oxide film forms initially as an amorphous variety of alumina that quickly transforms into a crystalline variety, gamma-alumina. These thin films, probably only a few nanometres thick, their thickness consisting of only a few molecules, inhibit further oxidation. However, after an incubation period the gamma-alumina in turn transforms to alpha-alumina which allows oxidation at a faster rate. Although many alloying elements in aluminium, including iron, copper, zinc and manganese have little effect on the oxidation process (Wightman and Fray, 1983), other alloys mentioned later exert important changes.

Figure 6.3 shows, approximately, the rate of thickening of films on aluminium and some of its alloys based on weight gain data by Theile (1962). The extremes are illustrated by the rate of thickening of Al-1.0 atomic %Mg at about 5×10^{-9} ms^{-1} that is more than 1000 times faster than the 1 atomic %Be alloy. Interestingly, Theile found that water vapour in an oxidising environment inhibited the rate of oxidation of Al (Figure 1.3). This finding seems to be in accord with shop floor experience of operating furnaces in which the incoming Al alloy charge is preheated by the spent furnace gas (flue gas necessarily containing much water vapour). Although Al melts would be normally expected to increase their hydrogen content in proportion to the water vapour in the environment, over a certain concentration of moisture the oxide film appears to become protective, inhibiting further uptake (but regrettably Cochran (1977) found this benefit does not extend to high Mg melts) so that high levels of hydrogen are not experienced in the melted metal from such furnaces.

The Al-Mg system is probably typical of many alloy systems that change their behaviour as the percentage of alloying element increases. For instance, where the aluminium alloy contains less than approximately 0.005%Mg the surface oxide is pure alumina. Above this limit, the alumina can convert to spinel, Al_2MgO_4. It is important to note for later reference that the spinel crystal structure is quite different from any of the alumina crystal structures. Finally, when the alloy content is raised to above approximately 2%Mg, then the oxide film on Al converts to pure magnesia, MgO (Ransley and Neufeld, 1948). These critical compositions change somewhat in the presence of other alloying elements.

In fact, the majority of aluminium alloys have some magnesium in the intermediate range so that although a pure alumina film forms almost immediately on a newly created surface, given time, it will usually be expected to convert to a spinel film.

The films have characteristic forms under the microscope. The newly formed alumina films are smooth and thin (Figure 2.10(a)). If they are distorted or stretched they show fine creases and folds that confirm the thickness of the film to be typically in the 20–50 nm range. The magnesia films are corrugated, as a concertina, and typically 10 times thicker (Figure 2.10(b)). The spinel films are different again, resembling a jumble of crystals that look rather like coarse sand paper (Figure 2.10(c)).

Rough measurements of the rate of thickening of the spinel film on holding furnaces show its growth to be impressively fast, approximately 10^{-9} to 10^{-10} $kgm^{-2}s^{-1}$. Although these speeds appear to be small, they are orders of magnitude faster than the rates of growth of protective films on solid metals. Because the oxide itself is fairly impervious, its rate of growth expected to be controlled by the rate of diffusion of ions through the oxide lattice, how can further growth occur quickly after the first layer of molecules is laid down?

It seems that this happens because the film is permeated with liquid metal. Fresh supplies of metal arrive at the surface of the film not by diffusion, which is slow, but by flow of the liquid along capillary channels, which is, of course, far faster. The structure of the spinel film as a porous assembly of oxide crystals percolated through with liquid metal, as coffee percolates through ground beans, may be an essential concept for the understanding of its behaviour.

We have already seen that progressive Mg additions to Al change the oxide from alumina, to spinel and finally to magnesia. A cursory study of the periodic table to gain clues of similar behaviour that might be expected from other additives quickly indicates a number of likely candidates. These include the other group IIA elements, the alkaline earth metals including beryllium, calcium, strontium and barium. The Ellingham diagram (Figure 1.5) confirms that these elements have similarly stable oxides, so stable in fact that in sufficient concentration, alumina can be reduced back to aluminium and the new oxide take its place on the surface of the aluminium alloy. The disruption or wholesale replacement of the protective alumina or spinel film can have important consequences for the melt.

In the case of additions of beryllium at levels of only 0.005%, the protective qualities of the film on Al-Mg alloy melts is improved, with the result that oxidation losses are reduced as Figure 6.3 indicates. Low-level additions of Be have been found to be important for the successful production of wrought alloys by continuous DC (direct chill) casting possibly because of a side effect of the strengthening of the film, causing it to 'hang up' on the mould or ladle, and so avoid entering the casting, as discussed earlier.

Attempts to measure the strength of films on Al alloys have been made from time to time (Kahl and Fromm, 1984, 1984; Syvertsen, 2006), but the measurements are not easy to make and the results seem of uncertain value at this time.

Strontium is added to Al-Si alloys to refine the structure of the eutectic in an attempt to confer additional ductility to the alloy. However, strontium has a significant effect on the oxidation behaviour of an Al-7Si-0.4Mg alloy as determined by Dennis et al. (2000). Strontium, as with magnesium, seems also to form a spinel, its oxide combining with that of aluminium to form $Al_2O_3 \cdot SrO$ alternatively written Al_2SrO_4. In addition, the resistance to tearing of the film is probably also increased, affecting the entrainment process. Because of this additional powerful effect on the oxide film, the action of Sr as an addition to Al alloys is complicated. It is therefore dealt with separately in Section 6.3.5.

Sodium is also added to modify the microstructure of the eutectic silicon in Al-Si alloys. In this case, the effect on the existing oxide film is not clear and requires further research. Sodium will have much less of an effect in sensitising the melt to the effect of moisture because it is less reactive than strontium. In addition, sodium is lost from the melt by evaporation because the melt temperatures used with aluminium alloys, typically in the range 650–750°C, approach its boiling point of 883°C. The wind of sodium vapour issuing from the surface of the melt will act to sweep moisture vapour and hydrogen away from its environment. Both the reduced reactivity and the vapourisation would be expected to reduce any hydrogen problems associated with Na treatment compared to Sr treatment, in agreement with general foundry experience.

However, Wightman and Fray (1983) found that all alloys that vapourise disrupt the film and increase the rate of oxidation. The additions they tested included sodium, selenium and (>900°C) zinc. The disruption of the film acts in opposition to the benefit of the wind of vapour purging the environment in the vicinity of the melt. Thus the total effect of these opposing influences is not clear. It may be that at these low concentrations of solute any beneficial wind of vapour is too weak to be useful, allowing the disruption of the film to be the major effect. However, the overall rate is in any case likely to be dominated by the reduced reactivity of Na with moisture compared with Sr.

Experience of handling liquid aluminium alloys in industrial furnaces indicates that the character of the oxide film is visibly changed when sodium, strontium or magnesium is added. For instance, as magnesium metal is added to an Al-Si alloy, the surface oxide on the melt is seen to take on a glowing red hue that spreads out from the point that the addition is made. This appears to be an effect of emissivity, not of temperature. Also, the oxide appears to become thicker and stronger. The beneficial effects found for the improved ductility of Al-Si alloys treated with sodium or strontium may be due not only therefore to the refined silicon particle size, as discussed in Section 6.3.5.

6.3.2 ENTRAINED INCLUSIONS

In aluminium alloy castings, the standard well known inclusions are isolated fragments of aluminium oxide. The fragments are too thick and chunky to be newly formed oxide, but are almost certainly particles carried over from the melting furnace. To be included in the alloy, they will have necessarily undergone an entrainment event so that they will themselves be wrapped in an envelope of new, thin oxide film. (Figure 2.3) If this entrainment was some minutes or hours before casting, there is a good chance that the defect can be regarded as extensively if not completely 'old' oxide, with its new envelope more or less welded in place over much of its surface as a result of continuing reaction with the entrained layer of air separating the wrapping from the granular oxide. Some sintering may also have occurred to supplement the welding process. Thus it will probably remain permanently compact, its welded patches inhibiting its relatively new envelope from re-opening in the mould cavity.

Such alumina inclusions are extremely hard. During machining, they are often pulled out of the surface, forming long tears on the machined face. The ease of pulling out is almost certainly the direct result of the inclusion being unbonded to the matrix as a result of its surrounding envelope acquired during its entrainment into the liquid. These rock-like inclusions cause the machinist additional trouble by chipping the edge of the cutting tool.

Carbides, and sometimes borides, are also reported for some Al alloys, but these appear to have been little researched. It seems likely in any case that the precipitation of such phases may occur on oxide bifilms, so, in a way, it is perhaps more of a bifilm problem than a carbon contamination problem. Only further research will clarify this situation.

Turning now to the subject of entrainment of the liquid surface, together with its oxide film, during surface turbulence: when the surface of the melt becomes folded in, the doubled-over films take on a new life, setting out on their journey as bifilms. The scenario has been discussed in some detail in Section 6.2 in the description of liquid metal as a slurry of defects.

As an overview of these complicated effects, Figure 6.5 gives an example of the kinds of populations of defects that may be present. This figure is based on a few measurements by Simensen (1993) and on some shop floor experiences of the author. Thus it is not intended to be any kind of accurate record; it is merely one example. Some melts could be orders of magnitude better or worse than the figures shown here. However, what is overwhelmingly impressive, and clearly shown, are the vast differences that can be experienced. Melts can be very clean (1 inclusion per litre) or dirty (1000 inclusions per cubic millimetre). This difference is a factor of a 1000 million. It is little wonder that the problem of securing clean melts has presented the industry with a practically insoluble problem for so many years. These problems are only now being resolved in some semi-continuous casting plants, and even in this case, many of these plants are not operating particularly well. It is hardly surprising therefore that most foundries for shaped castings have much to achieve. There is much to be gained in terms of increased casting performance and reliability. The mechanics of cleaning the melt, eliminating the oxides, and then ensuring that the oxides are not re-introduced before casting, is dealt with in detail in Chapter 10 and the following chapters.

6.3.3 GRAIN REFINEMENT (NUCLEATION AND GROWTH OF THE SOLID)

There are many reasons given for refining the grain size of castings. Yield strength and toughness are increased; microsegregated phases are more evenly distributed at grain boundaries—the susceptibility to hot tearing is reduced. The Al alloys are well behaved in this respect, so that the grain refining action is now thought to be relatively well understood and appears to be under good control. Even here, however, everything is not what it seems because the immense importance of bifilms, strongly influencing these processes, has been mainly overlooked.

The addition of titanium in various forms into aluminium alloys has been found to have a strong effect in nucleating the primary aluminium phase. It is instructive to consider the way in which this happens.

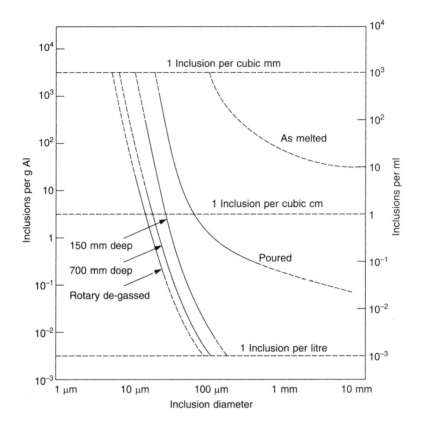

FIGURE 6.5

An example of inclusion content in an Al alloy.

Sigworth and Kuhn (2007) described a well-accepted theory in which the addition of Ti to Al alloys raises the melting point by 5°C. At the same time, the first phase to form is rich in Ti as a result of the relatively rare condition in which the partition coefficient is more than unity, causing solute to be preferentially absorbed by the growing solid rather than being segregated ahead. They point to elegant micrographic studies revealing Ti-rich dendritic shapes in the centres of refined grains as evidence of the correctness of the theory. All this experimental evidence is certainly in line with expectations from the phase diagram (Figure 6.6), but does not explain *refinement*.

The fundamental reason for the nucleation of grains appears to be the prior existence of $TiAl_3$ nuclei. There is no doubt that $TiAl_3$ is an active nucleus for aluminium because $TiAl_3$ is found at the centres of aluminium grains, and there is a well-established orientation relationship between the lattices of the two phases (Davies et al., 1970). These nuclei are stabilised by the huge undercooling of hundreds of degrees Celsius (Figure 6.6). At such high undercoolings, the $TiAl_3$ phase might even form by homogeneous nucleation. It seems quite probable that this could happen in the melt during the Ti addition, when the local Ti concentrations around the dissolving Ti addition start as extremely high Ti levels, well into the $TiAl_3$ undercooled region of the phase diagram before dispersing and falling to lower levels at which the $TiAl_3$ would be ineffective. Equally probably, the $TiAl_3$ particles are likely to be added ready-made in the grain refining addition.

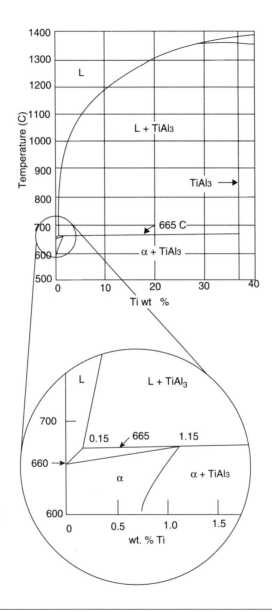

FIGURE 6.6

Binary Al–Ti phase diagram.

TiAl3 in the liquid metal at a Ti concentration above about 0.15 wt% would be expected to be stable (Figure 6.6). However, many Al alloys contain Ti but at levels lower than this, so there is the danger that the TiAl3 can disappear into solution in the melt. This is probably the mechanism for the 'fade' of the grain refinement effect with time.

Results of several researchers shown in Figure 6.7 illustrate that the effect of titanium in the grain refinement of aluminium starts at concentrations well below 0.10Ti. How titanium can be effective at concentrations anything lower

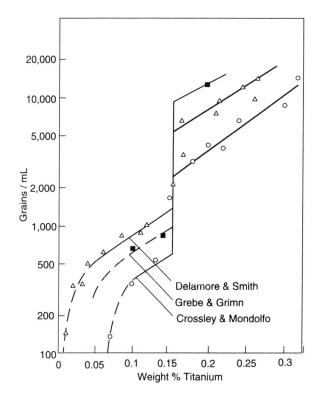

FIGURE 6.7

Increase in grain refinement with increasing titanium addition, especially at the peritectic 0.15Ti as found by several researchers.

than 0.15% has remained a mystery until the epoch-making research by Schumacher and Greer (1993, 1994). These researchers carried out their studies on an amorphous aluminium alloy as an analogue of the liquid state. Nucleation is more easily observed because the kinetics of reaction are 10^{16} times slower. Using TEM (transmission electron microscopy) they observed that $TiAl_3$ was present as adsorbed layers on titanium boride (perhaps more accurately, titanium diboride) (TiB_2) crystals, and so its existence was stabilised at lower levels of Ti than would be expected from the phase diagram, and it was thereby effective in nucleating aluminium for far longer periods as a result of the near-insolubility of TiB_2.

Effective grain refinement, however, seems to require more than simply the nucleation of new grains. A second important factor is the suppression of their growth. If growth is fast, grains can grow large before others have chance to nucleate. Conversely, if growth rates can be suppressed, greater opportunity exists for more grains to nucleate. For complex systems, where many solutes are present, the rate of growth of grains is assumed by Greer and colleagues (2001) to be controlled by the rate at which solute can diffuse through the segregated region ahead to the advancing front. They use a *growth restriction parameter* Q based on the concentration head of the front, defined as

$$Q = m(k-1)C_o \qquad (6.2)$$

Although they note that this relation should be modified by the rate of diffusion D of the solute, these factors are not well known for many solutes. They therefore assume that the rates of diffusion are fairly constant for all solutes of interest in aluminium (this is not far from the truth for the substitutional solutes shown in Figure 1.4(a)) Thus the

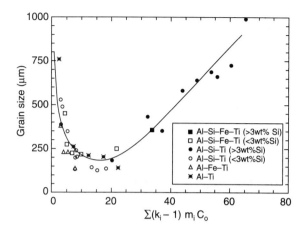

FIGURE 6.8
Effect of the growth restriction factor on the grain size of various Al alloys (Greer, 2001).

potentially more accurate summation of the effects of different solutes in solution, weighted inversely by their diffusivities as proposed by Hodaj and Durand (1997), is neglected in favour of a simple summation of Q values. The effect is shown in Figure 6.8. Initially, grain size clearly decreases with increasing total values of Q. With negligible values of growth restraint grains are seen to grow to nearly 1 mm in diameter. As restraint is increased grain size reduces to nearly one tenth of this, but can get no smaller. After this minimum, increasing Q now results in grain size returning to 1 mm.

The subsequent apparent growth of grains with increasing Q above about 20 is thought by Greer (2002) to be the result of the special effect of Si in 'poisoning' the grain refinement action of Ti at Si contents over 3%. The higher Q data are defined only by alloys with Si contents above 3%. For other solutes at high Q, particularly Cu, it is thought that the grain size remains small. Anyway, it seems that the attainment of fine grain size in Al-Si casting alloys has fundamental limitations, the attainable sizes being 5 to 10 times larger than those in some other casting alloys and in most wrought alloys.

Sigworth and Kuhn (2007) reported practical tests in Al-Si, Al-Si-Cu, Al-Cu, Al-Mg and Al-Zn-Mg alloys. They find a more confusing overall picture in which the grain refining response is different for each alloy system. With today's powerful Al-Ti-B refiners, they suggest there is no reason for large additions of soluble titanium in most alloys. In fact, they suggest it is more accurate to consider that we grain refine with boron, not titanium. The recommended addition is 10–20 ppm of boron, preferably in the form of Al-5Ti-1B or Al-3Ti-1B rod. Lower dissolved titanium levels provide better grain refinement.

However, copper-containing Al-Si casting alloys are the exception in their work. In alloys such as 319 or 355, they find it is best to have a minimum of about 0.1% Ti, which is of course in line with our earlier discussion in this section, but Chen and Fortier (2010) recommended that Ti should be near but not exceed 0.10% for all Al-Si alloys (see also Section 5.2.2). Clearly, we still have much to learn about optimising grain refinement.

Melt cleaning by grain refinement

We cannot leave the subject of grain refinement without adding a major caution to the reader. It is certain that the grain refining additives act to precipitate and grow on oxide bifilms which then become heavy and sediment to the bottom of the melt. The newly cleaned melt will then result in castings with improved properties. If the sediment is stirred in an (unfortunately misguided) effort to maximise the amount of grain refiner transferred to the casting then the bifilms with their precipitates of titanium-rich compounds will enter the casting, and properties are likely to be impaired. Thus it must be kept in mind that the major benefits of grain refinement are usually only marginally or even negligibly associated with the refining of grains, but mainly the result of the cleaning of the melt from oxide bifilms. This important aspect is discussed more fully in Section 9 on properties. The effect was elegantly demonstrated by Nadella, Eskin and Katgerman (2007) in which billets of continuously cast high strength 7075 alloy were found to crack on cooling. In an effort to reduce cracking, grain refiner was added to the metal flowing in the launder on its way into the mould, resulting in excellently grain refined ingots but still cracked. However, when the grain refiner was added to the ladle, permitting bifilms time to sediment, the ingots were grain refined but crack-free.

Other nucleation and growth effects are happening during the solidification of many Al alloys as a result of the many solutes that are present, both intended and unintentional.

6.3.4 DENDRITE ARM SPACING (DAS) AND GRAIN SIZE

In Figure 5.36, we can see the close link between freezing time and dendrite arm spacing (DAS), and in Figure 2.47 the close link between DAS and improved tensile properties. This is the reason, of course, for the strong motivation to chill castings to increase their strength and ductility.

In general, DAS is controlled by freezing rate alone (alloying has a minor effect that can usually be ignored). However, there is an interesting benefit to DAS from the action of grain refinement in removing bifilms from suspension. This occurs because the thermal conductivity of the melt increases when the bifilm barriers to heat flow are removed. Thus heat can be extracted more quickly from the casting, and DAS is consequently reduced.

In contrast, grain size is controlled mainly by the chemistry of grain refining additions, plus any mechanical action that will fragment grains, leading to an effect sometimes called grain multiplication. The combined and contrasting effects of DAS and grain size on mechanical properties are dealt with in Chapter 9.

6.3.5 MODIFICATION OF EUTECTIC Si IN Al-Si ALLOYS

The *modification* of Al-Si alloys is the refinement of the Si eutectic phase by the addition of a small amount of a *modifier* such as Na or Sr. Figure 6.9 illustrates the dramatic change in structure which can be achieved.

When Al-Si alloys are solidified the eutectic silicon is seen on polished sections to consist of coarse, sharp-edged plates (Figure 6.9(a)). These have usually been thought to be detrimental to mechanical properties, being assumed to act as crack initiators. For this reason, for alloys containing more than about 5–7% Si, the addition of sodium or strontium to the melt has been favoured to refine the eutectic silicon phase, converting, as if by magic, the coarse Si flakes into an attractively fine eutectic (Figure 6.9(b)). Modification usually benefits the mechanical properties. However, life is rarely so simple. This straightforward scenario is beset with contradictions and complications. We shall try to unravel some of these problems here as a salutary and instructive exercise in the influence of bifilms on behaviour of melts and cast products. The patience of the reader is requested while we plough through what appears to be an essentially simple mechanism, but clouded by a long history of misconceptions.

Hyper-eutectic Al-Si alloys

In *hyper-eutectic* Al-Si alloys, the primary silicon forms as compact polyhedral shapes. Here the nucleant is most definitely an aluminium phosphide, most probably AlP (although there is the possibility that the nucleus is actually AlP_3 as suggested by Dahle and co-workers, 2008). The particle has an epitaxial relation with Si as is clear from the images recorded by Arnold and Presley in 1961 (Figure 6.10).

The study by Cho (2008) confirmed that an aluminium phosphide (assumed here for simplicity to be AlP) is the preferred nucleant for Si, suggesting that when P is added to be present in sufficiently high concentration the AlP probably nucleates homogeneously in the melt. As small particles, the precipitating Si can wrap completely around them, therefore taking up a compact form.

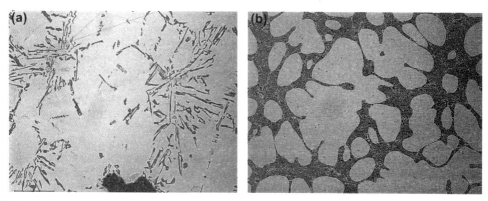

FIGURE 6.9

(a) An unmodified Al-Si eutectic in Al-7Si alloy; (b) the alloy modified by Sr addition.

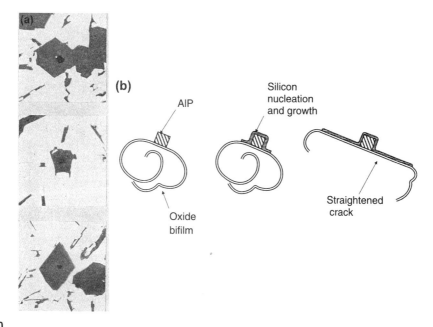

FIGURE 6.10

(a) Primary Si particles nucleated on AlP showing the clear crystallographic relationship (Arnold and Presley, 1961) and (b) the aluminium phosphide particle nucleating on an oxide bifilm. The subsequent nucleation and growth of Si straightens the bifilm, effectively extending and flattening the crack, thus reducing properties in this unmodified condition.

However, of course, at the same time the presence of a copious quantity of oxide bifilms will ensure that plenty of AlP particles will also precipitate on bifilms as favoured substrates for most intermetallics. The Si that precipitates on the nuclei attached to the oxides will not be able to wrap around completely because at least one side of the AlP will be firmly in contact with the oxide bifilm. Thus the Si will wrap around as far as it can, but subsequently proceed to grow out along the bifilm as a second favoured substrate, straightening the bifilm crack somewhat during its progress, and thereby progressively reducing mechanical properties. It seems that the oxide bifilm is not a sufficiently good substrate to encourage *nucleation* of the Si phase (although it seems just possible that this may occur if P concentration is sufficiently low), but contact with the Si reduces the overall energy of the combination sufficiently to encourage *growth*. The final alloy microstructure is now mixed, with some Si forming as particles wrapped around free-swimming AlP particles but much Si continuing to grow as plates on the substantial population of bifilms. This is typical of a hyper-eutectic Al-Si structure. Figure 6.11(c) shows the compact primary Si particles with surrounding Si primary platelets.

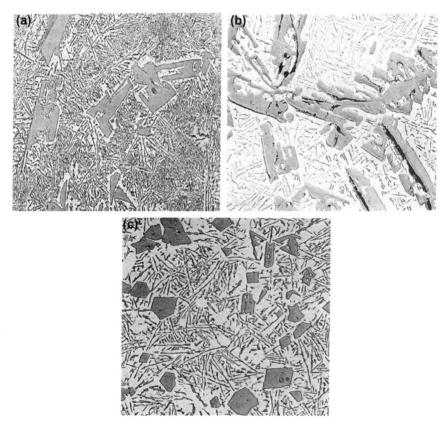

FIGURE 6.11

(a) The microstructure of a hyper-eutectic alloy, showing a mix of primary particles and unmodified eutectic silicon with relatively few AlP nuclei. (b) The cracks associated with growth on oxide bifilms can be clear, especially if solidification occurs unfed or with high hydrogen. (c) The hyper-eutectic alloy modified with additional phosphorus, as can be seen from the AlP nuclei in the centres of refined primary particles.

Please take note of this new observation that appears to be an important universal principle for new phases that grow freely in the liquid: *the new phase takes on the morphology of its substrate*. This disarmingly simple principle helps to explain many features of cast microstructures. Thus primary Si can adopt two quite distinct forms.

1. a *compact polyhedral particle* when it grows on AlP *particles*, and
2. a *plate-like form* when it grows on planar features such as *bifilms*.

(We shall see similar behaviour in other intermetallics and in the Fe-C eutectics.)

Hypoeutectic Al-Si alloys

Part of the action of modification was first explained by Flood and Hunt in 1981. They interrupted the solidification of unmodified and Na-modified Al-Si eutectic alloy by quenching partway through solidification to study the form of the growth front. Sections of their castings are seen in Figure 6.12. In the case of the unmodified alloy, they found the growth front appeared ragged, with some nucleation of eutectic phase apparently ahead of the freezing front, but after the addition of sodium, a smooth planar growth of the eutectic front was found. The effective length of solidification front was now a factor of 17 times shorter than the irregular front. Thus for the same given quantity of heat extracted by the mould over a given area, Flood and Hunt suggested that the interface in the case of the planar front would advance 17 times faster, and therefore have a much finer structure than the unmodified alloy.

The smoothing action of Na on the eutectic front has been confirmed by Crossley and Mondolfo (1966). In their excellent thoughtful, classic paper, they conclude that AlP as a nucleant is neutralised, and observe that instead of Si crystals taking the lead, growing ahead of the freezing front of the eutectic in unmodified alloys, but after modification,

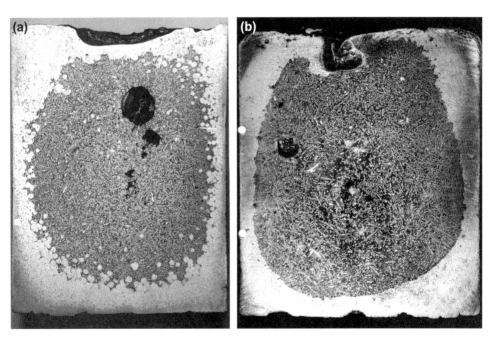

FIGURE 6.12

(a) A laboratory sample of unmodified Al-12Si alloy showing nucleation of eutectic ahead of the general solidification front; (b) the alloy modified with Na, showing suppression of nuclei in the melt, resulting in the advance of a planar solidification front. (Flood and Hunt, 1981).

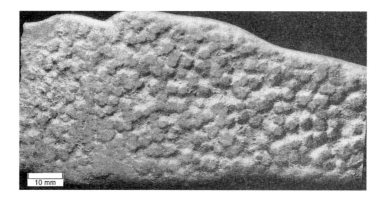

FIGURE 6.13

Thin casting of Al-7Si alloy modified with Sr, its unfed condition reveals its cellular growth form.

Al dendrites take the lead, implying that the eutectic now grows at a lower temperature. This clear thinking and accurate reporting was well ahead of its time.

Smoothing of the solidification front by Na to give a uniform and strong casting skin is used to good effect in most gravity die casting (permanent mould) shops. As the operator takes each casting from the die, he will check it for any sign of the drawing-in of the casting surface at a hot spot. This local collapse occurs because of the sucking-in of the surface that in places will have liquid near to the surface as a result of the ragged form of the freezing front, thinning and weakening the outer surface of the casting. On finding such a 'draw' or 'sink', the operator will add sodium to the melt. The local collapsing of the surface is instantly cured, only to return some time later, after the sodium has evaporated from the melt. However, the operator is watching for this event, and when he judges that a sink has become almost unacceptable, he adds sodium once again, straightening the freezing front and thus strengthening the solidifying wall of the casting, and the surface perfection is renewed.

Interestingly, in contrast with the *planar* front resulting from the Na addition, the form of the freezing front for the Sr-modified eutectic alloy is usually *cellular* in the freezing conditions commonly found in castings. This fairly smooth modified front has been studied by Anson et al. (2000). The intermediate (cellular) condition between extremely ragged (roughly dendritic) and perfectly smooth explains the intermediate performance of Sr. The cellular pattern is seen occasionally on the surface of Sr-modified castings that have suffered some degree of poor feeding, encouraging the loss of some residual liquid from around the cell, and so giving slight depressions on a cast surface that outline the cell boundaries (Figure 6.13).

The cellular structure of the front may lead to problems of feeding Sr-modified castings as suggested by McDonald and co-workers (2004). The cellular growth creating possible feeding problems is contrasted with the smooth front resulting from modification by Na. Surface-initiated porosity is favoured by modification with Sr because of weak patches of the casting surface between the cells, whereas modification with Na creates a smooth, uniformly strong cast surface.

Nucleation of Si

Flood and Hunt appreciated that the ragged interface in the unmodified condition could be explained by nucleation of silicon ahead of the general solidification front, but they admitted that the nature of the nucleus was at that time unknown.

After the realisation that oxide bifilms were common in liquid Al alloys, it was clear that Si crystals could form on oxide films as is seen in Figure 5.28, and reported by Jorstad (2008). The detailed mechanism of their formation is crucial to our understanding of Al-Si alloys. Current evidence indicates that the bifilms may not provide a sufficiently favoured substrate to *nucleate* Si, but they are sufficiently effective to provide a favoured substrate for its *growth*.

The paper by Cho and co-workers (2008) on the effect of Sr and P on Al-Si alloys presented excellent experimental data, confirming the nucleation of Si on AlP nuclei. However, these authors overlooked the probability that the phosphides will themselves nucleate on oxide bifilms (in common with many other intermetallics that appear to favour precipitation on oxide bifilms). In this case, if Si particles now form on the phosphides as illustrated in Figures 6.10 and 6.11(c), then the subsequent continued growth of the Si will naturally follow the oxide substrate on which the phosphide particle is sitting, illustrated schematically in Figure 6.10(b).

Although at this time these authors dismissed the possibility of a double oxide substrate because no central crack can be discerned in the particles (as in Figure 6.11(b) by Ghosh and Mott, 1964), it has to be kept in mind that the central crack will only obligingly reveal itself if opened by shrinkage or gas. In a well fed and well degassed sample, the crack will remain firmly closed, and essentially invisible, as in Figure 6.11(a). It may appear mischievous to make an apparently perverse recommendation for the use of deliberately poorly fed castings for such research to make bifilms clearly identifiable, but it would help to clarify many issues. (The alternative use of higher gas levels to open and reveal bifilms will be less reliable because of the difficulty of control of hydrogen levels, and the rapidity of the changes of hydrogen content as a result of its rapid rate of diffusion either in or out of the sample. The use of shrinkage pressure, equivalent to a tension in the liquid, to open bifilms is subject to fewer uncertainties.)

A potentially important observation can be discerned in Figure 6.14(a) in the paper by Cho and colleagues (2008). A transmission electron microscope (TEM) image of P-rich particles surrounded by Al$_2$Si$_2$Sr are significantly, perhaps, arranged on an impressively straight line. This has all the hallmarks of being an oxide bifilm. This is an interesting possibility because bifilms are rarely imaged successfully in the TEM. The line is also characterised by fine pores (probably fragments of the interior unbonded interface of the bifilm) and by clear discontinuities in the strain bands, as would be expected across the plane of a crack. Because the original AlP precipitates occur before the general solidification of dendrites, it suggests that the alignment is the result of the AlP particles having formed on the planar substrate of an oxide bifilm (corroborating the increasing evidence that nearly all intermetallics seem to favour formation on oxide bifilm substrates).

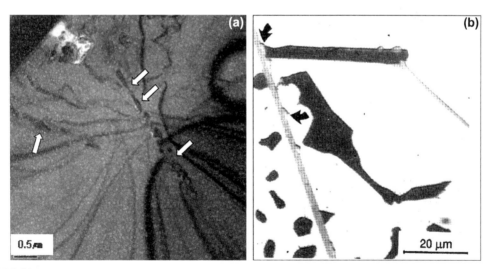

FIGURE 6.14

(a) TEM image by Cho (2008) of Al2Si2Sr phase containing P-rich particles (arrowed) and (b) SEM image by Pennors (1998) showing a row of AlP particles (arrowed) apparently on a β-Fe particle.

Courtesy F H Samuel (2009).

An optical image by Pennors (1998) from a quite different study shows a lineup of AlP particles on a bifilm, implied by Figure 6.14(b). The presence of the beta-Fe particle implies the prior location of an oxide bifilm. One can infer that the AlP particles arrived first, occupying one side of the bifilm and the beta-Fe particles arrived later, occupying the other side and straightening the bifilm, which all the time remains invisible. The alpha aluminium phase grows around the whole assembly later.

In many Al-Si alloys, P is already present in small concentrations. Because of the relatively low concentration of P, no free particles of AlP would be expected apart from those already precipitated on bifilms, favouring the subsequent growth of Si on bifilms, thus forming the typical plate-like primary Si particles that we know as the *unmodified* Al-Si structure (Figure 6.9(a)).

Ludwig and colleagues (2013) quantify the P addition and finds outcomes in line with our expectations. At low levels of P addition (0.2 to 0.5 ppm), the melt undercools 4.5°C and recalesces, achieving a fine, modified Si eutectic. At intermediate levels (2 to 3 ppm), the planar freezing front transitions to nucleation ahead of the front, creating equiaxed grains. At 20 ppm the undercooling is only 1.5°C, and the freezing is characterised by coarse flakes of Si and a great increase in nucleation rate of eutectic grains ahead of the front. Growth now takes place close to the normal 577°C temperature.

The formation of the *modified* Al-Si eutectic occurs as a result of the primary Si platelet formation being prevented. Cho (2008) found that AlP can be deactivated (i.e. poisoned) by the addition of Sr to the melt; the mechanism of deactivation being the precipitation of Al_2Si_2Sr on its surface. Thus all the microscopic nuclei of AlP attached to bifilms would then be deactivated, preventing any formation of Si on bifilms. Pandee and co-workers (2014) found exactly similar action by Sc in Al-Si alloys; the modification effect being the elimination of nucleant AlP by its conversion to the inactive ScP. (Note the closeness of Sr and Sc in the periodic table of the elements.)

(I had originally thought that the bifilms themselves acted as nucleating substrates, and the addition of Sr or Na caused modification by deactivating the bifilms as nuclei. However, the finding by Cho that the precipitation of Al_2Si_2Sr on AlP causes the deactivation of the AlP nucleant now appears to me to be convincing. Thus the concept of a hierarchy of nucleating substrates, first AlP, then oxide bifilms, then some lower temperature nucleus for the eutectic, now requires to be modified because bifilms appear to act only weakly or not at all as nucleation sites. Nevertheless, of course, the bifilms remain active as important growth substrates.)

When sufficient Sr has been added so there are no longer phosphide nuclei or oxide growth substrates that can operate to form Si particles or plates, a third process now comes into play. Si is forced to grow as a eutectic, probably nucleated on a boride such as CrB as suggested by Felberbaum and Dahle (2011), in the regions of significant undercoolings on or near the mould walls or near the melt surface. Because the easy nucleation mechanisms are now denied to Si, it is now forced to grow at a significantly lower temperature. At these greater 'undercoolings', the Si now takes the form of a regular, classical, fine, coupled eutectic, the *modified* Al-Si eutectic, as is usually observed (for instance, Glenister and Elliott, 1981).

Shamsuzzoha et al. (2012) researched Al-17Si alloy, finding that an addition of 3%Ba converts the eutectic to a modified form. Although the authors go to some lengths to invoke the effect of the solution of Ba in Si as some kind of explanation for the effect, it seems far more likely that Ba, an adjacent neighbour in the Periodic Table of the Elements, is simply acting precisely analogously as Sr. The expectation therefore is that the 3% level of Ba could be well over the top. Ultimately, however, the fascinating result is an alloy of excellent hardness, strength and toughness (160 HV, UTS 475 MPa and 5% elongation) and unusually high stiffness (85 GPa). The high Si alloys in the totally modified form would also be expected to be easily machinable. They clearly represent an untapped but impressive commercial opportunity.

Al-Si phase diagram

The *modified* or *non-modified* scenario can be described in terms of the Al-Si phase diagram in Figure 6.15. For normal melts of Al-Si alloys, which would all be expected to contain a generous population of bifilms, the eutectic freezing point would be 577°C, and the growth of the Si particles would follow the bifilms, forming large irregular plates. Interestingly, the eutectic composition appears to be not well defined, as might be expected with the melts freezing on rather variable, uncontrolled substrates. The diagram is based on the composition being 11.5%Si, although other works (Brandes and Brook, 1992) suggest 12.6%Si. This unmodified eutectic is marked 'U' in the figure.

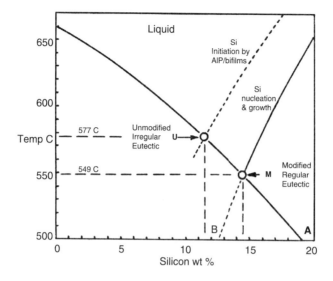

FIGURE 6.15

The Al-Si phase diagram showing the two eutectics, one at high temperature forming Si platelets on bifilms in suspension in the liquid and the other at low temperature as a classic coupled eutectic.

On addition of Sr, deactivating the AlP nuclei on the bifilms by coating them with Al_2Si_2Sr prevents Si nucleation so that the melt continues to undercool to the point at which the modified eutectic forms at point M. This point is located at $14.5 \pm 0.5\%Si$, as identified from the maximum fluidity clearly recorded by many studies of fluidity of modified Al-Si alloys (see Chapter 3), indicating an undercooling in the region of 28°C. This figure is subject to significant uncertainty because of the uncertainties in the compositions, and might therefore have a value somewhere between extremes of about 12–33°C. Whatever its exact value, it is a significant undercooling. In thermal analysis techniques, the undercooling is used to monitor the effectiveness of a modifying addition.

As an aside, it is probable that non-modified Al-Si alloys would also record a maximum fluidity in the region of 14.5% Si because the bifilms will be compacted during this period of rapid flow and will have little time to unfurl in the fairly rapid freezing associated with a fluidity test casting (Figure 3.18).

Although the reader with scientific interests will know that eutectic growth has been assumed to occur in a regime below the liquidus, or the extrapolation of the liquidus, at some point in a region skewed to the right under point U, in view of the new form of the phase diagram in which two eutectics are now clearly separated, it seems likely that eutectic growth actually takes place in the region MAB which is simply under both extrapolated liquidus lines. No complex argument to explain a skewed region appears to be necessary.

The consistent picture that emerges is of a succession of nucleation and growth processes operating at progressively lower temperatures. Summarising:

1. With adequate P addition, AlP particles may form freely in the melt and will nucleate Si particles (Figures 6.10 and 6.11(c)).
2. At the same time (and at more normal lower P concentrations), AlP particles will nucleate on oxide bifilms if present in suspension in the melt. These will form a favoured growth substrate for Si (Figures 6.10(b) and 6.11(c)), the growth starting at the AlP particle on the film; the eutectic freezing at U.
3. The addition of Sr coats the AlP nucleus with Al_2Si_2Sr and deactivates its nucleating potential. Such a surface mechanism, concentrating the action of the modifier onto microscopic areas, would explain why only parts per million of Sr in solution can be effective.

4. Thus neither AlP nor bifilms can operate to nucleate Si particles or grow Si plates. The melt will continue to cool without precipitation of Si until a new nucleus finally becomes effective to nucleate the eutectic phase at M. Felberbaum and Dahle (2011) found that borides such as CrB can nucleate the eutectic. The eutectic grows cooperatively with the Al phase on a substantially planar front (if modified with Na as in Figure 6.16(b)) or on a cellular front (if modified by Sr). The Si now takes on the form of a fine, fibrous eutectic seen deeply etched in Figure 6.17 and in its more familiar optical micrographic structure as a fine, rod-like or coral-like eutectic phase (Figure 6.9) that we know as a *modified* structure.

For (4), after modification the Si grows as a fine, fibrous eutectic, the inter-phase distance being controlled by diffusion in the liquid. A feature of all such eutectics that contain a phase that grows in a crystalline, faceted fashion is that this phase wishes to grow in a single direction, but is forced to continuously modify its growth direction because of the close proximity of its neighbours. The result is a fibrous, 'coral' structure resulting from the necessity to continuously generate growth defects, usually twins, to accommodate the enforced changes in direction (Figure 6.17). Thus such eutectic phases will naturally be expected to contain a high density of twin and other defects. This has been often reported, for instance as described by Steen and Hellawell (1975) and Song and Hellawell (1989), and in the absence of other explanations was once thought to explain the phenomenon of modification, even though no one could quite understand why. Contrary to all previous authors, we can conclude that the high density of growth twinning is not *a cause*

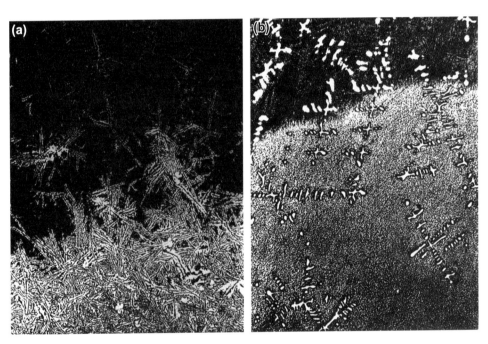

FIGURE 6.16

(a) Growth front of the high temperature unmodified Al-Si eutectic showing Si growth as platelets in the liquid phase; (b) the relatively compact planar front of the modified alloy growing as a classical coupled eutectic. (The eutectic front is seen growing through an open mesh of Al dendrites because this composition near the commonly accepted 12%Si is far from what appears to be a real location of the eutectic at nearer 14.5%Si.)

Courtesy Shu-Zu Lu and John Hunt (2010).

FIGURE 6.17

A magnified image of a quenched interface of the coupled eutectic (growing downwards in this image) illustrating the typical coral form of the silicon.

of the coral morphology but a natural *consequence*. The heavily twinned coral structure contrasts with the internal perfection of the primary crystals that grow freely, unrestricted in liquid ahead of the front.

Note that in contrast to the unmodified eutectic in which the Si is in the form of discontinuous particles which have had to repeatedly re-nucleate on separately swimming oxides, large regions of the modified eutectic nucleate only once in some undercooled region usually near to the mould wall, and becomes subsequently a continuously growing phase. Its main distinguishing feature is therefore its *continuous growth* leading to the development of *continuous lengths* of branching coral of Si. It grows as a 'classic' eutectic.

Formation mechanisms (1) to (2) explain the plate-like morphology and comparative crystallographic perfection of plate-like Si because at this stage of growth Si follows the extensive planar substrates which are sufficiently flexible to provide minimum mechanical constraint, so that the crystal grows essentially freely in the liquid. The growth of hyper-eutectic Si on AlP nuclei is another example of separated growth on particulate substrates. The eutectic morphologies that grew separated and necessarily not crystallographically related have been traditionally known as *irregular* or *anomalous* eutectics in contrast to *regular* or *normal* eutectics. The first has second phase particles that are irregular, whereas the normal eutectic, enjoying the coupled growth of both Si and Al controlled by diffusion, is highly ordered and predictable.

Practical application of Sr modification

In practice, the choice between the use of sodium or strontium for modification depends on the circumstances. The sodium is usually lost fairly quickly from a melt as a result of evaporation because the sodium is close to its boiling point. For instance, for a 200 kg crucible furnace a Na addition is lost within a time of the order of 15–30 min. Strontium, on the other hand, is a normal, stable, alloy in liquid Al alloys. Although it is slowly lost by oxidation at the surface, and sometimes can soak away into refractory furnace linings until they saturate, the alloy will usually survive several re-meltings.

The addition of strontium to the melt is an expensive option, but is taken in an effort to improve the ductility of the alloy. However, the results are not always straightforward to understand, and are accompanied by a number of problematical factors.

Effect of modification on mechanical properties

The precipitation and growth of Si on the originally crumpled and compact bifilm will force the straightening of the bifilm because the Si, as a diamond cubic lattice, has a distinctly favoured growth morphology, taking on the form of a plate. This is the reason that unmodified alloys are generally associated with poor properties—because the Si has become attached to the bifilm—so that its planar growth necessarily straightens the bifilms, forcing them to unravel, unfolding them to become serious planar cracks that can lower properties. The Si particle has always been assumed to be brittle, but the observed cracking behaviour is certainly associated with the presence of their internal bifilm cracks incorporated during their growth.

After modification the bifilms in the liquid no longer act as substrates, so they are no longer straightened, and remain in their original, crumpled, compact state. The mechanical properties now remain reasonably high. This appears to be the mechanism whereby modification tends in most cases to improve mechanical properties. To re-state this conclusion in different words, modification appears only to raise the properties of those alloys whose properties are already low because of the presence of compact oxides, but are lowered further by the growth of unmodified Si. Modification is a technique for avoiding the worsening of poor properties. We shall discuss the considerable evidence for this later.

Sr modification and porosity

Much confusion has existed over whether strontium can be successfully added without the deleterious effect of hydrogen pickup. Amid the confusion, the hydrogen already in solution in the strontium master alloy addition has often been blamed for this problem. However, we can, with some certainty, dismiss this minute source as negligible. What alternative possibilities remain?

Some have doubted that Sr leads to any increase the gas content of Al-Si melts. However, there are good fundamental reasons why Sr would be expected to lead to an increase of hydrogen in the melt, given the correct conditions of water vapour or other sources of hydrogen in the immediate environment. The enhanced reactivity of Sr will lead to enhanced reaction with environmental water vapour, leading to the oxidation of Sr and the release of hydrogen into the melt (Eqn (1.2)).

Even so, it is certainly the case that some foundries experience no problems because they happen to enjoy conditions for minimal uptake of hydrogen because of multiple factors:

1. low hydrogen content, usually associated with low humidity, of the local environment of the melt;
2. protection of the melt from its environment (as for instance in an enclosed low-pressure casting unit);
3. when undisturbed, the melt may continue to be protected from the environment by its oxide and create little additional surface oxide by reaction with water vapour or hydrocarbon gases with the associated release of hydrogen. Dennis and colleagues (2000) found that Al-7Si-0.4Mg alloy with 250 ppm Sr exhibited a short-term increase in the rate of oxidation, but over several hours exhibited a longer term dramatic decrease in rate. Thus melts subjected to repeated disturbance of the surface will react rapidly, whereas undisturbed melts will become practically inert. The stabilisation of the oxide by moisture as seen in Figure 1.3 is a further factor to strengthen this behaviour. However, disturbance of the melt surface by continuous bubble de-gassing would be expected to undermine such benefits.

The elegant research led by Gruzleski (1986) and Dinayuga (1988) employed minimal disturbance of the surface of the melt, and so minimised pickup of hydrogen, whereas Zhang et al. (2001) find a major uptake of hydrogen when regularly disturbing the surface by repeated sampling of their melt after Sr addition.

In an industrial furnace the stirring and ladling actions would be expected to fracture the protective oxide and so allow further reaction, encouraging the ingress of hydrogen in a furnace open to the atmosphere. In these conditions, the addition of strontium will usually be accompanied by an increase of hydrogen content of the melt. The hydrogen naturally increases with increasing strontium content, temperature, time and the presence of environmental water. This makes it practically impossible to add strontium without an increase in porosity in most foundry melting systems. In the experience of the author a single 0.05 wt% addition to a 1000 kg holding furnace caused the gas level to rise to such a high level that gas porosity caused all the castings during that shift to swell, solidifying oversize because of their generous pickup of gas, and had to be scrapped. It took 3 days of waiting for the melt to return to a castable quality. (The arrival of rotary de-gassing has, thankfully, eliminated such lengthy waits nowadays.) The absorption of hydrogen can, of course, be reduced by reducing the time available for absorption, for instance, by the casting of the whole melt immediately after the strontium treatment. Treating the strontium as a late addition in this way has been adopted successfully (Valtierra, 2001).

Strontium is, however, generally used with success in low-pressure casting furnaces where the melt is transferred immediately after treatment into an enclosed furnace that excludes any environmental moisture. It may then be held indefinitely, provided that the pressurising gas that is introduced into the furnace from time to time is dry or inert, and that the furnace lining is already saturated with strontium from previous additions.

So far we have examined the role of Sr in the increase of hydrogen content of the melt, leading to increased porosity. There appear to be additional factors at work.

A mechanism for the additional enhancement of porosity because of Sr has been put forward by Campbell and Tiryakioglu (2009). Briefly, their argument is as follows: After an Sr addition, the bifilms are no longer favoured substrates for the precipitation of Si. Thus the bifilms are no longer encapsulated inside the Si plates, but remain floating freely in suspension in the melt. These are now available for:

1. blocking interdendritic channels, reducing the permeability of the dendrite mesh practically to zero, as convincingly demonstrated by Fuoco and co-workers (1997, 1998) Figure 6.18(a). It is easily envisaged that piles of bifilms will be sucked into the dendrite mesh, where they will accumulate to block interdendritic channels. Furthermore, the bifilm at the base of the heap, experiencing the reduced pressure because of shrinkage, will open, effectively decohering from its other half because of its lack of bond, to create the start of a shrinkage pore (Figure 6.18(b)).
2. The precipitation of hydrogen into the central crack, expanding the bifilm into a pore (recall that pores are volume defects and cannot therefore be initiated by solidification alone; they have to be initiated from an entrainment defect such as a bubble or bifilm). Thus the freely floating bifilms can initiate the creation of hydrogen pores. Lui et al. (2003) affirm that oxides are responsible for pore formation in Sr modified alloys. Zhang (2001) describes how fissure-like pores gradually swell to become rounded after Sr addition, effectively describing the expansion of relatively large bifilms by the absorption of gas. Other authors only report the development of rounded pores immediately after the addition of Sr, indicating that their population of bifilms had a distribution of much smaller sizes, and thus less evidently initiating as cracks.

These bifilm actions leading to impairment of feeding, leading in turn to shrinkage problems, and their effectiveness as initiation sites for both shrinkage and gas pores, simply add to the already serious uptake of hydrogen possible under some conditions as we have seen. Thus the association of porosity with Sr modification has many aspects which have confused and resisted attempts at simplistic explanations for many years. The previous discussion, like much of foundry science, is seen to be non-trivial, and may yet be far from complete. It is hardly piety, but simply an admission, that at times it is necessary to respect the natural complexity of the real world.

Outside of the Al-Si system, the mechanism whereby Sr deactivates bifilm substrates for Si appears to be a standard mechanism applicable to other precipitates in other alloy systems. The approach explains for instance how the refinement of the structure of the Al alloys containing Mg_2Si precipitates is modified by the addition of lithium metal (Hadian, 2009).

Mixed experience with Sr

It is possible that modification, with its fine structure of the silicon phase in the Al-Si eutectic, may have better properties than the coarse unmodified structure. There are many results of published work that cite the benefits to strength and ductility following modification which have been thought to be the result of the finer Al-Si eutectic. However, despite eutectic refinement, some researchers have reported only mediocre improvements. Some have reported no benefit. A few have reported reductions in properties. This confusion of experience requires some examination.

Different foundries require massively different levels of Sr addition to achieve a useful measure of modification. This variable performance between foundries can be understood in terms of the deactivation by Sr of AlP and possibly oxide bifilms as favoured substrates for Si. Those with higher residual P contents, or possibly greater surface area of oxide per volume of melt, will require more Sr for deactivation.

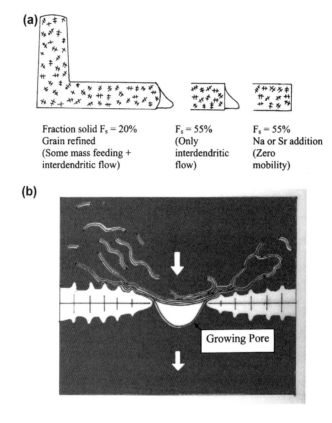

FIGURE 6.18

(a) Results by Fuoco (1997/8) revealing that the small amount of interdendritic flow at 55% fraction solid was completely stopped by eutectic modification; (b) the probable reason being the release of bifilms that would choke flow and have the potential to initiate porosity.

Conventional foundries might typically require 300–500 ppm Sr to achieve modification. This is the situation in many small foundries and research laboratories where the processing of the melt in small batches makes melt quality practically impossible to control. Furthermore, probably all castings and test pieces will have been cast under gravity, mostly using relatively poor filling system designs, and thus their material will be expected to be impaired with new oxide films. For such poor melt quality, the action of Sr in preventing the straightening of a large proportion of large oxide bifilms is a major advantage, and benefits are likely to be experienced.

On the other hand, a good casting operation may achieve relatively low levels of oxides and so require treatment with much lower levels of Sr, perhaps in the region of 50 ppm.

For a more ideal casting operation, taking for instance the counter-gravity filling of sand moulds to cast engine blocks, such a system may have very few new films because of the relatively quiescent handling, and the possibility of gentle counter-gravity filling. Using such a process, Hetke and Gundlach (1994) found their castings suffered a reduction in strength and ductility after Sr addition, and porosity was significantly increased. Similarly, Byczynski and Cusinato (2001) re-confirmed this finding in the Cosworth Process foundry operated by Ford/Nemak in Windsor, Ontario. Almost certainly these near-ideal operations do not need to discourage the straightening of bifilms because there are few bifilms to straighten. In fact, these better foundries find that Sr impairs the mechanical properties of their castings, almost certainly because, with no favourable effects of Sr, only its unfavourable effects are now experienced. Any increase in the hydrogen content would cause the few bifilms to open by hydrogen precipitation, thus suffering an impairment mechanism because of inflation (instead of the other impairment mechanism of crystallographic straightening) and so increasing microporosity and reducing properties.

The pickup of hydrogen is more serious in sand foundries, as described previously, because there is a significant influx of hydrogen from the sand binder during mould filling, and the slower solidification gives greater time for the inflation of the bifilms by hydrogen diffusion. The hydrogen level will rise during the flow of melt into the mould because of the reaction with the sand binder in the same way that hydrogen increases in greensand moulds as observed by Chen (1994). In heavy sections, there will be several minutes for the hydrogen to diffuse into the casting sections. From Figure 1.4(a) the coefficient of diffusion of hydrogen is seen to be close to 10^{-6} $m^2 s^{-1}$. From Eqn (1.5), for a typical time around 100 s, the average diffusion distance for hydrogen in aluminium is close to 10 mm. Thus the gas will easily penetrate the thickest sections of castings such as automotive cylinder blocks. The degradation of the sand binder raises the hydrogen level as the melt enters the mould, so increasing the damaging effect of the heightened gas level with maximum effectiveness.

In permanent mould foundries, the hydrogen influx from the mould is reduced (although it remains from any cores of course) and the time for expansion of the bifilms is reduced. Thus the significant disadvantages of Sr are greatly reduced, allowing the beneficial aspects to dominate. Thus Sr may now show a net benefit, and will particularly show a benefit if the melt has a poor quality (i.e. has a high residual P and high density of oxide bifilms). In this case, the amount of the addition will have to be high to be effective for such a poor melt, but the overall effect should be positive.

In summary, the most likely explanation of the action of Sr is that most operators using poorly designed gravity filling systems will benefit because the new bifilm defects introduced by pouring are prevented from forming plate-like silicon particles. Some will enjoy a useful net benefit if the increase of hydrogen can be controlled. In contrast, those operators using a process such as Cosworth Process will have few new films, and so will automatically achieve a high level of mechanical properties despite the automatically higher hydrogen level and longer freezing time in the sand mould. This high level of properties cannot be raised much further because there are practically no defects to deactivate, and will only suffer negatively from the problem of extra porosity. The refinement of the eutectic, much sought after by metallurgists and assumed to be the main mechanism of property enhancement, appears to have little effect either way. As we have noted before, the properties are not controlled so much by the microstructure, but are mainly under the influence of the defect population.

It seems inescapable therefore that the really important quality requirement should perhaps be the absence of bifilms. This means really clean metal and excellent designs of melt handling. If such conditions were achieved expensive additions of Sr would achieve nothing, and could be abandoned.

Summary of the modification mechanism

We can summarise the evidence, laying out the steps of the logic for the central role played by oxide bifilms; this is a long and detailed list of explanations of the different facets of the modification phenomena.

1. Bifilms are to be expected in our Al alloys because of the way our metals are handled in our cast houses and foundries.
2. The bifilm has two important morphologies: (a) a compact original state in which the bifilm has been crumpled by internal turbulence in the liquid, becoming a fairly harmless defect and (b) a second condition in which it may become straightened to become a serious planar crack.
3. All intermetallics studied so far (including Si, alpha-Fe, beta-Fe and Mg_2Si etc.) appear to precipitate and grow on bifilms. The cracked morphology of the intermetallic cannot be explained by solidification (which is incapable of creating such volume defects because of the effectively unbreakably powerful interatomic forces that hold the melt, the solid, and the interfaces in close atomic contact). The cracks can only originate by entrainment mechanisms (i.e. introduced externally into the liquid), explaining their necessary association with bifilms.
4. The known nucleation sites for Si are AlP particles. The AlP nuclei are most probably formed on oxide bifilms.

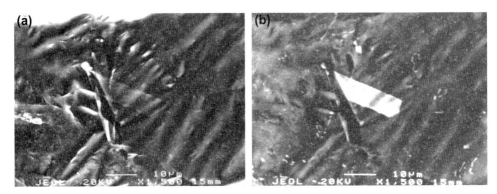

FIGURE 6.19
SEM image of an iron-rich particle nucleated on the underside of a thin alumina film imaged by (a) secondary electrons and (b) back-scattered electrons (Cao, 2000).

5. The Si particles are intermetallic with a diamond cubic crystal structure that forces its growth morphology to take up a plate-like form, attaching to the bifilm and therefore straightening the crumpled substrate into a flat planar bifilm. Thus the central bifilm crack now takes on the form of a large flat crack that has a serious effect on properties. The rather straight cracks sometimes observed in the centrescentres of primary Si and beta-Fe particles are thereby explained, together with the *apparent* 'weakness and brittleness' of these particles which in reality are expected to be strong and crack resistant (see also later in Figure 6.19).
6. The dilute addition of Sr (or Na) deactivates the AlP as a favourable nucleant for Si, and may consequently also deactivate oxide bifilms as favoured substrates.
7. After modification, early and easy nucleation of Si particles on bifilms can no longer occur. Thus the bifilms remain in suspension in the melt in their crumpled form (they are no longer straightened out by the plate-like growth of Si) so that properties are not degraded.
8. Having removed the oxide bifilms as nuclei, the next available opportunity for Si to form is as a eutectic at a significantly lower growth temperature, possibly nucleating on particles such as CrB. The eutectic spacing is now controlled by diffusion and so is significantly finer and controlled by the rate of heat extraction. Its coral-type morphology is a consequence of dense twinning, forced by the continuous change of direction of the Si fibres as a result of the constraint of neighbours.
9. The non-straightening of the bifilms (the retention of the bifilms in their original fairly harmless compact form) by the addition of Sr is expected to be the major benefit to mechanical properties, particularly increased strength and ductility. (The finer spacing of the eutectic is expected to be a minor contribution to any increased strength, as illustrated by the experiments of Cho and Loper (2000) who demonstrate with Bi additions how properties are unaffected by the modification of the structure of an A356 alloy.)
10. The addition of Sr has the potential disadvantage of increasing porosity in the casting. This effect is the result of (a) enhanced the reactivity of the melt, resulting in the danger of increased hydrogen pick-up in some circumstances and (b) the bifilms, no longer encased in Si, remain free-floating in the melt, where they can block interdendritic feeding and initiate porosity, although it is predicted that faster freezing and lower gas-content moulds can reduce these disadvantages.

Non-chemical modification

The bifilm hypothesis would predict that Al-Si melts completely free from oxide bifilms would be automatically modified without any necessity for additions. This is an interesting and testable proposition that appears in fact to have good support. Jian and colleagues (2005) used ultrasonics to drive bifilms out of suspension, causing modification without the aid of any chemical additions. Similar observations at the University of Windsor have been noted using electromagnetic stirring to centrifuge Al-Si melts clear of bifilms, producing a beautifully modified eutectic (Sokolovski, 2002). Wang et al. (2007), using what appears to be extremely clean Al-Si produced by direct electrolysis, found that a fully modified structure can be retained up to 16%Si. Above this level of Si, these authors find that primary Si is formed, but this does not seem to be a fundamental limitation, but was probably the result of the introduction of oxides because the additional Si was added using a master alloy containing 50%Si which would be certain to contain generous quantities of oxide.

It seems true that for very clean Al-Si alloys primary Si will be unable to precipitate, and the structure will therefore be automatically modified without the intervention of Na or Sr.

The experiments by Li and Xia (2014) are also fascinating. These researchers investigated the effects of high temperature melt treatment on Al-20%Si alloy in a scanning calorimeter. When increasing the temperature from 750 to 1000°C, they noticed a small exothermic peak at 951°C. Subsequently, they found the *eutectic* Si had been modified from a coarse flake to a fine coral form (although some finer but blocky *primary* Si is also present as would be expected from the equilibrium phase diagram but perhaps not expected compared with other non-equilibrium research results in this field). The authors suggested the mechanism was a heredity effect from the breakup of Si–Si atomic clusters in the melt when subject to high temperature. However, no evidence is presented in corroboration. It is tempting to assume that the effect is due to the change of oxide constituting the bifilms from gamma to alpha alumina, thus changing the effectiveness of the substrate, either discouraging the precipitation of AlP or discouraging the subsequent growth of Si. Similar effects of high temperature treatment of the melt are reported by Wang et al. (March 2014) and have been commented on by me (Campbell, September 2014).

6.3.6 IRON-RICH INTERMETALLICS

In addition to Si, it seems that most intermetallics also find oxides to be favoured substrates for growth. The iron-rich intermetallics alpha-iron ($Al_{15}Fe_3Si_2$) and beta-iron (Al_5FeSi) are included in this list.

Several authors have concluded that beta-iron first nucleates on AlP particles (Cho et al., 2008). If so, it would be expected, as in the case of Si nucleation, that the AlP particles would be already sited on oxide bifilms. Pennors and colleagues (1998) used elegant metallography to show AlP in a straight line, apparently attached to a beta-iron platelet (Figure 6.14(b)). Because the AlP appears to precipitate first, it seems reasonable to conclude that AlP must have aligned itself against a bifilm because no other extensive planar solids are known to be present at that stage of solidification. The beta-iron will arrive later, sited on the other side of the bifilm, and so, perhaps, in no way connected to the AlP. It seems most probable therefore that both AlP and beta-iron form on bifilms, but these are independent and unrelated events. They only appear to be related when the AlP forms on one side of the bifilm and the beta-Fe forms on the opposing side; because the bifilm is usually not easily seen, it appears that the AlP particles are sitting on the beta-Fe.

Cao and Campbell (2000) discovered that βFe plates in Al-Si alloys precipitated on the wetted outside surfaces of bifilms. Later work by these authors in 2003 indicated that a wide variety of oxide substrates might be effective, including αAl_2O_3, γAl_2O_3, $MgAl_2O_4$ and MgO. During the early stages of the growth of the precipitate, the βFe particle is sufficiently thin that it can follow the folds of the bifilm. On a fracture surface, the iron-rich phase can be clearly seen through the thin oxide film that represents one half of the bifilm (Figure 6.19). At this early stage, it is faithfully following the undulations of the oxide film. However, as the βFe particle thickens, the particle becomes increasingly rigid, taking on its preferred crystalline plate-like form, and so forcing the film to straighten. Finally, the bifilm is often seen as a crack aligned along the centre of the βFe particle (Figure 6.20(b)), or along the matrix/particle interface if the βFe happened to nucleate only on one side of the bifilm (Figure 6.21). The cracks are not always seen because the individual halves of the bifilm are usually extremely thin and are often tightly closed together. It usually requires some opening mechanism to come into play before the cracks are easily seen. Thus gas can inflate the bifilm, or shrinkage can pull it apart, as can stress applied during or after freezing.

The weakening effect of the straightened cracks seriously reduces the mechanical properties of the metal. *This* is the mechanism whereby iron is the most damaging impurity in Al alloys; it weakens the alloy as a result of its effectiveness in straightening oxide bifilms. It is important to remember that such inclusions as the βFe particle are probably extremely strong and resistant to fracture, in common with many intermetallics. As bears repeating, the observed cracks are not the

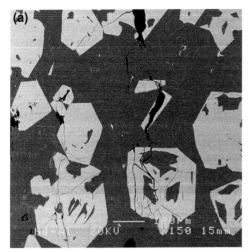 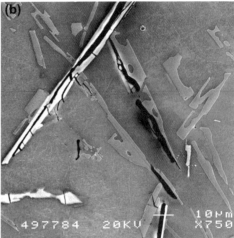

FIGURE 6.20

Fe-rich particles in an Al-12Si alloy growing on oxide bifilms, showing (a) alpha-Fe with its cubic symmetry wrapping around and sealing in situ compact bifilms, thereby maintaining properties; and (b) beta-Fe with its monoclinic crystal structure, and silicon particles with their diamond cubic structure, growing as platelets, flattening bifilms, thereby straightening cracks, and reducing properties. (Cao, 2000).

result of intrinsic brittleness but the result of the presence of the oxide bifilm. The strong effect of Fe in reducing the properties of Al alloys is additionally heightened because although the iron content in the alloy is low, the size of both alpha- and beta-iron inclusions is disproportionately large because their volume is increased fivefold by the content of aluminium, the formula having a 5:1 ratio in both $Al_{15}Fe_3Si_2$ and Al_5FeSi, respectively.

Wang and colleagues (2009) observed the formation of beta-Fe particles directly in a solidifying Al alloy using synchrotron radiation. They noted that the beta-Fe nucleates at a temperature between 550° and 570°C. This is a surprisingly large range for a well-characterised crystal, and strongly suggests that the problem lies not with the crystal but with the variability of its messy and fragmented oxide substrates, each one of which will be different.

Figure 6.20(b) demonstrates one of the convincing predictive successes of bifilm theory. When two oxide films originally impinge to form a bifilm, the film sizes will be random, never perfectly matched, so one film will always have a larger area than the other. Thus one film, which we call side 'A', will have to settle against its smaller partner 'B' by forming a series of small transverse folds on its side of the bifilm. When the βFe particles nucleate on A, they will only be able to grow as far as the nearest transverse fold which will form a barrier to further progress. Thus this side will have a series of short crystals separated by transverse cracks. Side B, the originally smaller side, will have no transverse folds and therefore will permit a continuous straight growth across its surface. The βFe particle is predicted therefore to be split by a central crack, but with transverse cracks on one side only, and a continuous crystal on the opposite side, exactly as seen in Figure 6.20(b).

Figure 6.21 illustrates different views of a bifilm, its presence inferred by the presence of a βFe plate. One side (note only one side) of the βFe plate, as a result of the plate having formed on only one side of the bifilm, has been inflated, growing away by gas precipitation or pulled away by shrinkage forces, opening a pore on one side of the βFe plate. Because it has been common to observe an association between pores and βFe particles, it has in the past been assumed that the βFe particles blocked the movement of feed liquid along interdendritic channels, and so caused shrinkage porosity. However, in view of the three-dimensional access routes for feed liquid and in view of the strong probability that pores probably cannot be formed without bifilms, any restriction of feeding is almost certainly less important than the presence of bifilms. An observation by Miresmaeili (2006) in which platelets of βFe were scattered and randomly

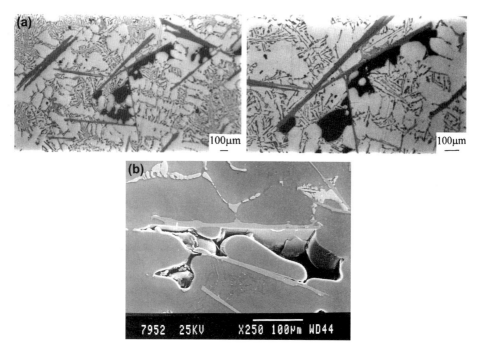

FIGURE 6.21

Platelets of βFe in an Al-Si alloy showing (a) pores opened by shrinkage or gas on only one side of each platelet; and (b) a 3-dimensional insight into the plate and pore geometry.

(a) Courtesy Cao (2000); (b) courtesy Samuel et al. (2001), Internat J Cast Metals Research.

oriented over a wide array of dendrites illustrated that the associated porosity at the sides of many of the plates were randomly oriented with respect to the dendrite direction and feeding direction. If the plates had been blocking the flow the pores should all have been sited 'down-stream' of the plates. Their random siting indicated that the association was not the result of blockage of feed metal, but more likely the result of the growth of a pore from the decoherence of the bifilm from the side of those plates formed on only one side of the bifilm.

Intermetallics with close-packed cubic symmetries (fcc, body-centred cubic) have such a high degree of symmetry that they often show little or no strong growth morphology; any direction of growth is as favourable as any other. The αFe phase has a hexagonal lattice which is close to cubic, so it can grow in practically any direction. For this reason, the αFe phase is usually seen to grow in a meandering fashion, following around the irregular form of the bifilm in its original crumpled state. In this way the compact, convoluted bifilm, with its associated central crack, is sealed in place, achieving a minimal impact on properties. This is why αFe as Chinese script is generally encouraged by metallurgists. On polished microsections the convoluted forms of αFe are often seen to contain a skeleton of convoluted cracks—the entombed trace of the originating crumpled bifilm (Figure 6.20(a)).

We might speculate that the form of Chinese script may arise, at least partially, from the original convoluted form of the bifilm, the αFe depositing on, and faithfully following, the original convolutions.

The conversion of the extended planar crack of the βFe particles (Figure 6.20(b)) into convoluted compact cracks inside the Chinese script particles (Figure 6.20(a)) is one of the key techniques for reducing the deleterious impact of iron impurity in Al alloys. There have been a very large number of research programmes to determine the best metallurgical solution to this problem. The common approach has been to use 0.5Mn addition for every 1.0Fe content (all in wt%).

Many other alloying solutions have been explored: Mahta and colleagues (2005) find the 1.0Fe level can be converted with 1.0Co or 0.33Cr. (Interestingly, this work indicates that primary Si in the form of compact nodules appears also to be stabilised by these additions, indicating that in addition to AlP there may be other effective nuclei for nodular Si based on transition metals.)

All of these alloying approaches to the control of βFe come with the built-in disadvantage that the Mn addition increases the total volume of unwanted intermetallic phases as pointed out by Crepeau in his 1995 review. Other elements have been used as listed by Crepeau. These include Be, Mo, Ni and S. This list is highly suspect; it suggests wildly different mechanisms. For instance, Ni may act as an alloying element similarly to the other transition metals. Mo seems unlikely to fit into this category and remains a mystery. Be will almost certainly not act in solution but as a significant strengthener of the oxide film on the melt, reducing entrainment of bifilms by a 'hanging up' mode during pouring, so reducing the total oxide bifilm content in the melt. Sulphur seems likely to be analogous to phosphorus and may provide or modify a nucleating substrate. Clearly, much more work remains to clarify this messy subject.

An alternative approach based on a further finding by Mahta may be important. The occurrence of the βFe phase appears to be particularly sensitive to the rate of solidification. Although more rapid freezing makes most features of the microstructure finer, there may be a reason for the βFe to be particularly sensitive. This is because for the particle to grow, the bifilm has to be mechanically unfurled as an extendable substrate against the viscous drag of the liquid. This means that the βFe can only form at relatively slow cooling rates. The αFe phase can also be suppressed by faster freezing as would be expected, but is less sensitive, requiring higher rates of freezing to reduce its formation.

It is not just Fe-rich intermetallics which are considered as exhibiting poor structure and properties. As just one instance of many, very many, Chen and colleagues (2013) investigated the intermetallics Al_7Cu_2Fe and T phase (AlZnMgCu) in high-strength Al alloys ascribed the loss of properties to these 'weak and brittle' phases leading to intergranular fracture along high angle boundaries and transgranular microvoid-induced fracture. The fallacy of such standard reasoning is examined critically in Section 6.3.7.

Si and Fe competition for sites

Caceres (2004) made a series of alloys of increasing Si and Fe contents. He found that as Si was increased the βFe particles were refined. Conversely, Liu (2003) noted that for some common Al-based casting alloys, higher Fe resulted in finer Si particles. This reciprocal relation between the formation of Si and βFe particles reflects that they are both competing for the same limited number of suitable substrates, as is clearly seen from their formation on bifilms in Figure 6.20(b).

Shabestari and Gruzleski (1995) found that Sr suppressed the precipitation of Si (presumably on bifilms) but that βFe particles still straightened and grew, but were slightly reduced in average length. This finding indicates that Sr is extremely effective in deactivating the precipitation of Si on the oxide substrate, possibly because of its action on the tiny AlP nuclei sitting on the oxide bifilms. The action of Sr is similar, but somewhat less effective for βFe, possibly because the βFe is nucleated directly on the oxide substrate, and this more extensive area may require deactivation by correspondingly more strontium. Mulazimoglu (1993) noted that Sr promotes the formation of more fragmented αFe particles in the form of a kind of Chinese script and coarsens the $CuAl_2$ phase in many copper-containing common Al alloys. His observations are again explained by the partial deactivation of the oxide substrates, resulting in fewer active substrates and consequently coarser precipitates.

Finally, it is significant that Caceres (2000) concluded that ductility of Al-Si alloys is controlled by the cracking of both Si and βFe particles. Both phases nucleate and grow on the oxide bifilms in suspension in the alloy. Thus both naturally contain similar cracks.

6.3.7 OTHER INTERMETALLICS

Miresmaeili and co-workers (2005), studying Al-7Si-0.4 Mg alloy with Sr additions, found Al_2Si_2Sr precipitated on both γAl_2O_3 and MgO films. In addition, alignments of the precipitates indicated an epitaxial relationship. The experimental technique was an admirable model of simplicity. The melts were poured into a small steel cup the size of an eggcup. The outside surface of the casting was then studied by scanning electron microscope (SEM). The intermetallics could easily be seen through the thickness of the oxide skin of the casting. Simple!

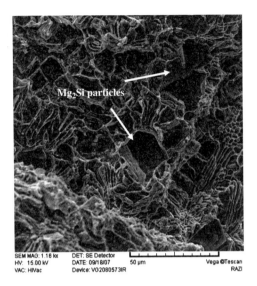

FIGURE 6.22

Fracture surface of an in situ Al-Mg$_2$Si intermetallic MMC showing the ductile failure of the upper intermetallic particle (Hadian et al., 2009).

There is some evidence that the CuAl$_2$ intermetallic in Cu-containing Al alloys also forms on oxide bifilms (Campbell, 2009). Thus the strong and tough Al-4.5Cu series of alloys, of which there are many, exhibit the CuAl$_2$ phase along grain boundaries. On careful examination, although it seems clear that the Cu segregates to the interdendritic regions around every dendrite arm, the CuAl$_2$ phase occurs only in some locations. It seems that the phase only precipitates when a bifilm is present; otherwise, the Cu merely remains in solution (Figure 6.23). (Although it is possible that the absence of the CuAl$_2$ phase is the result of a locally lower Cu segregation, this seems unlikely because the interdendritic segregation appears to be especially uniform.) A further important observation is that, probably as a result of its cubic symmetry, the CuAl$_2$ phase does not appear to have a strong straightening effect on the bifilm. Thus in contrast to Al-Si alloys in which the main alloying agent, Si, straightens the bifilms to reduce properties, this does not happen in Al-Cu alloys, explaining an important contribution to their excellent behaviour. (Another important contribution is its high Cu content creating the high precipitate density after heat treatment of course.)

Excess Ti in Al-Si alloys, often specified to be in the range 0.13–0.18%Ti, leads to the formation of TiAl$_3$-type intermetallics that form at temperatures well above the liquidus temperature of the alloy. These crystals can take up to 37%Si in solution, making a mixed intermetallic, usually written Ti(AlSi)$_3$. They take either a block or flake morphology (possibly as a result of the different sizes of bifilm on which they may nucleate). For optimum effect with modern grain refiners in most Al-Si alloys, including the popular Al-7Si types, Chen and Fortier (2010) recommended that 10–20 ppm B should be targeted, whereas the Ti level should be near, but not exceed 0.10%. The reduction of Ti reduces the content of the Ti-rich intermetallics.

The fracture resistance of intermetallics

The conventional description of intermetallics as brittle and weak does not stand up to critical appraisal. Although mentioned several times previously, it is worth devoting some space to some fundamental facts. It should be common knowledge that intermetallics are generally strong; in fact micrometre-sized intermetallic particles will be expected to be extremely strong. The great strength, hardness and crack resistance of carbides and borides are used worldwide for cutting tools for machining hard metals. It is also common knowledge that intermetallic alloys are currently being developed for turbine blade production and other ultra-high performance uses which require at least some toughness.

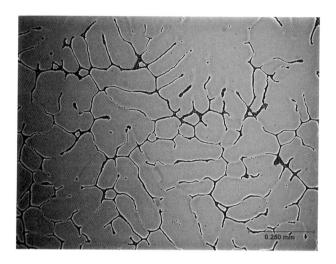

FIGURE 6.23

The as-cast Al-4.5Cu alloy showing $CuAl_2$ phase at many but not all interdendritic boundaries even though Cu segregation can be visually discerned in all boundaries.

Furthermore intermetallics are even likely to be ductile. There is evidence for this in occasional observations of intermetallics that have been plastically deformed (for instance, the Mg_2Si intermetallic particle observed by Hadian, 2009, Figure 6.22). Although carbides are renowned for their apparent hardness and brittleness, Yoshimi and co-workers (2014) studied two carbides of quite different structures, (Mo,Ti)C and $(Mo,Ti)_2C$, during high temperature deformation of Mo alloys. They observed significant plastic deformation of both carbides without any formation of cracks.

Ductile behaviour in hard crystals is well-known in physics; we need to divert to describe a simple but revealing experiment which deserves to be more universally known. A crystal of common salt is usually brittle when crushed in air, but when deformed in water becomes strong and ductile as a result of the dissolving away of surface defects that can lead to failure (Mendelson, 1962). All strong crystals deposited and grown in a solvent, and stressed while remaining in their solvent, are expected to be similarly strong and ductile. This is the case for all intermetallics grown in situ in metal matrices. It is to be expected therefore in Si and βFe particles formed during the solidification of Al alloys. Their apparent brittleness is therefore a deception arising from their formation on bifilms and therefore appearing to be cracked. We have all lived with this deception for years, and clearly it will take much unlearning.

Incidentally, we need to emphasise this behaviour is not to be expected of MMCs made by *mechanical mixing*; i.e. stirring intermetallic particles into a melt of the matrix. Such introduced particles naturally entrain the surface film on the melt during entry to the melt, as though stepping into a paper bag when entering. They are therefore effectively isolated from the matrix by the surrounding film and its associated layer of air. The entrained particle does not contact the melt (Figure 2.3) and effectively does not know that it is submerged in a metallic melt because it only experiences the layer of surrounding air. Such particles are thus always associated with, if not actually surrounded by, bifilm cracks. The attainment of ductility in such mechanically mixed MMCs is thereby fundamentally limited as seen in the work of Emamy et al. (2009). These submerged particles are equivalent to our grains of salt in air and are consequently truly brittle.

6.3.8 THERMAL ANALYSIS OF Al ALLOYS

The cooling curve of a solidifying alloy can be informative, usually in several ways. This is a potentially large subject, and although useful for many alloys (particularly cast irons), we shall deal only with Al alloys as an example of features that can be observed. For a variety of cooling curves the reader is recommended the excellent book by Backerud, Chai and Tamminen (1990) which gives clear curves for all common Al casting alloys. Readers are also recommended to the excellent review by Stefanescu (2015). In this short section, we shall only consider an Al-7Si alloy as a particularly simple example. This is shown in Figure 6.24(a).

In the first edition of this book it was assumed, in common with standard interpretations, that the cooling curve of temperature T versus time t shows the undercooling that is necessary to give the driving force to nucleate the alpha-Al dendrites. Strictly, this is an error. Figure 6.25(a) illustrates that the equilibrium freezing point is, of course, not shown on the cooling curve, but when inserted will often be found to lie above the dip in the curve. It is immediately clear therefore that undercooling will have occurred down to the nucleation start as found by construction from the first inflection of the first derivative curve and by extension of the alpha phase portion of the cooling curve. Only a single nucleation might occur at this point, so the dip in the cooling curve might only, in principle, correspond to the undercooling required to accelerate the growth of the nucleated solid. Thus, again emphasising in principle, the dip normally attributed to the undercooling required for nucleation may only correspond in fact to a condition of delayed heat evolution because of

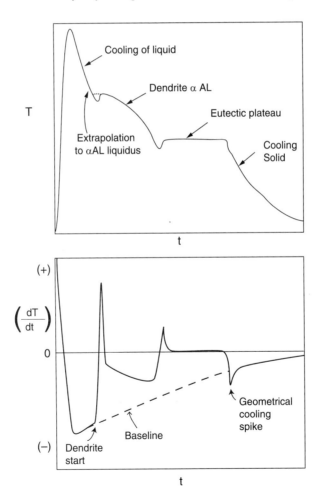

FIGURE 6.24

The thermal analysis of a simple Al-7Si alloy.

266 CHAPTER 6 CASTING ALLOYS

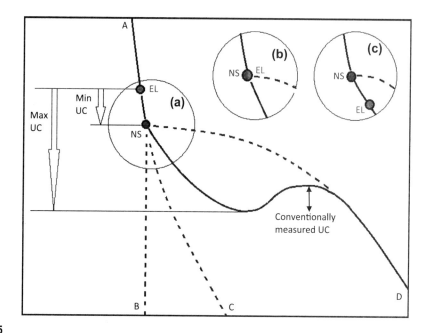

FIGURE 6.25

A cooling curve showing how the key measurement is sometimes incorrectly assumed to be the obvious dip in the curve. (a) The correct undercooling required below the equilibrium liquidus (EL) temperature to the nucleation start (NS) temperature; (b) illustrates zero undercooling condition; (c) shows nucleation on a highly favoured substrate before EL. The latter is common, for instance, for Si precipitation in Al-Si alloys and graphite in cast iron, where bifilms are present in the liquid.

slow growth. Perhaps more probably, however, is that additional nucleation is occurring during the downward portion of the dip, only stopping after the bottom of the dip when the rate of release of heat from the growth of the nucleated phase overcomes the rate of cooling, so that the temperature recalesces (literally 'reheats').

By the provision of effective grain refiners the initial minimum and dip undercoolings can both be reduced practically to zero (Figure 6.25(b)), and so can be used to assess the state of grain refinement of the melt prior to casting. Similarly, the undercooling necessary to initiate the eutectic phase, and the steady-state growth temperature can monitored to assess the state of modification of the eutectic. This 'plateau' is usually not so level in more complex alloys as a result of continued segregation during the course of the eutectic solidification, resulting in the 'plateau' taking on an increasing downward slope.

There has been occasional speculation among researchers that a sufficiently favoured substrate might cause the alpha phase to appear even above its equilibrium temperature. This seems to me to be quite possible, and is shown in Figure 6.25(c). It would be useful to identify such events.

The differential curve dT/dt versus time t is used by Backerud and others to show how the area between the cooling curve and the baseline can yield the heat evolved during the formation of that particular phase that is solidifying at the time. Thus if the heats of formation of the phases is known, the relative volumes of phases in the solidifying structure can be found. In this way, for instance, the quantity of Fe-rich phases in Al-Si alloys can be fairly accurately assessed.

Later developments of this technique to use the cooling curve to determine the quantity of the different phases occurring during freezing has been carried out by workers at the University of Windsor, Ontario, using a Newtonian method (Emadi, Whiting et al., 2004). More recently still, Chinese workers (Xu and Liu et al., 2012) have developed a clever technique employing extrapolations of the cooling curve itself to construct a baseline. These recent quantitative techniques agree excellently and are to be recommended.

A curious feature often observed at the end of freezing on a cooling curve is a 'spike' of cooling mysteriously diving down below the baseline. This does not appear to have been ever explained, but is simply the result of the rapid acceleration of freezing of the sample because of the geometrical effect of the residual melt dwindling to zero. In contrast with the rest of the thermal analysis record, it has nothing to do with any metallurgical reaction. It is the same effect as causes the 'reverse chill' in grey cast iron. It is quantified in Section 5.1.2 and Figure 5.10.

6.3.9 HYDROGEN IN Al ALLOYS

We turn now to the presence of hydrogen in aluminium. Hydrogen behaves quite differently to other solutes.

The aluminium-hydrogen system is a classic model of simplicity. The only gas that is soluble in aluminium in any significant amounts is hydrogen. (The magnesium-hydrogen system is similar, but rather less important in the sense that the hydrogen concentration increases only by a factor of about 1.4 on freezing, compared with a factor of about 20 for aluminium, so that dissolved gas in liquid magnesium is usually much less troublesome. Other systems are in general more complicated as we shall see later.)

Figure 6.26 is calculated from Eqn (1.4) illustrating the case for hydrogen solubility in liquid aluminium. It demonstrates that on a normal day with 30% relative humidity, the melt at 750°C should approach about 1 mLkg^{-1} (0.1 mL 100 g^{-1}) of dissolved hydrogen. This is respectably low for most commercial castings (although perhaps just uncomfortably high for aerospace standards). Even at 100% humidity, the hydrogen level will continue to be tolerable for most applications. This is

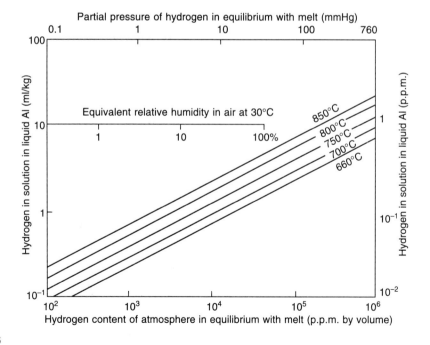

FIGURE 6.26

Hydrogen content of liquid Al shown as increasing with temperature and the hydrogen content of the environment as hydrogen gas or as water vapour.

the rationale for de-gassing aluminium alloys by doing nothing other than waiting. If originally high in gas, the melt will equilibrate by losing gas to its environment (as is also illustrated by the copper-based alloy in Figure 1.2).

Further consideration of Figure 6.26 indicates that where the liquid aluminium is in contact with wet refractories or wet gases, the environment will effectively be close to 1 atm pressure of water vapour, causing the concentration of gas in solution to rise to nearer 10 mLkg^{-1}. This spells disaster for most normal castings. Such metal has been preferred, however, for the production many non-critical parts, where the precipitation of hydrogen as dispersed porosity can compensate to some extent for the shrinkage on freezing, and thus avoid the problem and expense of the addition of feeders to the casting. Traditional users of high levels of hydrogen in this way are the permanent mould casters of automobile inlet manifolds and rainwater goods such as pipes and gutters. Both cost and the practicalities of the great length to thickness ratio of these parts prevent any effective feeding.

Raising the temperature of the melt increases exponentially the solubility of hydrogen in liquid aluminium. At a temperature of 1000°C, the solubility is over 40 mLkg^{-1}. However, of course, if there is no hydrogen available in its environment, the melt will not be able to increase its gas content, no matter what its temperature is. This self-evident fact is easy to overlook in practice because there is nearly always some source of moisture or hydrogen, so that, usually, high temperatures are best avoided if gas levels are to be kept under good control. Most aluminium alloy castings can be made successfully at casting temperatures of 700–750°C. Rarely are temperatures in the 750–850°C range actually required, especially if the running system is good.

A low gas content is only attained under conditions of a low partial pressure of hydrogen. This is why some melting and holding furnaces introduce only dried air, or a dry gas such as bottled nitrogen, into the furnace as a protective blanket. Occasionally, the ultimate solution of treating the melt in vacuum is employed (Venturelli, 1981). This dramatically expensive solution does have the benefit that the other aspects of the environment of the melt, such as the refractories, are also properly dried. From Figure 6.26, it is clear that gas levels in the melt of less than 0.1 mL/kg are attainable. However, the rate of de-gassing is slow, requiring 30–60 min because hydrogen can only escape from the surface of the melt, and takes time to stir by convection, and finally diffuse out. The time can be reduced to a few minutes if the melt is simultaneously flushed with an inert gas such as nitrogen.

For normal melting in air, the widespread practice of flushing the melt with an inert gas from the immersed end of a lance of internal diameter of 20 mm or more is only poorly effective. The useful flushing action of the inert gas can be negated at the free surface because the fresh surface of the liquid continuously freshly exposed by the breaking bubbles represents ideal conditions for the melt to equilibrate with the atmosphere above it. If the weather is humid the rate of re-gassing can exceed the rate of de-gassing.

Systems designed to provide numerous fine bubbles are far more effective. The free surface at the top of the melt is less disturbed by their arrival. Also, there is a greatly increased surface area, exposing the melt to a flushing gas of low partial pressure of hydrogen. Thus the hydrogen in solution in the melt equilibrates with the bubbles with maximum speed. The bubbles are carried to the surface and allowed to escape, taking the hydrogen with them. Such systems have the potential to degas at a rate that greatly exceeds the rate of uptake of hydrogen.

Rotary de-gassing systems can act in this way. However, their use demands some caution. On the first use after a weekend, the rotary head and its shaft will introduce considerable hydrogen from their absorbed moisture. Thus it is to be expected that the melt will get worse before it gets better. Thus de-gassing to a constant (short) time is a sure recipe for disaster when the refractories of the rotor are damp. In addition, there is the danger that if a vortex is formed at the surface of the melt, it may carry down air, thus degrading the melt by manufacturing oxides faster than they can be floated out. This is a common and disappointing mode of operation of a technique that has reasonable potential when used properly. The simple provision of a baffle board to prevent the rotation of the surface will suppress the vortex formation at reasonable rates of rotation. However, it is common to see such high rates of stirring that additional vortices are generated by the baffle board, making a bad situation worse.

High rates of loss of hydrogen during rotary de-gassing are entertaining because of the mass of flashes of blue flames of hydrogen burning as it is released from the melt surface. If there are no blue flames there is probably no hydrogen in the melt.

Despite these proper considerations surrounding the de-gassing of melts, a major effect of rotary de-gassing may not be the de-gassing. The action of greatest importance is most probably the floating out of the major oxide films resulting from the skins of charge materials. These oxides are fractions of square metres in area. It is essential to eliminate them if castings of any integrity are to be achieved. It seems likely that these large area films will be eliminated within the first few minutes. At that point the major benefit of rotary de-gassing may be over.

The danger that subsequently arises in continued operation is the generation of millions of small oxide bifilms from the bursting of small de-gassing bubbles as they arrive at the surface of the melt. As they release their contents of inert gas plus any flushed out hydrogen, the waft of fresh air across the newly opened surface will oxidise the surface. Thus as the sides of the opened bubble collapse together, they will create a new oxide bifilm. Thus the original massive bifilms are replaced by millions of small bifilms. The compositions are likely to be different too; the old films will be mainly spinels but the new films will be relatively pure alumina. Only later will the alumina films convert to spinel if some Mg is available in the alloy. This behaviour of rotary de-gassing is not fully researched and understood, so an optimum practical procedure is not easily recommended at this time.

One technique that has yielded good quality metal is the use of rotary de-gassing together with the use of a flux. Once again, rather than any significant effect on hydrogen removal, the action seems to be important for the wetting and elimination of oxide bifilms, or possibly the gluing shut of bifilms (keeping them permanently closed and therefore resistant to forming pores or cracks). Here too the recommended operational details are sketchy. The big Al producers of the world are known to have carried out their research, but keep the details a closely guarded secret.

The rate of pickup of hydrogen

When dealing with the rate of attainment of equilibrium in melting furnaces, the times are typically 30–60 min. This slow rate is a consequence of the large volume to surface area ratio. We call this ratio the 'modulus'. Notice that it has dimensions of length. For instance, a 10-tonne holding furnace would have a volume of approximately 4 m^3, and a surface area in contact with the atmosphere of perhaps 10 m^2, giving a modulus of $4/10$ m $= 0.4$ m $= 400$ mm. A crucible furnace of 200 kg capacity would have a modulus nearer 200 mm.

The values around 300 mm for large bodies of metal contrast with those for the pouring stream and the running system. If these streams are considered to be cylinders of liquid metal approximately 20 mm in diameter, then their effective modulus is close to 5 mm. Thus their reaction time would be expected to be as much as $300/5 = 60$ times faster, resulting in the approach towards equilibrium within times of the order of 1 min. This is the order of time in which many castings are cast and solidified. We have to conclude, therefore, that the melt will continue to equilibrate actively with its environment at all stages of its progress from furnace to mould.

There are methods available of protecting the liquid by an inert gas during melting and pouring which are claimed to reduce the inclusion and pore content of many alloys that have been tested, including aluminium alloys, and carbon and stainless steels (Anderson et al., 1989). It requires to be kept in mind that such techniques only protect the free surface of the melt. Hydrogen may continue to rise if moisture is present in the furnace refractories.

Ultimately, however, the gas in solution is normally not a problem to the castings of most metals and alloys (steels might be an exception because the exact mechanism of hydrogen embrittlement is not yet understood). If hydrogen can be persuaded to remain in solution, it appears to be normally perfectly harmless; it simply acts as an ordinary solute in solution, and therefore might provide some strengthening. Retaining the hydrogen in solution, but reducing oxide bifilms by such techniques as appropriate handling and treatment, and the elimination of pouring by, say, counter-gravity filling of moulds, hydrogen control becomes largely irrelevant.

6.4 COPPER ALLOYS

The ease of reducing copper from its ore has ensured copper and its alloys have enjoyed a long history, lasting thousands of years. Even so, copper and its alloys continue to be important at the present day for corrosion resistant cast products from domestic water fitting to high-strength marine applications in ships.

Pure copper castings are used for applications in which maximum thermal conductivity is required; a demanding application being blast furnace tuyeres. Similarly, pure copper is used in applications where maximum electrical conductivity is required, such as heavy electrical switch gear and bus bars (although dilute additions of cadmium or chromium are often used to increase strength without reducing conductivity too much). 'Pure' copper and its dilute alloys have a reputation for being particularly difficult to cast, the castings often suffering hot tears and porosity. This is almost certainly the direct result of poor casting technique, leading to the entrainment of surface oxides. The reader is recommended the sections of this work on casting manufacture for solutions to this common problem. The answer, as most often for castings, lies in the correct casting practice (for instance, the alloy composition and other metallurgical aspects being relatively unimportant).

The common brasses are red brass (70Cu30Zn) and yellow brass (60Cu40Zn), although many more complex and stronger brasses based on these alloys are also common.

Bronze alloys were traditionally solely alloys with tin (for instance bell metal is almost universally 77Cu23Sn), but nowadays aluminium bronze (5–10Al) and silicon bronze are important and widely used. Al bronze with additions of Fe, Ni and Mn is used for ships' propellers.

Gunmetals are alloys in which Zn has been added to a tin bronze. For instance 'Navy Gun Metal' was 88Cu8Sn4Zn. If lead is also added the alloy is known as 'leaded gunmetal'. One of the most famous of these is *ounce metal*, so-called because it was made up of a pound of copper, an ounce of zinc, an ounce of tin and an ounce of lead, also known as 85-5-5-5 metal. The lead was useful because of its action to seal pores in these very long freezing range alloys. Leaded brasses and bronzes are used for plain bearings, the lead- and tin-rich phases probably acting as high pressure lubricants. However, lead has been phased out of domestic water fittings, and is being steadily eliminated from nearly all foundry alloys as a result of the toxicity of its fumes and contamination of sand moulds during casting. Much exploratory work has been expended to investigate Si-containing alloys as a replacement in domestic water fittings (Fasoyinu, 1998).

6.4.1 SURFACE FILMS

Pure liquid copper in a moist, oxidising environment, causes water molecules to break down on its surface, releasing hydrogen to diffuse away rapidly into its interior. This behaviour is, of course, common to many liquid metals. The oxygen released in the same reaction (Eqn (1.2)), and copper oxide, Cu_2O, may be formed as a temporary intermediate product, but is also soluble, at least up to 0.14% oxygen. The oxygen diffuses and dissipates more slowly in the metal so no permanent film is created under oxidising conditions unless the solubility limit is exceeded. If the solubility limit is exceeded at the melt surface if the rate of arrival exceeds the rate of solution, a surface film of Cu_2O will build up. However, this will dissolve away if the rate of input of oxygen falls below the rate of dissolution of the film. In reducing conditions no film of any kind is expected. Thus pure liquid copper can be free from film problems in many circumstances. (Unfortunately, this may not be true or in the presence of certain carbonaceous atmospheres that may create a carbonaceous film, as we shall see later. Carbon-based films on copper and its alloys do not appear to have been researched so far.)

Certain Cu alloys, particularly the various aluminium bronzes that contain typically 5–10 wt% Al are quite a different matter. The great reactivity of Al with oxygen in the atmosphere and the high melting temperatures of these alloys combine to create conditions for the growth of a thick, tough and tenacious oxide that can give major problems if allowed to enter the casting.

6.4.2 GASES IN COPPER-BASED ALLOYS

Copper-based alloys have a variety of dissolved gases and thus a variety of possible reactions. In addition to hydrogen, oxygen is also soluble. These elements in solution (denoted by square brackets) can produce water vapour according to the reversible reaction:

$$2[H] + [O] = H_2O \tag{6.3}$$

Thus water vapour in the environment of molten copper alloys will increase both hydrogen and oxygen contents of the melt. Conversely, on rejection of stoichiometric amounts of the two gases to form porosity, the principal content of the pores will not be hydrogen and oxygen but their reaction product, water vapour. An excess of hydrogen in solution will naturally result in an admixture of hydrogen in the gas in equilibrium with the melt. An excess of oxygen in solution will result in the precipitation of copper oxide.

6.4 COPPER ALLOYS

Much importance is often given to the so-called *steam reaction*:

$$2[H] + Cu_2O = 2Cu + H_2O \tag{6.4}$$

This is, of course, a nearly equivalent statement of Eqn (6.3). The generation of steam by this reaction has been considered to be the most significant contribution to the generation of porosity in copper alloys that contain little or no deoxidising elements. This seems a curious conclusion because the two atoms of hydrogen are seen to produce one molecule of water. If there had been no oxygen present, the two hydrogen atoms would have produced one molecule of hydrogen, as indicated by Eqn (1.3). Thus the same volume of gases is produced in either case. It is clear therefore that the real problem for the maximum potential of gas porosity in copper is simply hydrogen. Depending on how much oxygen is present in solution, dissolved hydrogen will produce either a molecule of water vapour or a molecule of hydrogen. The volumes of gas are the same in either case.

(However, as we shall see in later sections, the presence of oxygen will be important in the nucleation of pores in copper, but only if oxygen is present in solution in the liquid copper, not just present as oxide. The distribution of pores as sub-surface porosity in many situations is probably good evidence that this is true. We shall return to consideration of this phenomenon later.)

Proceeding now to yet more possibilities in copper-based materials, if sulphur is present in solution, then a further reaction is possible:

$$[S] + 2[O] = SO_2 \tag{6.5}$$

and for the copper-nickel alloys, such as the monel series of alloys, the presence of nickel introduces an important impurity, carbon, giving rise to an additional possibility:

$$[C] + [O] = CO \tag{6.6}$$

Systematic work over the past decade at the University of Michigan (see, for instance, Ostrom et al. (1981)) on the composition of gases that are evolved from copper alloys on solidification confirms that pure copper with a trace of residual deoxidiser evolves mainly hydrogen. Brasses (Cu-Zn alloys) are similar, but because zinc is only a weak deoxidant the residual activity of oxygen in solution gives rise to some evolution of water vapour. Interestingly, the main constituent of evolved gas in brasses is zinc vapour because these alloys have a melting point above the boiling point of zinc (Figure 1.6). Pure copper and the tin bronzes evolve mainly water vapour with some hydrogen. Copper-nickel monels with nickel above 1% have an increasing contribution from carbon monoxide as a result of the promotion of carbon solubility by nickel.

Thus when calculating the total gas pressure in equilibrium with melts of copper-based alloys, for instance inside an embryonic bubble, we need to add all the separate contributions from each of the contributing gases.

The brasses represent an interesting special case. The continuous vapourisation of zinc from the free surface of a brass melt carries away other gases from the immediate vicinity of the surface. This continuous out-flowing wind of metal vapour creates a constantly renewed clean environment, sweeping away gases which diffuse out of the melt, carrying them from the alloy surface and preventing contamination of the local environment of the metal surface with furnace gases or other sources of pollution. For this reason cast brass is usually found to be remarkably free from gas porosity.

For alloys with more than 20%Zn, there is sufficient zinc vapour to burn in the air with a brilliant flame known as zinc flare. Flaring may be suppressed by a covering of flux. Similarly, aluminium is often added to brasses to reduce the loss of zinc, probably as a result of the formation of a dense alumina film. However, the beneficial de-gassing action is thereby suppressed, raising the danger of porosity, mainly from hydrogen.

The boiling point of pure zinc is 907°C. But the presence of zinc in copper alloys does not cause boiling until higher temperatures because, of course, the zinc is diluted (strictly, its activity is reduced). Figure 6.27 shows the effects of increasing dilution on raising the temperature at which the vapour pressure reaches 1 atm, and boiling occurs. The onset of vigorous flaring at that point is sufficiently marked that in the years before the wider use of thermocouples foundrymen used it as an indication of casting temperature. The accuracy of this piece of folklore can be appreciated from Figure 6.27,

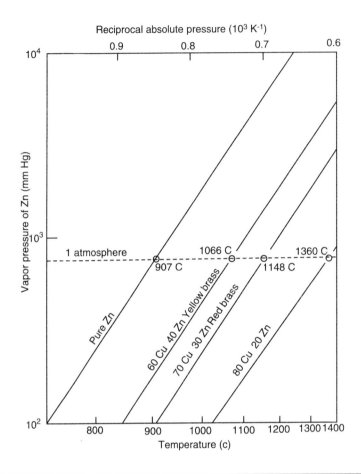

FIGURE 6.27

The vapour pressure of Zn and some brasses.

in which it is clear that the flaring temperatures increase in step with the increasing copper contents (i.e. at greater dilutions of zinc), and thus with the increasing casting temperatures of the alloys.

Around 1% of zinc is commonly lost by flaring and may need to be replaced to keep within the alloy composition specification. In addition, workers in brass foundries have to be monitored for the ingestion of zinc fumes.

Melting practice for the other copper alloys to keep their gas content under proper control is not straightforward. Following are some of the pitfalls.

One traditional method has been to melt under oxidising conditions, thereby raising the oxygen in solution in the melt in an attempt to reduce gradually the hydrogen level. Before casting, the artificially raised oxygen in solution is removed by the addition of a deoxidiser such as phosphorous, lithium or aluminium. The problem with this technique is that even under good conditions the rate of attainment of equilibrium is slow because of the limited surface areas across which the elements have to diffuse. Thus in fact little hydrogen may have been removed. Worse still, the original oxidation has often been carried out in the presence of furnace gases, so raising oxygen and (unwittingly) hydrogen levels simultaneously (Eqn (1.2)) high above the values to be expected if the two dissolved gases were in equilibrium. The addition of deoxidiser therefore may leave hydrogen at near saturation.

The further problem with this approach is that the deoxidiser precipitates out the oxygen as a suspension of solid oxide particles in the melt, or as surface oxide films. Either way, these by-products are likely to give problems later as non-metallic inclusions in the casting, and, worse still, as bifilm initiation sites to assist the precipitation of the remaining gases in solution, thus promoting the very porosity that the technique was intended to avoid. In conclusion, there is little to commend this approach.

An additional problem should be noted with respect to the relatively common practice of deoxidation with phosphorus. Gunmetals typically require only 0.01%P and Sn bronze typically 0.2%P (French, 1957). If excess is used a liquid Cu_3P film forms on the top of the melt. If this is poured with the alloy, it penetrates the mould leading to impaired surface finish and possibly entrained phosphide liquid.

A second reported method is melting under reducing conditions to decrease losses by oxidation. Hydrogen removal is then attempted just before casting by adding copper oxide or by blowing dry air through the melt. Normal deoxidation is then carried out. The problem with this technique is that the hydrogen-removal step requires time and requires the creation of free surfaces, such as bubbles, for the elimination of the reaction product, water vapour. Waiting for the products to emerge from the quiescent surface of a melt sitting in a crucible would probably take 30–60 min. Fumes from the fuel-fired furnace would be ever-present to help to reverse any useful de-gassing. Clearly, therefore, this technique cannot be recommended either!

Marin and Utigard (2010) described a variant of this technique in which they directed a flame of mixed O_2 and CH_4 (methane) on to the surface of the melt. Higher ratios of CH_4/O_2 gave higher reducing conditions that raised the rate of deoxidation of the copper. However, they noted that hydrogen is simultaneously increased by this practice, eventually becoming so high that the melt 'boils' with the evolution of water vapour by reactions as in Eqn (6.3). Boiling may now flush the melt to low levels of hydrogen, but the surface turbulence of the boil may now increase reaction with the air, increasing oxygen once again. All this resounds of lack of control.

A less dramatic and more straightforward technique involves a simple cover of granulated charcoal over the melt to provide the reducing conditions. This is a genuinely useful way of reducing the formation of drosses (dross is a mixture of oxide and metal, so intimately mixed that it is difficult to separate) as can be demonstrated from the Ellingham diagram (Figure 1.5), the traditional free energy/temperature graph. The oxides of the major alloying elements copper, zinc and tin are all reduced back to their metals by carbon, which preferentially oxidises to carbon monoxide (CO) at this high temperature. (The temperature at which the metal oxide is reduced, and carbon is oxidised to CO, is that at which the free energies for the formation of CO exceed that of the metal oxide, i.e. CO becomes more stable. This is where the lines cross on the Ellingham diagram.)

However, even here, it is as well to remember that charcoal contains more than just carbon. In fact, the major impurity is moisture, even in well-dried material that appears to be quite dry. An addition of charcoal to the charge at an early stage in melting is therefore relatively harmless because the release of moisture, and the contamination of the charge with hydrogen and oxygen, will have time to be reversed. In contrast, an addition of charcoal at a late stage of melting will flood the melt with fresh supplies of hydrogen and oxygen that will almost certainly not have time to evaporate out before casting. Any late additions of anything, even alloying additions, introduce the risk of unwanted gases.

Reliable routes for melting copper alloys with low gas content include:

1. Electric melting in furnaces that are never allowed to go cold.
2. Controlled use of flaring for zinc-containing alloys.
3. Controlled dry environment of the melt. Addition of charcoal is recommended if added at an early stage, preferably before melting. (Late additions of charcoal or other sources of moisture are to be avoided.)

In summary, the gases and vapours which can be present in the various copper-based alloys are:

Pure copper	H_2, H_2O
Brasses, gunmetals	H_2, H_2O, Zn, Pb
Cupro-nickels	H_2, H_2O, CO, (N_2?)

Various tests have been proposed from time to time for copper-based alloys. Dion (1979) recommended a test for oxygen by simply dipping a carbon rod into the melt. If the oxygen is high, the rod 'sings' with the rapid evolution of microscopic bubbles of CO, producing a vibration that can be sensed by hand. This test can detect oxygen in solution down to a limit of about 0.01% at which the rod stops singing. The rod needs to be at the temperature of the melt because a cold graphite rod evolves gas. This raises the potential problem, of course, of the probability of moisture in the graphite, although it is possible that the test itself will assist to eliminate this providing the rod is reasonably dry. If the rod is dry it will also contribute to the flushing out of any hydrogen.

The total sum of gases has often been assumed to be testable using the Reduced Pressure Test exactly similar to that used for Al alloys. Studies of the test are reported by numerous workers, including Matsubara (1972) and Ostrom (1974, 1976). The test is also described in the ASM Handbook 2008 vol. 15. As with experience in Al alloys, the test has proved controversial and confusing for the same reason: the test is also highly sensitive to the presence of oxide bifilms. In reality, the test would be a better test of oxides than a test of gas, but with intelligent use, should give a useful indication of both.

Sub-surface pinholes in copper base alloys have been widely researched. One of the more thorough reports is by Fischer (1988). The phenomenon appears to be exactly analogous to that observed in Al alloys. The pores arise as a result of a combination of factors, naturally leading to widespread confusion in the literature. The key factors include the following.

1. The gases in solution in the melt (of which, of course, there are several in copper alloys).
2. The gases released by the mould binder material and which diffuse into the casting surface.
3. The build-up of gases ahead of the advancing freezing front, raising local concentration of the total gas content after a distance of 1–2 mm (i.e. just sub-surface).
4. The presence of a population of oxide bifilms in suspension in the melt as a result of poor melting practice or poor casting practice. The bifilms act as initiation sites and are themselves pushed ahead of the front, so being in precisely the correct location for maximum effect of pore generation. The variation of difficulty of unfurling of the bifilms will give a spectrum of pore initiation times, giving a mix of round and dendritic pore morphologies corresponding to early and late arrivals respectively (see Figures 7.40 and 7.41).

6.4.3 GRAIN REFINEMENT

There has been much practical work carried out on the grain refinement of Cu-based alloys, but, it seems, little fundamental research so far.

It was Cibula in 1955 who used iron and cobalt borides, plus zirconium carbide and nitride, to grain refine bronzes and gunmetals. As is usual with such work, hot tears were found to be reduced although porosity became more evenly dispersed or at times concentrated as layer porosity. Finer grains also appeared to be associated with reduced ductility and a lack of pressure tightness. These apparently perverse results are almost certainly the result of a high oxide bifilm population in his cast material which is only to be expected from the awful casting techniques in general use at that time. Techniques such as those described later in this book should greatly assist to reduce these defect levels, and reverse such effects of loss of ductility and pressure tightness.

With similar poor casting methods, Couture and Edwards (1973) used Zr master alloys to successfully grain refine a variety of gun metals, finding improved strength and hot tear resistance, but confirmed Cibula's finding of a drastic reduction in elongation and pressure tightness.

Gould (1960) studied the grain refinement of pure copper, and found additions of Li and Bi to be effective, but many additions that were tried were of no use.

Clearly, copper alloys would benefit from additional work to understand the fundamentals of grain refinement. In addition, casting with improved filling system designs should help to remove the current confusion relating to loss of ductility and pressure tightness that has accompanied early work.

6.5 CAST IRON

Cast irons are nature's gift to foundrymen. They melt at accessible temperatures, requiring relatively low energy, they run like water, are strong enough for many engineering applications and are relatively insensitive to poor casting techniques and so yield adequate castings despite sloppy and out-dated casting practices.

The commercial exploitation of simple grey irons was a major factor triggering the start of the industrial revolution, but the current spectrum of different cast irons is now dauntingly broad. It stretches from grey irons (so-called because of their fracture surface is coloured grey from the graphite flakes that provide the fracture path) through compacted graphite irons, spheroidal graphite (known as ductile) irons, to white irons (the apparently brittle iron carbides providing a bright, white fracture surface) for wear resistance. Highly alloyed irons are commonly used for heat and oxidation resistance. Special heat treatments, particularly austempering, can provide high-performance irons with strengths approaching 1 GPa—the traditional reserve of high-strength steels.

For cast irons, carbon is the key alloying element. In pure binary Fe-C alloys, the minimum carbon content for a cast iron to distinguish it from a steel is close to 2%, but the liquidus temperature at this carbon content is high at 1290°C and the long freezing range of about 150°C and limited graphite content would make feeding difficult. At a carbon content of approximately 4.3%, the alloy is of eutectic composition and the freezing temperature is comfortably low at 1150°C with negligible freezing range. This is also close to the composition at which the expansion—because of graphite precipitation—almost exactly counters the contraction resulting from austenite solidification, conferring on eutectic irons the enormous benefit of requiring little if any feeding. Beyond the eutectic composition, the liquidus temperature rises once again and the precipitation of primary graphite prior to general solidification at the eutectic temperature means that graphite flotation in these hyper-eutectic irons can become a problem, limiting carbon contents usually to a maximum close to 4.5%.

With silicon as the next most common element in grey irons, equivalent structures can be achieved with reduced carbon, trading off C against increased Si. It is usual to quantify this relation by an *effective* carbon content, known as the carbon equivalent (C_E) or sometimes the carbon equivalent value (CEV). This is approximately

$$C_E = \%C + \%Si/3$$

A typical engineering quality of grey iron would consist of around 3.2C and 2.0Si, to give a C_E close to 3.9%C, just slightly hypoeutectic. In general, grey irons contrast with ductile irons which have a C_E commonly of 4.2–4.4%.

In the days of making elaborate ornamental cast iron ware, or for certain wear-resistance applications, 1–2% of phosphorus was often added. This addition increased fluidity of the melt (Figure 3.13) and the hard phosphide phases added wear resistance to such applications as brakes. The %P acted similarly to the Si in its replacement of effective carbon, so that their combined effect gave approximately:

$$C_E = \%C + (\%Si + \%P)/3$$

The reasonable cost, huge availability and attractive machinability of the graphitic irons, especially grey irons, guarantees the prosperous future of these alloys (despite all criticisms of its weight compared to Al alloy castings for instance).

Having said all this, it is true that cast irons have always been and continue to be the most used casting alloys, the most researched, but the least understood. This is why this Section 6.5, perhaps unfortunately, is effectively pressurised into being the most speculative section in this book. For want of any other coherent explanation, I present my own approach to an understanding of the microstructures of irons in terms of the mechanisms of inoculation and nodularisation. Until disproven, it is offered as no more than a potentially useful working hypothesis. In the meantime, we can look forward to the outcome of researches that might confirm the approach, or suggest an improved hypothesis.

6.5.1 REACTIONS WITH GASES

Before moving on to describe the formation of porosity from gases in solution or reacting with its environment, it is worth emphasising that most porosity I have seen in most castings, including cast irons, is mostly air bubbles entrained by the poor design of filling system. Such clumps of porosity, rather than uniformly dispersed pores, are easily identified as filling system defects.

Moving on now to reactions leading to porosity, as for copper-based alloys, the production of iron-based alloys is also complicated by the number of gases that can react with the melt and that can cause porosity by subsequent evolution on solidification. Again, it must be remembered that all the gases present can add their separate contributions to the total pressure in equilibrium with the melt. We shall deal with the gases in turn.

Oxygen is soluble and reacts with the high carbon content of cast irons. CO is the product, following Eqn (6.6). Carbon dioxide is seen to be blamed by many experimenters who note bubbles in cast irons usually in association with oxide (usually silicate) slags. It has been assumed that the high carbon content of the iron has reacted with the oxygen in the slag to create bubbles of CO. However, this is almost certainly an error. The slag is more usually not the result of carryover from the ladle, but is created in situ in the mould by surface turbulence. The bubbles are much larger than could be generated by reaction because the rate of reaction is limited by the rate of diffusion which is necessarily rather slow, producing, as for the dispersed microporosity of Al alloys, bubbles of maximum diameter perhaps 0.5 mm. It would require the coalescence of 1000 of these bubbles to make the 5 mm diameter bubbles typical of those seen in irons. Thus it is certain that the bubbles are not the product of a diffusion-controlled reaction but are simply *air bubbles* introduced by the turbulence involved with the poor filling systems usually employed. Naturally, they will contain some traces of carbon monoxide. However, with a well designed naturally pressurised filling system, both slag and so-called CO bubbles will disappear.

CO is still expected to be an important gas for the creation of porosity observed from time to time in cast irons, simply because of the high availability of both carbon and oxygen. However, it needs to be kept in mind that such bubbles will be expected to be fine, usually submillimetre in size, and evenly distributed.

CO can be encouraged to evolve from cast iron as a froth of bubbles, known as a carbon boil. Such an evolution of CO can be induced in molten cast iron, providing the silicon is low, simply by blowing air onto the surface of the melt (Heine, 1951), thus reinforcing that, if not already obvious, oxygen can be taken into solution in cast iron even though the iron already contains high levels of carbon. During subsequent solidification, in the region ahead of the solidification front, carbon and oxygen are concentrated still further. It is easy to envisage how, therefore, from relatively low initial contents of C and O, they can increase together so as to exceed a critical product $[C] \cdot [O]$ to cause CO bubbles to form in the casting. The equilibrium equation, known as the solubility product, relating to Eqn (6.7) is:

$$[C] \cdot [O] = kP_{CO} \quad (6.7)$$

We shall return to this important equation later. It is worth noting that the equation could be stated more accurately as the product of the activities of carbon and oxygen. However, for the moment we shall leave it as the product of concentrations, as being accurate enough to convey the concepts that we wish to discuss.

Hydrogen is soluble, as in Eqn (1.3), and exists in equilibrium with the melt, as indicated in Eqn (1.4). Although a vigorous carbon boil would reduce any hydrogen in solution to negligible levels by flushing it from the melt, such techniques are not usual in iron melting. Thus any hydrogen will tend to remain in the melt. Even so, although some hydrogen might be present, it does not generally appear to lead to any significant problems in irons. This is probably the result of the hydrogen solubility being rather low, in contrast to steels, where higher melting temperatures lead to the possibility of high hydrogen levels in solution.

In addition, of course, the rate of diffusion of hydrogen is astonishingly high. Thus hydrogen might be quickly lost from a melt, probably only in a few minutes, if the melt has the benefit of a dry environment. Similarly, in the presence of moisture or hydrogen, the hydrogen content of a melt could rise quickly. It clearly pays to work with furnaces and their backup refractories that are always kept hot.

Nitrogen is also soluble in liquid iron. The reaction follows the normal law for a diatomic gas:

$$N_2 = 2[N] \quad (6.8)$$

and the corresponding equation to relate the concentration in the melt $[N]$ with its equilibrium pressure P_{N_2} is simply:

$$[N]^2 = kP_{N_2} \quad (6.9)$$

As before, the equilibrium constant k is a function of temperature and composition. It is normally determined by careful experiment. As for hydrogen, in general, it seems that nitrogen is probably not an important source of porosity in irons. Exceptional circumstances might include the reactions involving high-nitrogen binders because the decomposition of amines they contain appear to lead to such problems as nitrogen fissures (see the Nitride Films section in this chapter). However, these seem to be as much surface nitride problems as nitrogen solution problems.

6.5.2 SURFACE FILMS ON LIQUID CAST IRONS

When cast iron is held at a high temperature (e.g. 1500°C) in a furnace or ladle lined with a traditional refractory material such as ganister, a fascinating sight can be witnessed. The surface of the liquid iron is seen to be continuously punctuated by the silent and mysterious arrival of bright circular patches. These suddenly appear and spread from nothing to their full size of several centimetres within about a second. The patches drift around, coalesce with other patches, and finally attach themselves to the wall of the vessel where they cool and add to the solidified rim of slag. These patches are droplets of liquid refractory, melted from the walls and bottom of the vessel. As the vessel is tips and empties, upstanding 'stalactites' on the base, and upward runs and drips on the walls can usually be clearly seen, marking the sites where the drops detached.

In common with all the components of molten metal systems, the slag will be changing its composition rapidly as it interacts with the molten metal. At high temperature its contents of iron, manganese and silicon will be reduced from their respective oxides and taken into solution in the liquid iron, whereas the remaining stable oxides, such as those of aluminium and calcium, will remain to accumulate as a dry slag, sometimes called a dross. These reactions will be explained further in the next section.

Such layers of slag on the surface of molten iron can be anything from 0.1 mm thickness upwards. (In the cupola, of course, the thickness is often around 100 mm or more.) It is not intended to consider such macroscopic surface-layer problems in this section. The following section considers only the microscopically thin surface film that, under certain conditions, will form automatically on the surface of the melt (no matter how good is the melting resistance of the lining material of the holding vessel).

Oxide films

Work by Heine and Loper at the University of Wisconsin, dating from 1951, has done much to explain the complex formation of surface films on cast irons. A slightly later study by Merz and Marincek (1954) is also illuminating. Based on these studies, we can explain the changes that occur as the temperature falls.

When the iron is at a high temperature, 1550°C, the Ellingham diagram indicates that CO is a more stable oxide than SiO_2. Thus carbon oxidises preferentially and is therefore lost at a higher rate than silicon, as is seen in Figure 6.28. Here the blowing of air on to the surface of a small crucible of molten metal serves to accentuate the effect. Silicon is observed to fall only after all the carbon has been used up. At this high temperature, no film is present on the melt—any silicon oxide, SiO_2, would be immediately reduced to silicon metal which would be dissolved in the melt, simultaneously forming CO which would escape to atmosphere.

At around 1420°C the stability of the carbon and silicon oxides is reversed. The exact temperature of this inversion seems to be dependent on the composition of the iron as pointed out by Merz and Marincek; de Sy (1967) reported a range of 1410–1450°C for the irons that he investigated, whereas the Ellingham diagram (Figure 1.5) predicts an inversion temperature for pure Fe-C alloys of about 1500°C. The agreement is, perhaps, as good as can be expected because of compositional uncertainties. Below approximately 1400°C, therefore, SiO_2 appears on the surface as a dry, solid film, rather grey in colour. This film cannot be removed by wiping the surface because it constantly reforms.

At a temperature of 1300°C, and in alloys that contain some manganese, it is clear from the Ellingham diagram that MnO is the least stable, SiO_2 is intermediate, and CO the most stable. Thus manganese is oxidised away preferentially, followed by silicon, and finally by carbon. The contribution of MnO to the film at this stage may reduce the melting point of the film, causing it to become liquid.

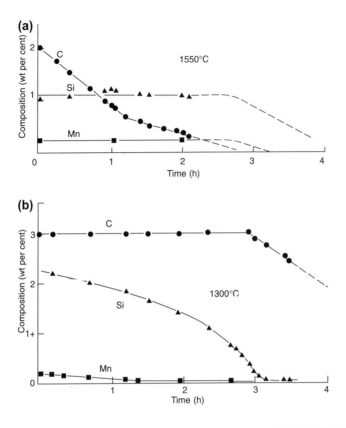

FIGURE 6.28

Change in composition of 3.6 kg of molten grey iron held in a silica crucible, whilst air was directed over its surface at the rate of 22 mL/s (a) melt at 1550°C; and (b) melt at 1300°C.

Data from Heine (1951).

At around 1200°C, iron oxide, FeO, contributes to the further lowering of the melting point at the ternary eutectic between FeO, MnO and SiO_2. If sulphur is also present in the iron, then MnS will contribute to a complex eutectic of melting point 1066°C (Heine and Loper, 1966).

The author finds that, in general terms, the previous considerations nicely explain his observations in an iron foundry where he once worked. For a common grade of grey iron, the surface of the iron was seen to be clear at 1420°C. As the temperature fell, patches of solid grey film were first observed at about 1390°C. These grew to cover the surface completely at 1350°C. The grey film remained in place until about 1280°C, at which temperature it started to break up by melting, finally becoming completely liquid at 1150°C.

When casting grey iron in an oxidising environment, the falling temperature during the pour will ensure that the surface film will be liquid at the most critical late stage of the filling of the mould. If the film becomes entrained in the molten metal, it will therefore quickly spherodise into compact droplets. The droplets are of much lower density than the iron and so will float out rapidly. On meeting the surface of the casting they will mutually assimilate, and be assimilated by, the existing surface liquid film, and so spread over the casting surface. The glassy sheen of some grey iron castings may be this solidified skin. The harmless dispersal of the oxide film in this way is the reason for the good natured behaviour of cast iron when cast into greensand moulds; it is one of the very few metal-mould combinations capable of exhibiting tolerance towards surface turbulence. Even so, there appears to be some experience indicating the irons cast without turbulence exhibit improved properties. This confusing situation requires to be resolved by future research.

Only on one occasion has the accumulation of liquid oxide at a casting surface given the author some problems. This was in a grey iron casting where a small amount of surface turbulence was known to be present just inside the ingate, because it was not easy to lower the velocity below 0.4 ms^{-1} at this point and was judged to be a negligible risk of any kind of internal defect. However, so much liquid surface was created at that location, and so many droplets of the entrained slag floated out at a point just down-stream, that the layer of surface slag accumulated at the down-stream location exceeded the machining allowance, scrapping the casting.

In **ductile irons**, in contrast to grey irons, the entrainment of the surface is nearly always a serious matter. The small percentage of magnesium that is required to convert the iron from flake to the spheroidal graphite type dramatically alters the nature of the oxide film.

Above 1454°C, Heine and Loper (1966) found that the surface of liquid ductile iron remains clear of any film. Below this temperature, a film starts to form, increasing in thickness to 1350°C, at which point the surface exhibits solidified crusty particles. By the time the temperature has reached 1290°C, the entire surface is covered with a dry dross. Magnesium vapour distils off through the dross because the molten iron is above the boiling point of magnesium. Presumably, the oxidation of the vapour to powdery MgO at the upper surface of the dross is a major contributor, causing the dross to grow quickly and copiously. The dross makes life difficult for the ductile iron foundryman, forming films, and agglomerating into dry, non-wetting heaps, that, if entrained, spoil otherwise excellent castings. Ductile iron is renowned for being difficult to cast cleanly, without unsightly dross defects. Surface films something like those on Al alloys are to be expected as in Figure 6.29.

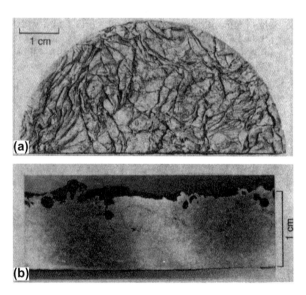

FIGURE 6.29

(a) Oxide skin on liquid Al-9Si-4Mg alloy wrinkled by repeated disturbance of the surface: and (b) a cross-section of the solidified metal. The appearance in both cases is extremely similar to graphite films on grey iron.

Courtesy of Agema and Fray (1990).

Oxide bifilms naturally in suspension in grey irons

De Sy (1967) has shown that liquid cast iron generally contains significant quantities of oxygen in solution in excess of its solubility. He concluded, on the basis of careful and rigorous experiments, that the undissolved fraction of oxygen was present as SiO_2 particles. Interestingly, by heating to 1550°C he confirmed the expectation that the SiO_2 solids dissolved because they became less stable than CO, but reappeared on cooling once again. Hartman and Stets report for irons containing Mg not only the presence of SiO_2 in suspension, but also the more complex iron-magnesium-silicate, olivine, $2(Mg,Fe)O \cdot SiO_2$. Hoffman and Wolf (2001) find a variety of oxides including SiO_2, FeO and MnO among others, but superheating and holding at high temperature eliminated many of these. In their elegant study of the thermodynamics Mampaey and Beghyn (2006) show how mainly SiO_2 together with some FeO forms in a typical melt when cooling from 1480° to 1350°C.

It seems reasonable to speculate that these oxides almost certainly would not be compact spheres, cubes or rods etc., but would most likely be in the form of *films*. Only films would have a sufficiently low Stokes velocity to remain in suspension for long periods of time associated with these experiments, and the long periods during which irons are held molten in holding furnaces.

In any case, of course, the *film* morphology is to be expected. During melting in the cupola as droplets of iron rain down, the natural folding-in of the surface film of SiO_2 of each droplet would ensure a natural population of SiO_2-rich double films (bifilms). Additional treatments or handling such as pouring actions, stirring in induction furnaces, and the oxide introduced from the surface of the charge (whether steel, pig or foundry returns) would increase this already large, natural population of oxide bifilms. Furthermore, it is well known that iron from electric furnaces is more liable to chill formation problems in thin sections than cupola iron, and this effect has been widely accepted as the loss of nuclei (in agreement with proposals made here), especially during extended time in holding furnaces or pouring systems.

Even if no silica bifilms are formed naturally in the iron as described previously, they will certainly be introduced during the process of inoculation, when Si-rich particles are added through the silica surface, entraining the surface silica as they penetrate and become submerged, as illustrated in Figure 2.3.

Thus there appear to be at least two quite different populations of oxide bifilms in suspension in liquid iron. (1) In grey irons, the silica-rich bifilms are a natural population in equilibrium with the melt, the amount of silica-rich phase being predictable by thermodynamics. (2) In ductile irons, the magnesia-rich bifilms are the result of mechanical accidents involving the turbulent entrainment of the surface film into the bulk liquid, resulting in non-reversible damage. We expand on this problem in ductile irons in this next section.

Bifilms in ductile iron

The population of oxide bifilms observed in ductile irons arises from the trauma of a turbulent filling system. In this case, the presence of Mg stabilises the magnesia, MgO, in the surface film, although Si might also contribute, thus forming a magnesium silicate $MgO \cdot SiO_2$ (also written as $MgSiO_3$). Both magnesium oxide and magnesium silicate are extremely stable, and when entrained, represent permanent damage to the liquid metal and subsequently to the casting. Although the bifilms are known to have an initially compact morphology as a result of the turbulence during their formation, and so are relatively harmless as cracks, the subsequent straightening of these bifilms by various natural processes such as the growth of dendrites, creating extensive planar cracks, is common, resulting in the development of brittleness in the form of plate fracture (Figures 6.30 and 6.31), as discussed in more detail in the next section.

The ductile casting industry has referred to entrained surface films observed on polished microstructural samples as 'dross stringers' (Figure 6.32). This name, based on their one-dimensional appearance on a polished two-dimensional section, has led to a comforting self-deception, concealing their real nature as extensive planar defects in three-dimensional space in the form of films floating about in a sea of metal. The occasional appearance of clusters of graphite nodules that have floated up and been trapped under such 'stringers' corroborates their real nature as films; nodules would not be trapped under one-dimensional 'strings' but naturally collect under two-dimensional films. Also, as we are now aware, if the film is solid (as it clearly is in this case), the entrainment process will fold them in dry side to dry side, thus forming a bifilm, a crack.

6.5 CAST IRON 281

FIGURE 6.30

Plate fracture in the feeder neck of a ductile iron casting (Karsay, 1980).

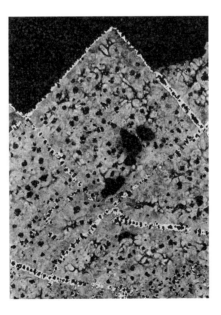

FIGURE 6.31

Polished microsection through the fracture (Barton, 1985).

Courtesy Casting Technology International.

282   CHAPTER 6 CASTING ALLOYS

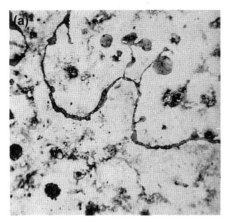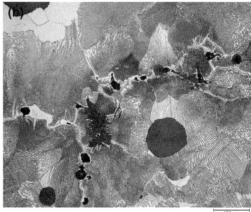

FIGURE 6.32

(a) So-called silicate 'stringer' in ductile iron (actually a visible silicate bifilm; (b) an alignment of mis-shapen nodules at a grain boundary in a pearlitic ductile iron, indicating the probable presence of an invisible bifilm.

The films appear on the liquid metal only at low temperature as we have seen, and seem to be mainly magnesium silicate, probably with a thick upper layer of solid MgO. If the ductile iron is cast at a low temperature, and if the surface is entrained, the creation of seriously damaging bifilms is guaranteed. Naturally, as the hot liquid iron cools during its passage through the running system it is likely to cool to the temperature at which the solid film starts to form, so that defects will be expected in most filling systems in which surface turbulence is not controlled. Once entrained, the defects can, of course, lead to a variety of additional problems. One of these serious problems is discussed later.

The magnesium silicate dross 'stringers' (actually, we should immediately stop calling them 'stringers' and call them 'bifilms') are particularly thick bifilms, and very clear when seen on a polished microsection as seen in Figure 6.32(a). Figure 6.32(b) probably also shows a bifilm in ductile iron, but the bifilm in this case is too thin to reveal its presence directly. We can be fairly sure a bifilm is present from the long line of particles of flaky graphite and mis-shapen spheroids, all typical of features that prefer to form on an oxide bifilm, and all apparently following a grain boundary, a common site for a bifilm. Elsewhere in this alloy, away from contact with the bifilm, the spheroids are beautifully formed.

Plate fracture defect in ductile iron

As with nitrogen fissures in grey irons, plate fracture in ductile irons has also never been satisfactorily explained. *Ductile* irons, should, of course, always exhibit a *ductile* mode of failure. Sometimes, however, a casting will exhibit poor strength and poor elongation to failure, with the fracture surface exhibiting large planar facets, the alloy appearing to consist of large embrittled grains. These unpredictable events give rise to serious concern that the material is not under the proper control that either the foundry or the customer would like to see. Everyone's faith is shaken. The question naturally arises, 'Is ductile iron a reliable engineering material?' These are the questions that should never arise, and that no one wishes to hear.

Following the description given by Karsay (1980) and Gagne and Goller (1983), the features of the plate fracture are large, flat, apparently brittle fracture planes, in ductile irons that in normal circumstances would exhibit only ductile failure. The planar facets appear to grow mainly vertically at right angles to the bottom surface of the casting (Figures 6.30). When viewed closely in cross-section (Figure 6.31), the planes are seen to be studded with small, irregularly shaped graphite spheroids, arranged with an accuracy almost resembling a crystal lattice. When polished and etched the planes are characterised by a matrix that is somewhat lighter than the rest of the casting. Karsay suggests that the colour difference may be the result of a higher Si content in this region, stabilising the ferrite. Finally, in this region, there is a high incidence of small inclusions that appear to be mainly magnesium silicates.

All these features are consistent with the defect being an oxide bifilm, probably a magnesium silicate, explaining the high Si content, the high inclusion content, and possibly the malformed spheroids as a result of local loss of Mg together with the natural tendency of graphite to form preferentially on bifilms. The planar form of the failure surface arises from the bifilm being pushed by the raft of austenite dendrites and organised into an interdendritic sheet, similar to that commonly seen in other alloy systems (Figures 2.43, 2.44(a) and 2.46). The vertical orientation is also understandable because of the greater rate of heat transfer from the base of the casting where gravity retains its contact with the mould, enhancing cooling, so that grains growing vertically from the base grow fastest and furthest. In addition, the magnesium silicate bifilm will necessarily be trapped at the mould wall (otherwise, dendritic straightening without some part of the film being anchored is not easily envisaged—bunches of film will only be pushed ahead if the film is not anchored) and buoyancy will encourage its vertical orientation, and so assist the advancing dendrite to straighten the film. Spheroids in interdendritic regions would then be revealed at the regular spacing dictated by the dendrite arm size. During solidification and cooling at the high temperature the bifilm probably disintegrates to some extent because of its surface energy tending to spherodise it. What remains are the changes in chemistry and numerous silicate fragments as inclusions plus remnants of the original central bifilm crack to encourage the direction of growth of the crack that finally causes failure.

Other features of plate fracture are its occurrence in slowly cooled regions, such as in a feeder neck. This may be the result of the lower rate of growth allowing the dendrites to straighten films more successfully (at high growth velocity, the drag resistance of films would resist dendrite growth, and resist film straightening).

Loper and Heine (1968) found that '*spiking*' observed on fracture surfaces of both white and ductile irons are similar and sometimes appears as oxidised facets. The occasional oxidation is easily understood if the bifilm connects to the surface and allows the ingress of air deep into the casting. Similar *internal oxidation* during heat treatment of irons is described in Section 9.10.

The less common appearance of plate fracture in irons of higher carbon equivalent and its reduction in resin-bonded sand moulds reported by Barton (1985) is probably not so much the result of a more rigid mould as he suggests, but an indication that the entrainment of the oxide film is less damaging in this more carbonaceous environment.

Heine and Heine (1968) described some fascinating observations and use telling language: 'when iron is damaged by melting and pouring so that it solidifies "spikey" it will not permit feeding even though proper sized feeders are provided'. This description acknowledges the possibility of deterioration of the melt by improper melting and handling and the consequent facetted nature of the freezing pattern. Furthermore, the presence of extensive oxides, probably from wall to wall in the casting, and bridging the necks of feeders, act to prevent the flow of feed metal. I am always impressed by intuition of good foundrymen for understanding the underlying scientific mechanisms.

In passing, it is worth reminding ourselves of the possibility that other non-oxide bifilms are to be expected in cast irons.

Nitride Films

Nitride bifilms probably form in cast irons giving rise to the 'nitrogen fissure' defects associated in the past with high nitrogen binders. Nitrogen fissures in grey iron castings are large cracks, often measured in centimetres, which appear to have been associated with the use of sand binders that contain high levels of nitrogen. They are an enigma that has never been satisfactorily explained. The blame is normally given to high-nitrogen binders that contain amines, whose breakdown probably contributes both nitrogen and hydrogen to the liquid iron. However, although such binders are known to be associated with fissure defects, their use does not always result in fissures.

It seems most likely that entrained bifilms, perhaps consisting of nitride films, are also required, so that the filling system may also be highly influential. Any involvement of the filling system has not previously been suspected, but would explain the current confusion in the results of studies carried out so far. Once entrained, the combination of high hydrogen and nitrogen pressure in the iron might be sufficient to inflate any nitride bifilms to some extent, opening the bifilms to revealing their presence as crack-like features. Further discussion is given in Section 7.2.3 under Nitrogen Porosity.

Carbon films (lustrous carbon)

The liquid film present on cast iron at low temperature in an oxidising environment has made iron easy to cast free from serious defects. This marvellous natural benefit of cast iron when cast into moulds made from sand bonded with clay and water must have played an important part in the success of the industrial revolution. In general (although acknowledging a number of infamous and tragic exceptions) the bridges did not fall into the river, and the steam engines continued to power machinery. Later, this benefit was to be extended to moulds made using one of the first widely used chemical binders: sodium silicate. This environmentally friendly chemical is still widely used today as a low cost sand binder for the production of strong moulds (despite a number of significant disadvantages that some foundries have been prepared to live with, but which are now being successfully overcome).

However, it is one of those ironies of history that the arrival of modern chemical binders based on resins was to change all of this.

Binders based on various kinds of resins; furan, phenolic, acrylic, polyurethane, etc. were heralded as the breakthrough of the twentieth century. Indeed, the new binders had many desirable properties, making accurate and stable moulds, with excellent surface finish, at good rates of production from simple low-cost equipment, and with good breakdown after casting.

However, when iron was poured into some of these early resin-bonded moulds, especially those based on polyurethane, a new defect was discovered. It became known as lustrous carbon. It had been occasionally seen, especially if the volatile additions to greensand had been high, but it was never so common nor so damaging. This shiny, black film resulted in casting skins wrinkled like elephant hide. Studies deduced from X-ray diffraction patterns (Draper, 1976) concluded that it was pyrolytic carbon (a microcrystalline form) that was deposited from the gas phase on to hot surfaces in the temperature range of at least 650–1000°C. Being a form of pyrolytic carbon, similar to carbon black, it was unlike the nicely crystallographic regularity of graphite.

The hot surface was originally assumed to be the sand grains of the mould, and somehow the deposit was pushed ahead of the advancing liquid front, to become incorporated into the surface as folds (Naro and Tanaglio, 1977). The explanation is clearly problematical on several fronts: in most instances, the sand surface is rather cold, and thus incapable of chemically 'cracking' (i.e. breaking down) the polymeric gases to precipitate carbon. Also, it is difficult to imagine how a film deposited on the complex and rough sand grains could be detached from its grip on these three-dimensional shapes before the arrival of the liquid metal and so be pushed into rucks and folds. After the arrival of the metal the film would be assisted in keeping its place by being held against the surface of the sand grains by the pressure of the liquid. Clearly, this explanation cannot be correct.

The only explanation that fits all the facts is that the graphitic film forms on the surface of the molten metal itself. Photographs of lustrous-carbon defects, particularly those seen on the fracture surfaces of parts that have suffered brittle failure, beautifully reveal their origin as the surface film on the liquid metal (Bindernagel, 1975; Naro and Tanaglio, 1977). The caption to the photograph of the oxide skin on an aluminium alloy (Figure 6.29) could be changed to read that it was a carbonaceous skin on a grey iron; to the unaided eye, the appearance of the two types of film is practically identical.

Part of the confusion that has surrounded the lustrous carbon film, claiming that it deposited on sand grains, dates from a misreading of the brilliant original work by Petrzela (1968). This Czech foundry researcher devised a test in which he demonstrated that the vapours released from coal tar and other hydrocarbon additions to moulding sands would decompose, depositing carbon as a shiny, silvery film on a metal strip, resistance heated to at least 1300°C. In his test, it happened that the sand was also heated to this temperature. Thus he observed carbon to be deposited directly onto the sand grains in addition to that deposited on the heated metal strip. The mistake of subsequent generations of researchers has been to assume that lustrous carbon always deposits onto sand grains, even though, at the instant that the metal is filling the mould, the sand is usually nowhere near the temperature at which a hydrocarbon vapour could be decomposed.

The reader is recommended to Petrzela's engaging, chatty and candid account. He was clearly one of our great foundry characters. His writing contains other fascinating asides to some of his observations on the release of carbon from hydrocarbons. He was possibly the first to describe a sooty deposit among sand grains that had a fibrous, woolly appearance. It has been observed many times since, although not previously explained. It is considered later, for instance in Figure 6.35.

Other work has studied the generation of lustrous carbon in greensand moulds. It is clear that the mould atmosphere can provide a hydrocarbon environment for the liquid metal if sufficiently high concentrations of hydrocarbons are added to the sand mixture. Such additives help the mould to resist wetting by the metal, and so improve surface finish, as appreciated in the original work of Petrzela (1968) and later by Bindernagel et al. (1975). Excess additions have sometimes been claimed to give lustrous carbon defects. One such defect is shown in Figure 6.33. However, it is certain that the defects form only if the surface turbulence can cause the film to be entrained. Otherwise, if the film is retained on the surface by careful uphill filling as in Figure 6.34, it is a valuable effect, significantly enhancing the surface finish of the casting. It is a pleasure to see iron castings that shine like new shoes.

The mechanism for the improvement of surface finish by the addition of hydrocarbons to the mould repays examination in some detail. The carbon film forms on the front of the advancing liquid. There is a suggestion by me (Campbell and Naro, 2009) that the carbon deposits on a precursor oxide film (Figure 6.34), on the grounds that the surface graphite, Kish graphite, appears to do the same (as noted by Liu and Loper, 1990) and flake graphite seems likely to form on oxides. Whether or not the film forms on a prior oxide film, it becomes trapped between the melt and the mould, and is held there by friction. Thus, as the meniscus advances, the film bridging the meniscus is forced to tear, splitting apart, but of course, immediately re-forming, as illustrated for the films on the advancing melts shown in Figures 2.2 and 6.34. The film is therefore continuously formed and laid down between the melt and the mould by the advancing metal, as though the advancing metal were rolling out its own track like a track-laying vehicle. The film forms a mechanical barrier between the metal and the mould. It is the mechanical rigidity of this barrier, helping to bridge the sand grains, that confers the improved smoothness to the cast surface. Thus, over the years, although there has been much talk about the action of surface tension bridging the gaps between sand grains (and this is clearly true to some extent) the main action in many alloys appears to be the result of the presence of the mechanically rigid surface film. The paper by Campbell and Naro (2009) beautifully illustrated the lustrous carbon film resembling a plate of steel, easily bridging sand grains and thus smoothing the surface of the casting (Figure 6.35(a)–(d)).

History now appears to have turned full circle because some resin binders for sands have recently been developed to yield iron castings with reduced incidence of lustrous carbon defects. At this stage, it is not clear whether the surface finish of the castings has suffered as a result. If better filling systems using bottom gating had been employed, the lustrous carbon would have been given its best chance to enhance the surface finish of the casting with no danger of an entrainment defect. Thus the new reduced lustrous carbon binders need never have been developed.

In lost foam castings using polystyrene foam, lustrous-carbon films cause troublesome defects. In this situation, the vapourisation of the polystyrene to styrene, and the subsequent decomposition of the styrene to lower hydrocarbons and eventually to carbon, deposits thick carbonaceous films on the advancing surface of the iron (Figure 4.9 shows the decomposition products). Gallois et al. (1987) found the film to consist of three main layers: (1) an upper lustrous multilayered structure of amorphous carbon; (2) an intermediate layer of sooty fibres consisting of strings of crystallites; and (3) a layer adhering strongly to the surface of the iron consisting of polycrystalline graphite enriched in manganese, silicon and sulphur. Clearly there has been some exchange of solutes from the iron into the film.

In sand castings the often-observed apparent decohesion of the carbon film from the metal surface and adhering to the surface of the mould (Figure 6.35) has previously led to much confused thinking, concluding that the lustrous carbon had formed on the mould. The decohesion of the carbon film from its original substrate had been difficult to explain because most if not all other films such as oxides strongly adhere to their originating matrix, as is demonstrated by anodised layers on Al alloys.

The presence of the woolly, fibrous carbon behind the film and among the sand grains in Figure 6.35(a)–(d) is illuminating and probably holds the explanation of this behaviour. Clearly the fibrous carbon forms after the film is put in place against the mould wall. Thus it seems that the fibrous material forms at perhaps lower temperatures or at lower concentrations of hydrocarbons. The dense array of fibres, interconnected between the film and the grains, explains how the film, originally on the metal, can become rather firmly attached to the mould. Further encouragement for the film to detach from the casting will arise from the enormously different thermal contractions of the substrate casting and the carbon surface film.

Graphite films that have grown on molten iron have been studied in the form of crystals formed on the surface of Fe-C alloys held in graphite crucibles, and so saturated with carbon before being allowed to cool (Sumiyoshi, 1968).

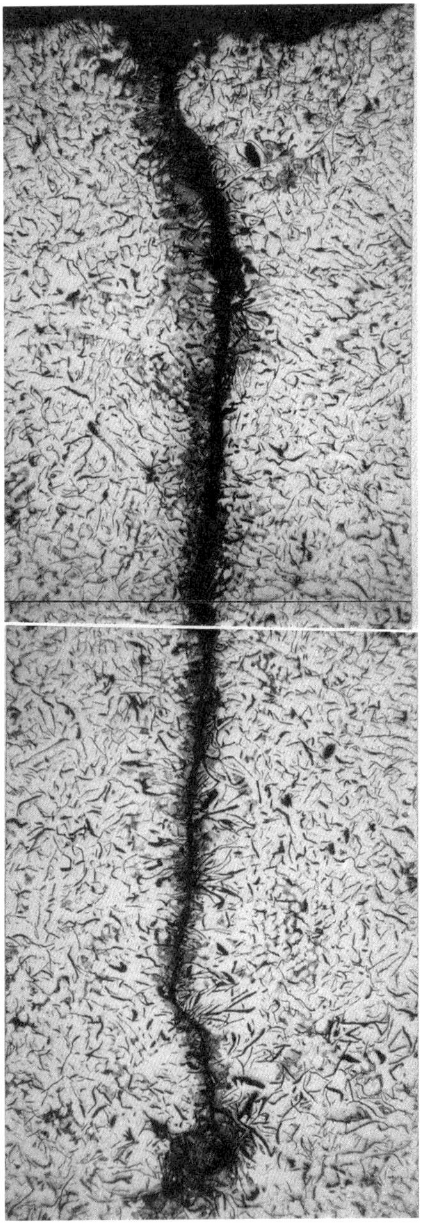

FIGURE 6.33

A folded-in lustrous carbon film, forming a carbon bifilm in grey iron.

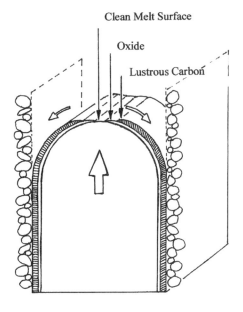

FIGURE 6.34
A schematic view of the advancing cast iron front causing the deposited films of oxide and carbon to be trapped between the mould and the metal, leading to the necessary splitting of the films at the advancing front.

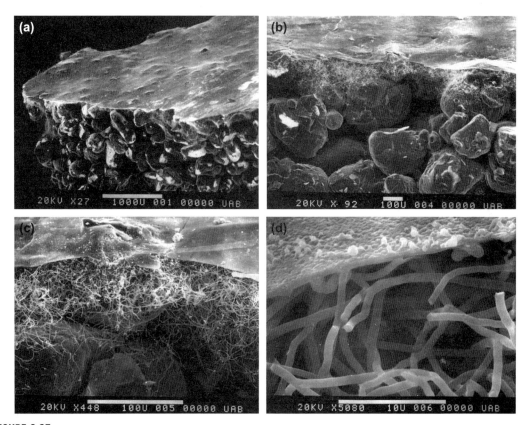

FIGURE 6.35
SEM images (a-d) illustrating the lustrous carbon film bridging sand grains.

These shiny black sheets that float on the surface are analogous to the Kish graphite that separates from hyper-eutectic cast irons during cooling, and are likely therefore to be growing on the underside of the surface oxide film. (For the scientifically minded, the graphite films in this research had interesting features. They were single crystals with numerous cracks along certain crystal directions, and hexagonal growth steps on the underside that showed how the film grew by gradual deposition of carbon atoms, probably onto ledges from emergent screw dislocations.)

The growth of graphite on melts saturated with carbon, as discussed previously, is easy to understand; but how does graphite grow in the case of lustrous-carbon films where the composition of the iron is far from saturated? Such films should go into solution in the iron! Every student of metallurgy knows that at the eutectic temperature the carbon in solution in iron has to exceed 4.3 wt% before free (Kish) graphite will precipitate. At higher temperatures, the carbon concentration for saturation increases, following the liquidus line for the solidification of Kish graphite in hyper-eutectic irons, as is clear from the Fe-C phase diagram. It seems unlikely that a graphite film can result from an equilibrium reaction. The explanation that follows is an example first mentioned in Section 2.5 relating to soluble films. It shows the effect relies on an interesting dynamic condition.

In an atmosphere containing hydrocarbons, if the rate of arrival of reactants at the free liquid surface is low, then both carbon and hydrogen can diffuse away from the surface into the bulk liquid. The free surface of the melt therefore remains clean.

However, in a highly concentrated environment of hydrocarbon gases the rate of arrival of reactants may exceed the rate of diffusion away into the bulk. Thus carbon will become concentrated on the surface (hydrogen less so, because its rate of diffusion is much higher) and may exceed saturation, allowing carbon to build up at the surface as a solid in equilibrium with the local high levels of carbon. Once formed, it would then take time to go into solution again, even if the conditions for growth and stability were removed. Thus it would appear to have a pseudo stability, with a life just long enough so that in some conditions the film could be frozen into the casting if a chance event of surface turbulence were to enfold the surface into the melt.

The folding in of the graphitic film is known to result in the familiar lustrous carbon defect (Figure 6.33). This is, of course, simply a bifilm crack lined with the lustrous carbon films. In heavy sections, such defects are not seen, probably because they will have time to go into solution before being frozen into the casting. In thinner sections of a ferritic matrix or mixed ferrite/pearlite matrix, the films can be seen to be partly dissolved as indicated by the layers of the higher carbon content pearlite on either side of the defect.

More speculatively, many irons are poured so turbulently that it is to be expected that huge numbers of graphitic films will be entrained. This raises a distinct possibility that many of the graphite flakes seen on a polished section of grey iron will not be formed as a result of a metallurgical precipitation reaction, but may be the remnants of entrained graphitic bifilms. The occasional appearance on microsections of what seem to be isolated large flakes amid uniform smaller flakes is suggestive of the bi-modal distribution to be expected if such a mixed source of graphite were present.

Recent research has indicated that the conditions for the growth of the graphite film on liquid metals are similar to the conditions required for the growth of diamond films. Reviews by Bachmann and Messier (1984) and Yarborough and Messier (1990) listed conditions for the growth of diamond as the breakdown of hydrocarbons and the presence of hydrogen. In the case of iron, the temperature is a little too high for diamond, and so would tend to stabilise the formation of graphite films. But for metals such as aluminium in a hydrocarbon environment, the conditions seem optimum for the creation of diamond on the metal surface. Prospectors and investors will be disappointed to note, however, that the rate of growth is slow, only 1 μm/h. Thus in the time that most liquid metal fronts exist while pouring a casting, the diamond layers, if any, will be so thin as to be a disappointing investment.

6.5.3 CAST IRON MICROSTRUCTURES

The forms of graphite in cast irons have been the subject of intense interest and huge research efforts over many decades, but a full understanding has been elusive. Readers are referred to the review by Loper (1999) for a wide-ranging synopsis covering many details not included in this study. Here, a rather different review is made of the literature, exploring the possibility of a unifying approach based on the hypothesis that oxide bifilms are present in liquid irons.

6.5 CAST IRON

We have seen how a comprehensive understanding of the microstructure of Al-Si alloys has been proposed in terms of bifilms, explaining both the mechanism of modification and the structures of hypo- and hyper-eutectic alloys. Nakae and Shin (1999), among many others, have drawn attention to the analogous features of Al-Si and Fe-C alloys. This section of the book is an extension of the bifilm hypothesis, apparently valuable to an understanding of the Al-Si system, to a possible understanding of the various morphologies of carbon in the form of graphite and carbides in the Fe-C alloy system.

Adopting an analogous phase diagram to that shown in Figure 6.15 for Al-Si, the equivalent for the Fe-C system is shown in Figure 6.36(a). The eutectic at 4.3%C and 1130°C seems fairly well established, although freezing point values up to 1150°C have also been commonly assumed. This seems to correspond to the formation of graphite as flakes on the silica bifilms (in detail, most probably nucleated not directly on the bifilms but on nuclei already formed on the bifilms). If easy growth on silica is avoided by elimination of the silica, formation of graphite then occurs at some lower temperature, growing as compacted or nodular forms as discussed later. The precise value of the lower temperature is not known at this time, but is probably in the region of 1100–1125°C, and will certainly be affected by the growth form as is clear from thermal analysis curves.

The eutectic composition is also moved to higher levels of carbon, so that the higher carbon levels typical of ductile irons may be at least partly the result of this effect, and not solely the result of the founder seeking greater graphite content to reduce feeding problems. Interestingly, the peak in fluidity, normally indicating the composition of a eutectic, are seen from the work by Porter and Rosenthal (1953) to be close to 4.5%C (Figure 3.11), even though this work was carried out with grey iron. This result indicates that rapidly flowing iron, which would contain be turbulently ravelling and retaining bifilms in their compact form, flows similarly to ductile iron which contains no bifilms, as might be expected.

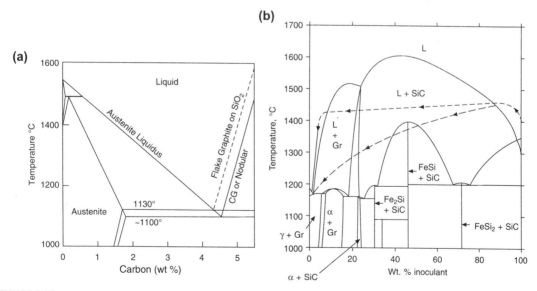

FIGURE 6.36

(a) Fe-C phase diagram showing (i) the high temperature eutectic in which graphite forms on silica-rich bifilms in suspension in the melt and (ii) the lower temperature eutectic of Mg-treated iron in which the eutectic can take different forms in the partial or complete absence of bifilms. (b) The Fe-FeSi phase diagram showing possible melting and mixing routes for a dissolving FeSi inoculant particle (Harding et al., 1997).

Graphite nuclei

Mizoguchi and co-workers (1997) have demonstrated that austenite is ineffective in nucleating graphite. In fact, they find that undercoolings below the liquidus of between 200 and 400°C are required to trigger nucleation by austenite. A more unfavourable nucleus than austenite would be difficult to imagine. The question therefore arises, 'what does nucleate graphite?' This question is all the more intriguing following the work by Mampaey and Xu (1997) in which they found that a single population of nuclei (although the nature of the nuclei remained unidentified) could explain both grey and ductile irons.

There is now a growing consensus that both flake and spheroidal graphite nucleate on similar if not identical nuclei (for instance, Warrick, 1966) composed of particles of complex oxides and sulphides. This was the conclusion reached in the first study following the development in the United Kingdom of the microprobe analyser (Jacobs, 1974). Jacobs and colleagues were the first to carry out an elegant study suggesting that within graphite nodules there is a central seed of a mixed sulphide (they suggested Ca and Mg sulphide) surrounded by a mixed (Mg,Al,Si,Ti) oxide spinel. They found matching crystal planes between the central sulphide, the spinel shell, and the graphite nodule, indicating a succession of nucleating reactions. This exemplary work has been confirmed a number of times, most recently by Solberg and Onsoien (2001).

Many confirmations of this general conclusion have because been made (for instance, Skaland, 2001) suggesting that the oxy-sulphide mix of the various elements will have a spectrum of lattice spacings ensuring that at least part of the compound will match graphite, and therefore possibly constitute a favoured substrate. As an example of a recent study, while working on preconditioning treatments for grey irons (treatments involving small additions of elements such as Al before inoculation—possibly to enhance the population of naturally occurring nuclei in uninoculated irons), Riposan et al. (2008) defined a three-stage model for the nucleation of graphite that differs in some details from that by Jacobs:

1. Small oxides (<2 μm) are formed in the melt (from preconditioner)
2. Complex sulphides (<5 μm) nucleate on and wrap around the oxides (the presence of Mn and S is necessary; many authors have reported the beneficial effects of S, for instance Chisamera et al., 1996)
3. Graphite nucleates on the sulphides (Ca, Sr, Ba in the sulphide assists nucleation), and, we shall suppose, will attempt to wrap itself around the sulphide particle.

Whether the details of the successive steps of nucleation follow exactly that described by Jacobs or by Riposan are not important for our understanding of the overall mechanism of inoculation. In the following discussion, the mechanisms proposed to explain the various morphologies of graphite are based on the possibility of

1. *nucleation* on specific oxy-sulphide nucleating particles which are effective for all types of graphite and
2. *growth* morphology of the graphite depends on the presence or absence of favourable growth substrates such as oxide bifilms for flake graphite and oxysulphide particles for nodular graphite (it is proposed in this work that the initial spherical growth on a nucleating particle appears to require stabilisation by plastic deformation of austenite at a later stage of growth.)

For a proper understanding of inoculation, it will be essential for the reader to keep in mind the separate actions of nucleation and growth.

6.5.4 FLAKE GRAPHITE IRON AND INOCULATION

The treatment of cast irons by the deliberate addition of material to aid the formation of graphite is generally called inoculation. Inoculation of cast irons is important to achieve a reproducible type and distribution of graphite to give reproducible mechanical properties and machinability.

Uninoculated iron is characterised by poor control of the graphite flake morphology. Flakes occur, but are relatively few in number, and uncontrolled in size. The relatively few opportunities for the carbon to precipitate lead to relatively large regions of the iron elsewhere being supersaturated with carbon. Thus iron carbide (Fe_3C) precipitation is likely in regions that are deficient in graphite nuclei. The mechanical properties of the iron are generally poor. In general, as will become clear during the progress of this account, it seems that some nuclei exist prior to inoculation, but their number and effectiveness cannot be relied on.

The history of inoculation

I am indebted to my good friend, Reginald Forrest, for an account of the history of the inoculation process.

Ross and Meehan were co-owners of the Ross Meehan Foundry based in Chattanooga, Tennessee, USA (a city later to be immortalised by Glenn Miller with his swing band hit 'Chattanooga Choo Choo'). Gus Meehan was an inquisitive and observant foundryman. He was intrigued by the known benefits of 'treating' the melt with floor sweepings and put the treatment on a more controlled basis in his foundry. The practice became 'standard' in the Ross Meehan foundry and customers were happy. By chance an English foundryman working in Newcastle, UK, learned about this development and corresponded with Meehan to exchange experiences. This was Oliver Smalley, another observant, smart and practical foundryman. They experimented with various mixtures and materials and finally decided on calcium silicide as a major component of their mix, together with ferrosilicon fines and graphite. The mixtures were a closely guarded secret. All this time, the consistency of the Ross Meehan cast iron castings was becoming widely known.

They decided to name their iron 'Meehanite' and later 'Meehanite Metal'. They patented their invention in 1922 and began a system of licencing foundries to produce iron castings from controlled (inoculated) liquid cast iron. This system, spearheaded by Smalley, who had the marketing flair, became progressively better controlled with measured response (wedge controls) and producing a spectrum of grades of grey iron known variously as Meehanite irons.

Smalley, the genius of marketing Meehanite, developed the concept of Meehanite as bridging the gap between steel and cast iron. Meehanite licencee foundries mushroomed in the United States and Europe; the issue of licenses being limited so far as possible to good, disciplined foundries who would follow the prescribed practices. Smalley realised that by reaching the end user (the engineers and designers) and convincing them that Meehanite was a high quality, consistent and reliable material and that castings produced by Meehanite licenced foundries were to be trusted—he tapped into a rich vein. This was perhaps one of the first examples of technological foundry marketing.

Gus Meehan was the origin of the Meehanite name that became so famous in the cast iron industry—but Smalley was the real driving force in its commercial exploitation. The Americans and Smalley eventually went their separate ways, splitting the operation: Meehanite Metal Corporation controlled the American and Far East operations and International Meehanite Metal Company based in the UK controlled European Licensee operations. Materials & Methods Limited was the birth child of International Meehanite Metal Company and produced and marketed all of the inoculants and, later, nodularisers used by the licenced Meehanite foundries (and later sold openly to any foundry).

The gradual introduction of the inoculation process from 1922 onwards, and its continued development to the present day (for instance, Skaland, 2001 and Hartung et al., 2008), was found to greatly increase the number of nuclei available, giving a copious crop of graphite flakes of good uniformity of size, with a reduced tendency to carbide formation, and a consequent benefit to the mechanical properties and machinability of the iron. These benefits are now freely available to all cast iron foundries.

The mechanism of inoculation

Clearly, inoculation was some kind of process to provide the nucleation of graphite particles. However, the detailed mechanism remained unknown.

What was known was that successful inoculants include ferrosilicon (an alloy of Fe and Si, usually denoted as an inaccurate shorthand, FeSi, but usually containing approximately 75 wt% silicon, so sometimes written Fe75Si), calcium silicide and graphite. These were added to the melt as late additions, just before casting. Additions designed to work over 15–20 min were used in a granular form, of size around 5 mm diameter, whereas very late additions (made to the pouring stream) were generally close to 1 mm. Late inoculation was favoured because the inoculation effect gradually disappeared; a process known as 'fade'.

Ferrosilicon is the normally preferred addition, and is known as a 'clean' inoculant. Calcium silicide is known to be a rather 'dirty' addition, almost certainly because the calcium will react with air to give solid CaO surface films (in contrast to FeSi that will cause liquid silicate films). The folding-in of CaO films during turbulent filling would create particularly drossy bifilm inclusions together with entrained air as porosity. The calcium silicide addition would probably have been much more acceptable with better-designed filling systems that reduce surface turbulence. (Ductile iron casters experience similar problems when using Mg as a noduliser.)

It is immediately clear that the common inoculant, FeSi, does not perform any nucleating role itself. This is because liquid iron at its casting temperature (perhaps 1350–1400°C) is well above the melting point of the FeSi intermetallic compound (1210°C), so that the whole FeSi particle melts (Figure 6.36(b)).

The evidence now suggests that the inoculants, originally a particle, continue to exist for some time as a molten high Si region in the liquid iron. Although the Si-rich region is liquid, and the iron is liquid, and the two liquids are completely miscible, the two nevertheless take time to inter-diffuse. This time is probably the fade time. The Si-rich region slowly dissipates in the melt, eventually disappearing completely. However, during its short lifetime, it provides a local environment with a high effective CEV. To get some idea of the scale and importance of this effect it is instructive (although admittedly not really justified, as we shall see) to calculate the carbon equivalent in one of these regions. For an iron of carbon content about 3%, assuming CEV = (%C) + (%Si/3), we have CEV = 3 + 75/3 = 28%C. Extrapolating the carbon liquidus line on the equilibrium diagram to an iron alloy with the huge level of 28%C predicts a liquidus temperature in the region of several thousand degrees Celsius. (This is actually not surprising in view that graphite itself has an effective melting point of more than 10,000°C.) Clearly, therefore, there seems good reasons for believing that the carbon in solution in the Si-rich regions is, in effect, enormously undercooled. It is a form of artificial *constitutional undercooling* (because the graphite is effectively *undercooled* as a result of a change in the *constitution* of the alloy).

Now, in reality, it is not appropriate to extrapolate the CEV beyond the eutectic value of 4.3%C. In fact, when this part of the equilibrium phase diagram is calculated, the liquidus surface is nothing like linear, as seen in Figure 6.36(b) (Harding, Campbell and Saunders, 1997). Even so, this figure shows the liquidus in the hyper-eutectic region to be very high, so that the essential concept is not far wrong. The path of the dissolving particle is marked on the figure, showing its gradual loss of silicon as the melted liquid region makes its way from right to left across the figure. Different paths can be envisaged for different rates of loss of temperature as are seen in the Figure. This slowly disappearing liquid 'package' has to pass through regions where it will experience undercooling of several hundred degrees Celsius, providing a massive driving force for the precipitation of graphite.

The large undercooling, creating such a substantial driving force in these liquid regions is almost certainly the reason why, over the years, so many different nuclei have been identified for the initiation of graphite. It seems that even nuclei that would hardly be expected to work at all are still coaxed into effectiveness by the extraordinary and powerful undercooling conditions that it experiences. Studies have shown that many particles that are found in the centres of graphite spherules, and thus appear to have acted as nuclei, whereas identical particles are also seen to be floating freely in the melt of the same casting, having nucleated nothing (Harding, Campbell and Saunders, 1997). This is understandable if the nuclei are not particularly effective. They will only be forced to act as nuclei if they happen to float through a region that is highly constitutionally undercooled.

Studies by quenching irons just after inoculation have revealed a complex series of shells around the dissolving FeSi particle. Hurum (1952, 1965) was the first to draw attention to this phenomenon, but it has been studied by several others since (for instance, Fredriksson, 1984). Although FeSi itself contains almost no carbon, the carbon in the cast iron diffuses into the liquid FeSi region quickly. Data from Figure 1.4(c) and Eqn (1.5) indicate a time of 1 s for an average diffusion distance $d = 0.1$ mm, but 100 s for $d = 1$ mm. The flow resulting from the buoyancy of the high Si melt, and the internal flows of metal in the mould cavity, will smear the liquid Si-rich region into streamers, reducing the diffusion distance to give shorter estimated times for the homogenisation of carbon. Thus the shell of silicon carbide (SiC) particles around a dissolving FeSi particle (Figure 6.37) appears logical as a result of the high undercooling in the part of the phase diagram where SiC should be stable (Figure 6.36(b)). It seems likely that the SiC nucleates homogeneously because of the high constitutional undercooling. In a shell further out from the centre of the dissolving inoculant particle, graphite starts to form. It may be that graphite does not simply nucleate homogeneously as a result of the generous undercooling but can also form in this region by the decomposition of some of the SiC particles.

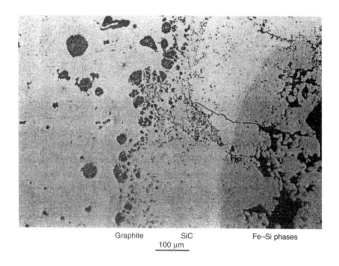

Graphite   SiC   Fe–Si phases
100 μm

FIGURE 6.37

Microsection of a dissolving FeSi particle in a ductile iron, quenched from the liquid state (Bachelot, 1997).

If all of this were not already complicated enough, there is even more complexity. In addition to the dissolving FeSi particle providing (1) a local solute enrichment of Si there will also be (2) a release of sundry complex inclusions including oxides and sulphides. Commercially available inoculants contain various impurities, and various deliberate additions that supplement the natural nucleating action in this way. These additions include group 1A elements of the periodic table, Mg, Ca, Sr and Ba, and often some rare earths such as La and Ce, that will react to create oxides and sulphides. At least some of these may be good heterogeneous nuclei for the formation of new graphite crystals (or perhaps new SiC crystals that may subsequently transform to graphite particles). Also, of course, these particles are provided exactly where they are needed, in the heart of the highly undercooled liquid region. These intentionally added particles will augment the naturally occurring population of nuclei already present in the melt. The overwhelming driving force explains the wide variation of successful nuclei which, in other circumstances, would be expected to be of only mediocre, if any, effectiveness.

This action of the inoculating material in providing a combination of copious heterogeneous nuclei together with good growth conditions explains the action of graphitisers such as ferrosilicon, and the importance of the traces of impurities such as aluminium, rare earths and sulphur that raise the efficiency of inoculation.

Ferrosilicon and calcium silicide are not, of course, the only materials that can act as inoculants. SiC is also effective, as is graphite itself. Both of these materials can be seen to provide similar transient conditions consisting of pockets of liquid in suspension in the melt in which high constitutional undercooling promotes the nucleation of graphite.

The chain of nucleating effects, oxides-sulphides-graphite, and only in the effectively supercooled regions, has the outcome that graphite particles exist in the melt at temperatures well above the eutectic. The prior existence of graphite particles in the liquid at high temperature, well above the temperature at which austenite starts to form is quite contrary to normal expectations based on the equilibrium phase diagram, but explains many features of cast iron solidification. The expansion of graphitic irons prior to freezing (the so-called 'pre-shrinkage expansion') has in the past been difficult to explain (Girshovich, 1963). The existence of graphite spheroids growing freely in the melt above the eutectic temperature has been a similar problem, seemingly widely known, and seemingly widely ignored, but now provided with an explanation, even though it would be highly desirable to have some additional confirmation as soon as possible.

Harding and co-workers (1997) pointed out that once nucleated in the regions of high driving force for initiation, the graphite particles attached to their nuclei now will emerge into the general melt where they will become unstable and

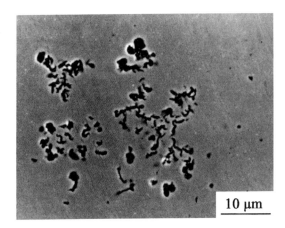

FIGURE 6.38
Coarsening of tips of graphite particles on emerging from the undercooled FeSi region (Benaily, 1998).

start to re-dissolve. Feest et al. (1980) found that although the Si-rich inoculant regions disperse relatively rapidly, the graphite which formed rapidly in these regions is slow to re-dissolve.

Although they were in the undercooled region, the graphite particles would have been expected to grow with extreme speed, thus adopting a thin and branching dendritic morphology. However, on leaving this rapid growth environment and entering a region where growth will suddenly be arrested, and re-solution starts, the dramatic change will be expected to result not only in the arrest of growth but the coarsening of the graphite dendrite tips. This effect is seen in material quenched from this region (Figure 6.38) by Benaily (1998).

The observations by Loper and Heine (1961) confirmed that graphite can form and survive in both hypo- and hyper-eutectic irons at 1400°C, well into the liquid range, high above the expected liquidus temperatures. Mampaey (1999) also confirmed that graphite forms in the melt before the appearance of austenite. (These observations are quite contrary to expectations based on the equilibrium diagram. However, of course, the equilibrium diagram is based on the assumptions not only of (1) equilibrium behaviour but also (2) perfectly uniform composition, neither of which applies during the inoculation of cast irons.)

Thus nuclei will have initiated graphite nucleation in the *constitutionally undercooled* pockets of liquid, but will emerge and start to re-dissolve in the open melt. However, if they happen to pass through other undercooled regions the graphite will experience sudden bursts of growth, followed by slow dissolution in the bulk of the melt. Finally, the graphite particle will approach the eutectic front, which will be *thermally undercooled*, and so enjoy stability and a final spurt of growth before being frozen into the advancing eutectic.

Given sufficient time, all the inoculant particles will have melted and dispersed, leaving no pockets of undercooling floating about in the melt. This is almost certainly the phenomenon known to all foundry personnel as 'fade' of the inoculation effect, occurring within a time of approximately 5–20 min.

Assuming that the nucleated graphite particles survive, whether their subsequent growth occurs in the form of flakes or spheroids is a completely separate issue, unrelated to the nucleation/inoculation treatment. This is a growth problem. We shall deal with growth separately.

Growth of graphite

Bearing in mind that many second phases and intermetallics precipitate on bifilms as preferred substrates, it seems reasonable to assume that these new graphite nuclei would also preferentially form on substrates provided by oxide bifilms. Having nucleated on the bifilm, the nucleus would in turn nucleate graphite. Figure 6.39 schematically shows a graphite nucleus formed on an oxide bifilm. As the bifilm by chance enters an undercooled region provided by a

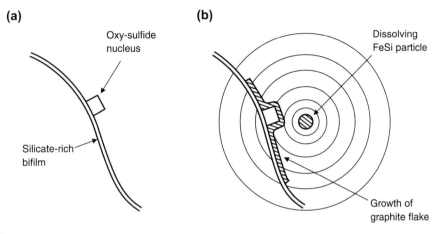

FIGURE 6.39

The mechanism of inoculation: (a) a bifilm in suspension in the melt, with oxy-sulphide nucleus (from trace impurities or preconditioners); (b) the bifilm floating into a constitutionally supercooled region, causing graphite to nucleate and grow on the bifilm.

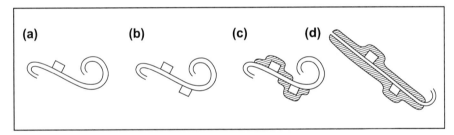

FIGURE 6.40

(a) A bifilm possibly supporting some initial nuclei; (b) additional nuclei provided by inoculant; (c) graphite forms on nuclei in supercooled regions; (d) graphite grows as fairly straight flake, straightening the bifilm to create a central crack.

dissolving FeSi inoculant particle, the nucleus experiences a massive driving force as a result of hundreds of degrees of effective undercooling, forcing graphite to nucleate around the nucleus, and forcing rapid growth.

The newly forming graphite is not able to grow completely around the nucleating particle because the particle itself has itself grown on the planar bifilm substrate so that at least one of its faces is inaccessible (Figure 6.40). The silica-rich bifilm will form a 'next best' substrate for graphite, so although insufficiently favoured to cause nucleation, it is sufficiently favoured to support the further growth of the graphite. Thus in grey irons the graphite extends across the bifilm, leading to the fairly flat morphology of flakes in grey iron. The flakes grow in regions ahead of the solidification front (i.e. slightly above the general eutectic freezing temperatures) because of the energetically favoured growth of graphite on the oxide substrates in suspension (Figure 6.41). The growth morphology of graphite, extending in the directions in its basal plane, would favour the straightening of the bifilm (Figure 6.40(d)). The bifilm would be expected to be extremely thin, possibly measure in nanometres, its minimal rigidity exerting

296 CHAPTER 6 CASTING ALLOYS

(10a) flake graphite; R = 1.2 μ/s

FIGURE 6.41

A straight graphite flake formed on a bifilm freely floating in the melt is overtaken by the coupled growth of the eutectic, and incorporated into the solid (Li, Liu and Loper, 1990).

negligible constraint of the advancing graphite crystal. The freedom from restraint would explain the development of relatively perfect crystals of graphite as observed growing in the liquid ahead of the coupled eutectic graphite (Figure 6.41).

The mechanism proposed previous explains the growth of flake graphite from nucleating particles introduced by inoculation. Particles that appear to be nuclei for the initiation of flakes have often been observed and seem likely to be a universal phenomenon in both flake and nodular irons (Rong and Xiang, 1991).

Although several experimenters have concluded that graphite nucleates on silica (Tyberg and Granehult, 1970; Gadd and Bennett, 1984; Nakae et al., 1991), this is probably an understandable error. In fact, it is far more probable that they were observing *growth* of the graphite on silica. Nakea (1993) goes further to identify the particular structural form of silica as cristobalite. It seems certain in fact that graphite nucleates on compact oxy-sulphide particles, but subsequently grows on silicon oxide bifilms, having the structure of cristobalite if Nakea is correct.

Eventually, the advancing solidification front will overtake those flakes growing on bifilms floating freely in suspension in the liquid. Thus eventually, these freely floating flakes will become incorporated into the solid as seen in the centre of the solidified coupled eutectic in Figure 6.41 by Li, Liu and Loper (1990). Thus it would be expected to be common to see grey irons with two separate populations of flakes: (1) those formed as primary particles by free growth in the liquid and (2) those formed by coupled growth with austenite at lower temperatures. The coupled growth mode is discussed later. A bi-modal distribution of graphite flakes is therefore to be expected in many microstructures. A bi-modal distribution is suggested in Figure 6.41 but is more clearly seen in Figure 6.42. Less obvious but important bi-modal distributions are probably common, as may be inferred from the work of Enright and colleagues (2000) using automated fractal analysis of microstructures. If lustrous carbon surface films on the liquid iron are also incorporated by turbulent entrainment events, it is conceivable that trimodal graphite forms can be present. No one has yet looked for such curious features so it is not known whether they exist.

Practical experience with inoculation

Goodrich (2008) attributes the *type C* iron (ASTM A247), characterised by large, very straight flakes, with some branching, to the result of the growth of the flakes in the liquid, unencumbered by the presence of austenite. He calls these 'proeutectic' flakes. They originate in suspension in the melt and are therefore capable of flotation to the upper regions of

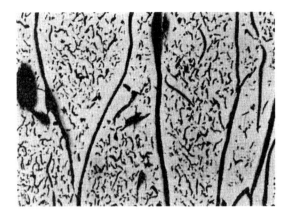

FIGURE 6.42
Two populations of flakes; one formed on bifilms and the other formed as a classical couple eutectic appearing as undercooled or coral form. (Hillert and Rao, 1967).

a casting. The more common *type A* graphite flakes are similar, displaying only minimal irregularity, suggesting a similar origin and behaviour in the melt. Loper and Fang (2008) use deep etching to reveal what they call 'pre-eutectic' flakes with elegant hexagonal symmetry, and apparently largely free from defects. For many other irons, the presence of a dense mesh of austenite dendrites constrains the size and shapes of flakes and prevents any significant buoyancy problems (Loper, 1999).

More usually in castings, the graphite flakes are seen to branch relatively frequently. In terms of the bifilm substrate, this is straightforwardly understood from the irregular structure of the bifilms. During their entrainment from the liquid surface into the bulk melt they tend not to entrain as nicely parallel double films, but as randomly folded, messy structures. Thus folds leading to parts of the double film at irregular angles to the main bifilm fold are to be expected, and would account for the branching of growing flakes.

Experience of variable performance is also to be expected. For instance, on a Monday morning, after melt has been held for the weekend, operators commonly find the iron has poor graphite structure, despite attempts to provide nuclei by inoculation. We may speculate that this is the result of the gradual floating out of the bifilm substrates. Similarly, iron heated to a high temperature suffers a similar degradation of graphite structure, almost certainly as a result of the dissolution of the bifilms because of the instability of SiO_2 above about 1450°C in the presence of carbon. It would be interesting to know whether the melt, after losing its silica-rich bifilms at high temperature, would regain its good solidified structure when cooled once again, because although de Sy reports that the silica reappears in the melt on cooling, without some kind of surface turbulence the form of the silica may not be a bifilm, nor even a film, but may be a compact particle. As such, it is not likely to be a good substrate for the development of a good flake structure. Even so, the process of inoculation, entraining the surface oxide during the act of submerging the inoculants particles, and the turbulence of the final pouring of the melt into the mould may address this problem, making the problem essentially invisible to those attempting to study the effect.

In agreement with the prediction that graphite grows on oxide bifilms, the effect of oxygen addition to the melt during the pouring of iron into the mould is demonstrated by a number of authors. For instance, Basdogan et al. (1980) and Chisamera et al. (1996) found oxygen to be highly effective in converting carbidic irons into beautifully 'inoculated' flake graphite irons. Liu and Loper (1990) found that oxygen was necessary to nucleate Kish graphite on the surface of grey iron melts. Larger quantities of Kish were formed at temperatures below 1400°C, below which SiO_2 is stable, but Kish was not observed in Si-free melts. Moreover, Johnson and Smart (1977) described a critical experiment in which they use sophisticated Auger analysis to prove that two or three atomic layers of oxygen (and interestingly, sulphur) are present on fracture surfaces of the matrix adjacent to graphite flakes (fractured and observed in high vacuum), but the surfaces of the hollows left by spheroidal graphite nodules exhibited no oxygen. This is behaviour consistent with a bifilm hypothesis, and with the assumption that the graphite itself may be strong, the presence of the bifilm gives it the appearance of weakness in tension. Briefly, if flake graphite formed on one side of an oxide bifilm, the fracture surface would necessarily travel along the centre of the oxide bifilm, revealing the oxide on both the graphite and the matrix, but oxides would be absent in the case of spheroidal graphite iron (as discussed later). Interestingly, if the graphite flake had nucleated on both sides of the oxide the fracture path would have passed through the flake, but would still have revealed the presence of oxygen, this time on both of the graphite faces. Johnson and Smart do not appear to have tested this condition.

Although all the previous discussion relates to silica-based bifilms, there is evidence that alumina-based, or possibly Al-containing Si-based bifilms (for instance based on mullite or other stable alumino-silicate compound) exist. Carlberg and Fredriksson (1977) found that cast irons based on Fe-C-Si exhibit fine graphite structures, whereas those based mainly on Fe-C-Al display coarse graphite flakes. Chisamera (1996) confirmed that conventional grey irons containing Al develop coarse graphite flakes.

The relatively poor mechanical properties of grey iron seems likely to be more to do with the presence of bifilm cracks down the centers of graphite flakes (or the sides of graphite flakes if graphite grows on only one side of the bifilm—the impression given that the flake has decohered from the matrix) rather than any intrinsic weakness of the graphite itself. A crack down the centrecentre of a flake is seen in Figure 6.43. In their studies of crack initiation and propagation in irons, Voigt and Holmgren (1990) reported many centreline cracks in graphite flakes plus some decoherence from the matrix. As is well known, graphite exhibits easy slip parallel to the basal plane. However, what is only recently uncovered from molecular dynamics simulations is that at right angles to its basal plane van der Waal forces result in a reasonable tensile strength approximately 0.65 GPa (Okamoto, 2013).

The previous discussion relates to those graphite flakes growing freely in the melt giving rise to large, randomly oriented flakes which tend to float or settle irregularly, creating what has been called in the past an 'anomalous' or 'irregular' eutectic; i.e. a eutectic whose phases are not regularly sized and spaced, in contrast to a classical eutectic whose spacing is regularly ordered and controlled by diffusion.

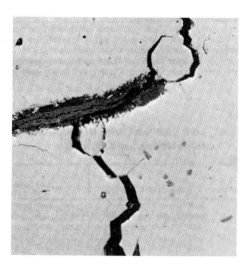

FIGURE 6.43

Graphite flake exhibiting a central crack (the solid state precipitation of surrounding temper graphite is also fractured off) (Karsay, 1971).

6.5.5 NUCLEATION AND GROWTH OF THE AUSTENITE MATRIX

The nucleation of the austenitic matrix of cast irons has, to the author's knowledge, never been systematically researched, although it is interesting that in irons with high Ti and S levels Moumeni and colleagues (2013) found a mixed Mn + Ti sulphide in the form of stars and ribs in the middle of austenite dendrites, strongly suggesting a nucleating role. Even so, it is not especially clear that the problem is at all important. For instance, if a fine austenite grain size could be obtained, would it be beneficial? The answer to this question appears to be not known. Moreover, in the section on steels the grain refinement of austenite is seen to be unsolved. Thus in all this disappointing ignorance, we shall turn to other matters about which at least something is known.

Only recently have two different teams of researchers revealed for the first time the growth morphology of the austenite matrix in which the graphite spherulites are embedded. Ruxanda et al. (2001) studied dendrites that they found in a shrinkage cavity, finding them to be irregular, each dendrite being locally swollen and mis-shapen from many spherulites beneath their surfaces (Figure 6.44). Rivera et al. (2002) developed an austempering treatment directly from the as-cast state that revealed the austenite grains clearly. The grains were large, about 1 mm across, clearly composed of many irregular dendrites, several hundred eutectic cells, and tens of thousands of spherulites. The dendrites from both these studies have some resemblance to the aluminium dendrite shown in Figure 5.33.

It seems fairly certain therefore that the growth of the austenite dendrites occurs into the melt in which there exists a suspension of graphitic particles. The particles hover with what seems to be neutral buoyancy despite their very different density. This arises because their small size confers on them such a low Stokes Velocity that they are carried about by the flow of the liquid: using Stokes relation it is quickly shown that a 1 μm diameter particle has a rate of flotation of only about 1 μms^{-1}, corresponding to a movement of the order of one dendrite arm spacing in a minute. Particles of 10 μm

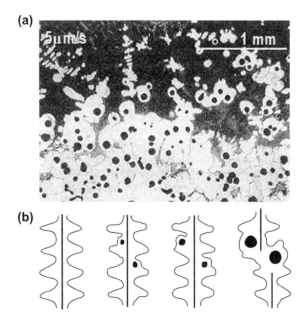

FIGURE 6.44

(a) The growth front of ductile iron (Li, Liu and Loper, 1990); (b) the distortion of dendrites as a result of the internal expansion of nodules.

After Hillert (1968).

diameter might have a more irregular form as in Figure 6.38. Thus despite their larger size, this would reduce their overall average density difference, and increase their viscous drag, so their flotation rate would hardly be higher. Thus many particles would have plenty of time to be incorporated into the dendrite structure.

Once trapped, the surrounding dendrite would be expanded and distorted by the continued growth of the graphite particle as in Figure 6.44, because, at these temperatures, the surrounding solid will be no barrier to the rapid diffusion of carbon to feed its growth. This micro-expansion of the dendrites to accommodate its content of graphite nodules translates of course to the macroscopic expansion of the whole casting, the expansion of the mould, and even the expansion of the surrounding steel moulding box, if any. Sub-microscopic rearrangements of atoms can accumulate to become irresistible forces in the macroscopic world. This is *mould dilation* leading to an increased demand from feeders, or, if not fed, to increased porosity in the casting.

6.5.6 COUPLED EUTECTIC GROWTH OF GRAPHITE AND AUSTENITE

So far, we have considered the separate nucleation and growth modes of graphite and austenite, as though these were unrelated. In fact the relation between the two is sometimes so poor to make any eutectic relation hardly discernible. This rather loose relation between the two major phases in grey iron castings has sometimes resulted in such eutectics being called 'anomalous' or 'irregular' as mentioned previously.

In this section, we move the focus from 'anomalous' to the truly regular, 'classical' eutectic form, in which austenite and graphite grow in a closely cooperative mode, known as coupled growth.

In the absence of suitable nuclei that have formed on oxide substrates in suspension in the melt, the carbon in solution will be unable to precipitate. Thus the melt will continue to undercool until the undercooling finally becomes sufficient to provoke precipitation on some other (less favourable) substrate. Only relatively few such nuclei will operate, activating in those parts of the melt that are especially cool, such as those regions close to the mould walls. The subsequent evolution of heat will inhibit other nuclei from becoming active.

Muhmond and Fredriksson (2013) found that if the Mn or S is too low to aid the formation of nuclei for flakes, then only an undercooled classical eutectic forms, outlining the austenite dendrites. Similarly, Moumeni et al. (2013) found that an addition of 0.35Ti suppresses the formation of graphite as flakes, and causes it to grow as a superfine 10 μm interdendritic graphite/austenite eutectic. Such behaviour is also clearly seen when the melt is cooled quickly.

At modest undercoolings, the coupled growth takes the form of rosettes, often called 'cells' (Figure 6.45). Thus it seems likely that a single initiation event on a nucleus, often sited on the mould wall, expands the coupled growth front as a hemisphere to form the rosettes, or cells (Figures 6.45 and 6.46(a)). The cells are beautifully regular structures, with inter-flake distances now strictly controlled by diffusion in the boundary layer immediately ahead of the advancing front. The rosette form seems to be a strictly coupled growth, not requiring the presence of bifilms. It seems probable that it forms in the pockets of undercooled liquid that would impinge from time to time on an outside surface, thus the constitutional undercooling and the thermal undercooling near to the wall of the mould would be additive to promote nucleation on some marginally favourable particle. The strongly undercooled (i.e. fine) graphite at the centres of some rosettes seen in Figure 6.46(a) seems to confirm this suggestion for some conditions, but Figure 6.46(b) suggests conditions more gently undercooled on this occasion. The even spacing is easily understood if nucleation is prolific, because those nucleation events that occur too near to a neighbour will be less favoured and may even re-melt as its neighbours give off their heat of formation as they grow.

At still lower undercoolings, after initiation, the growth of the coupled eutectic is probably so fast that it will spread sideways to cover large undercooled regions at the mould wall. It will subsequently proceed on a substantially planar growth front, growing away from the wall. Thus only *growth* can now occur. It is the growth phenomenon that dominates the structure we call coral graphite (Figure 6.17). This fine, more highly undercooled eutectic, sometimes designated Type D or E according to ASTM specification A247, seems in general to have been avoided for general engineering castings. This is possibly because the inter-flake diffusion distance is now so small that only ferrite can be formed, limiting the strength of such irons in the as-cast or heat treated conditions.

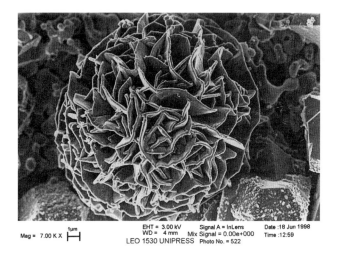

FIGURE 6.45

A rosette of flake graphite, expanding to form a 'cell' or 'eutectic grain'.

Courtesy of Fraz, Gorney and Lopez (2007).

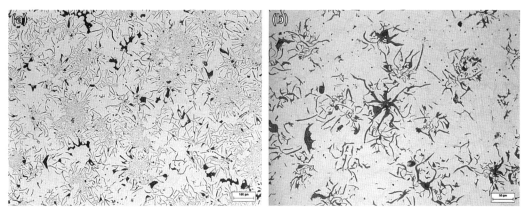

FIGURE 6.46

An array of eutectic graphite grains (cells) in a grey iron (a) rapidly solidified at a high undercooling to give a fine graphite spacing at the center of the cell; (b) slower freezing showing a coarser growth.

Courtesy Serge Grenier QIT 2010.

During coupled growth, flakes have to continually realign their growth direction because of the intrusion of their neighbours into their growth space. Because the growth direction of graphite is mainly parallel to the basal (0001) plane, this means that the crystal has to develop faults to allow it to change direction. This explains the 'coral' type of graphite morphology which is highly faulted (Zhu et al., 1985). We would expect therefore that types D and E graphite would be highly faulted, containing high defect densities, whereas rosette (or cell) graphite would represent an intermediate case as a result of its larger spacing. Flake graphite, as mentioned previously, would be expected to contain the least faults, growing while floating in the liquid, experiencing no significant restraint, and resulting in a nearly perfect crystal.

For interested readers, Nakae and Shin (1999) presented beautiful micrographs to illustrate in detail the close similarity between the coral forms of Fe-C and Al-Si alloys (even though they fail to mention, and appear possibly not even to have noticed, the coral growth of Fe-C shown by their work).

6.5.7 SPHEROIDAL GRAPHITE IRON (DUCTILE IRON)

When excess magnesium is added to the melt, the oxide bifilms are completely eliminated. In the case of silica bifilms the silica will be reduced by magnesium to (1) silicon metal which will dissipate into solution in the matrix and (2) solid magnesium oxide that will precipitate probably on the pre-existing nuclei that originally sat on the films, possibly augmenting these original particles. The reaction is simply:

$$SiO_2 + 2Mg = Si + 2MgO$$

The sudden and total loss of bifilms means that only solids remaining in suspension in the melt now are the original particulate nuclei, possibly augmented by additional MgO. If sulphur is also present in the melt, the MgO likely to contain a component of MgS. These compact nuclei are now the only nucleation sites available for the precipitation of graphite. The precipitating graphite grows over the compact nucleus, wrapping completely around it so as to form a compact initiating morphology.

The disappearance of the bifilms and the initiation of spheroids are shown schematically in Figure 6.47. The 'wrapping around' process (Figure 6.47(d)) may consist of re-nucleation of many separate microscopic grains of graphite on favourable fragments of the oxy-sulfide surface. The growth mode was originally proposed by Hillert and Lindblom (1954) to be an addition of carbon atoms to spiral growth steps generated by <0001> oriented screw dislocations (Figure 6.48). In this way, the radial structure of graphite nodules develops from the graphite grains growing radially out from the compact nucleus to form the familiar approximately spherical nodule (Figure 6.49). Lacaze and co-researchers (Theuwissen et al., 2012) found recent evidence that growth might occur instead by a repeated nucleation from the radial grain boundaries, so that growth occurs by rapid propagation in the 'a' direction.

Johnson and Smart (1977) used the sophisticated and respected perturbation analysis by Mullins and Sekerka to suggest that interfacial energies are of importance in spherodising graphite nodules up to a diameter of perhaps 50 nm, after which the spherical form can no longer be stabilised. Thus, much speculation by earlier authors that interfacial energies may be important in defining the shape of spheroidal graphite seems irrelevant.

In his review, Stefanescu (2007) concluded that all the evidence points to nodules initially growing freely in the liquid, subsequently developing a shell of austenite, and finally contacting and becoming incorporated into an austenite dendrite. A minor modification of this development may be envisaged in which the graphite nodule does not grow a shell of austenite until it contacts an austenite dendrite. At that moment, a shell of austenite would be expected to wrap itself

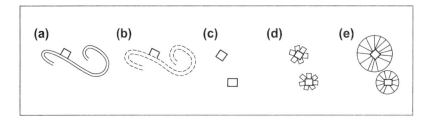

FIGURE 6.47

(a) The melt with a population of bifilms with sundry attached nuclei from impurities or preconditioning additions; (b) the elimination of the silica-rich bifilms by Mg; (c) the survival of the existing nuclei plus possible additional nuclei from inoculation; (d) the nucleation of graphite, wrapping completely around nuclei, particularly if they happen to pass through supercooled regions; and (e) growth of spheroids.

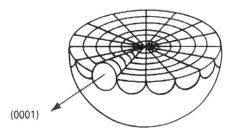

(0001)

FIGURE 6.48

The probable structure of a graphite nodule (Stefanescu, 1988).

FIGURE 6.49

Graphite nodules in an austempered iron indicating nucleation on a small central inclusion followed by radial growth (Hughes, 1988).

rapidly around the nodule. Painstaking metallography would be required to clarify this detail. Anyway, whatever the finer details of the encapsulation process, the shell of austenite seems a key feature associated with the growth of spheroids.

The separate nature of the growth problem can be appreciated from a close look at the graphite structure around some central nucleating particles. The structure in graphite spheroids close to the nucleating particle is sometimes seen to be highly irregular (Figure 6.50). The graphite form in this region appears chaotic, as perhaps might be expected if the effective undercooling leading to dramatically fast initial growth were to be in some kind of dendritic form (Figure 6.38). Clearly, after a very short growth distance, whatever original crystallographic orientation the graphite might have enjoyed with the oxy-sulphide nucleus is quickly lost in the rapid, chaotic growth. However, after a further small distance, the graphite organises itself, and develops its nicely ordered radial grains typical of a good spheroid. Thus the organisation of the growth takes time to develop and is a macroscopic phenomenon. There is a strong analogy with the planar growth condition of a metal under conditions of low constitutional undercooling.

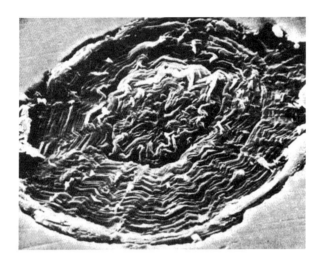

FIGURE 6.50

The chaotic growth structure of a graphite spherule, cathodically etched in vacuum, and viewed at a tilt of 45° in the SEM (Karsay, 1985, 1992).

Reprinted with permission of the American Foundry Society.

The spherical growth form almost certainly has at least some contribution from macroscopic influences. To influence the roundness of the growth form, a mechanism cannot be on an atomic scale, but must act on the scale of the spheroid itself. Such mechanisms might include (1) a low constitutional undercooling condition in the surrounding liquid when in the free-floating state, encouraging a smooth interface analogous to the planar growth of the metal at low driving force for growth, or (2) a mechanical constraint imposed on the expanding sphere when surrounded by solid, but plastically deforming, austenite. It is just possible that (3) some adsorption on the surfaces of the growing crystal, limiting growth directions, may be important. There are no shortages of theories on this issue, and facts are hard to establish.

We shall attempt to evaluate here what seems to me to be perhaps the most likely mechanism: the mechanical constraint by the surrounding solid.

The spherical morphology of the graphite nodules may be encouraged by the mechanical constraint provided by the nodule having to force its growth against the resistance provided by its surrounding shell of austenite (Roviglione and Hermida, 2004). Several studies have clearly revealed the deformation of austenite dendrites by the growth of internal nodules (Figure 6.44). This lumpy dendrite morphology has been attributed to various mechanisms, all of which are likely to contribute to some degree:

1. Ruxanda and colleagues (2001) and Stefanescu (2008) assumed the protrusions to be the natural growth shapes arising from cooperative growth of austenite and graphite by diffusion from the liquid.
2. Buhrig-Polackzed and Santos (2008) indicated in a schematic illustration that the contact between nodules already surrounded by austenite shells and the austenite dendrites results in the mutual assimilation of the two sources of austenite, to create a local bump on the dendrite. (Some subsequent surface smoothing driven by surface energy would be expected to occur rapidly.)
3. Deformation of the dendrite by plastic flow, locally expanding the surrounding solid to accommodate the increasing volume occupied by the graphite has to be important. This effect appears to have been generally overlooked, but appears to be important and worthy of examination, as discussed later.

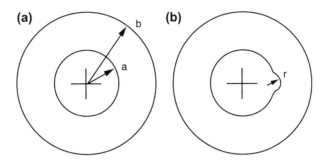

FIGURE 6.51
(a) A thick spherical shell expanding plastically because of internal pressure; (b) a perturbation radius r is not favoured because higher local pressure is required, so that sphericity is encouraged.

The pressure developed in a thick spherical shell (Figure 6.51) deforming plastically because of internal pressure is quantitatively expressed by (Chadwick, 1963)

$$P = 2Y \ln b/a \quad (6.10)$$

Where P is the internal pressure, Y is the yield stress and a and b are the internal and external radii of the shell. The logic is as follows: if a perturbation to the spherical graphite shape were to occur, having a necessarily smaller radius, r, the pressure to plastically extend the growth at this location would (according to Eqn (6.1)) be increased (Figure 6.51(b)). Thus growth of the extension of smaller radius would be discouraged because additional pressure would be required to stabilise the perturbation. The easier spherical growth mode, simply expanding the uniform radius a would therefore be encouraged. It seems therefore that there is some *qualitative* justification for believing that mechanical forces stabilise the spherical growth mode of the nodule.

However, it is useful to ascertain whether there is *quantitative* justification for this mechanism. If we take Y to be approximately 6 MPa (Campbell, 1967) for austenite at the melting point of iron, and $a = 2$ nm and $b = 20$ nm, we find $P = 30$ MPa approximately. Even at values of $a = 20$ μm and $b = 200$ μm, P is of course unchanged at 30 MPa because the ratio a/b is the same, indicating that there is a substantial restraining pressure, approximately five times the yield stress, on the growth of the nodule during most of its life.

With regard to the possible asymmetric effect of a perturbation of radius r, taking $r = a/2$ to $a/10$ locally increases P to approximately between 35 and 60 MPa, respectively. Thus a rounding effect from mechanical smoothing of the forces to expand the austenite shell appears to be important. Although a creep model rather than the above plastic model might give a somewhat more accurate result, the previous result can be relied upon to give us an order-of-magnitude estimate of the effect. I have found creep models and plastic flow models to give very similar answers provided the rates of deformation are similar (Campbell, 1968a). Even so, clearly, more work is required to confirm this preliminary indication.

In general agreement with this conclusion, Jiyang et al. (1990) used colour etching to reveal the austenite shells around graphite. They found that if the shell formed quickly and completely, the nodule developed as a sphere, whereas slow-developing or non-enveloping shells led to mis-shapen nodules, in agreement with our previous logic.

As part of their work in the metals treatment industry, Lalich and Hitchings (1976) observed that some nuclei for nodules were far from round. They noted that the graphite originally wrapped completely around such curious shapes (in agreement with our assumptions in this book) but as growth of the nodule advanced, so the nodule quickly became progressively more round as would be expected from the effect of mechanical constraint encouraging smoothing described here.

In passing, it may be significant that on the addition of Mg causing dissolution of the silica-rich oxide bifilms, any residual gas trapped between the films is expected to be released. In this way, it seems possible that clouds of fine bubbles, consisting mainly of argon, will be released into the melt. It seems likely that some Mg vapour will also diffuse into the bubbles. It is not easy to define the sizes of such bubbles with any accuracy. For instance, a bifilm of 100 µm square and average 1 µm gas gap would yield a pore of approximately 20 µm diameter. A similar bifilm of average 0.01 µm gas gap would form a pore approximately 5 µm diameter. A very small bifilm of 10 µm square and 0.01 µm average spacing would create a 1 µm diameter pore. Thus it seems that a fog of bubbles in the range of approximately 1–20 µm is to be expected.

It is intriguing, but not perhaps relevant, that there is a theory proposing that nodules nucleate from Mg bubbles in suspension in the melt (Gorshov, 1964; Itofugi, 1996). However attractive this hypothesis might be to explain graphite coatings inside pores in solidified castings, as a result of the reduction in strain energy involved, any strain energy relief in the liquid state is zero, and the reduction of surface energy to encourage such precipitation in the liquid state seems negligible. Furthermore, the successful incorporation of solid particulate oxide nuclei into the bubble depends critically on further reduction in interface energies. This is unlikely for particles of oxide formed by precipitation in situ in the liquid which will be in atomic contact with the melt (i.e. will be well 'wetted', being a necessary condition for the particle to be a nucleating agent). Such particles will be energetically rejected by bubbles. Thus although a mechanism for the presence of extremely fine bubbles may be provided by the present analysis, it appears to be irrelevant to the formation of graphite nodules. It does not offer support to the gas bubble nucleation hypothesis.

A further interesting aside can be noted. The transformation of the planar cracks sandwiched inside the bifilms into clouds of fine bubbles that may float and escape from the alloy is the essence of the process by which apparently brittle grey iron becomes ductile. (Ductile iron only becomes embrittled once again, as noted later, if oxide bifilms are re-introduced by turbulence by handling of the melt or poor filling system design of the casting.)

Finally, curious observations such as those reported by Yamamoto and colleagues (1975) in which flake iron is converted into nodular iron by simply purging the melt with fine bubbles of nitrogen, argon or carbon dioxide become explicable. From experience in the light metals industries, it is known that purging with gases can eliminate bifilms from melts. Thus spherodisation of the graphite appears to be achievable via a purely mechanical route, replicating the condition achieved chemically by the addition of Mg. However, the observations by Aylen et al. (1965), in which the graphite structure of pig iron is refined by bubbling nitrogen through the melt, is less easily explained. Perhaps nitride bifilms are active, or possibly sufficient oxygen contaminates the nitrogen to create thin silica films. Such observations are tantalising, and will not be understood without incisive research.

Mis-shapen spheroids

The presence of ill-formed spheroids, particularly if present in large numbers, is widely known to be associated with the reduction in mechanical properties of nodular irons. Hughes (1988) describes how a good ductile iron can achieve at least 90% nodularity but less good irons can fall to as low as 50% or less and suffer reduced properties. The 50% or so component of flakes or other non-nodular shapes does not particularly affect properties such as proof strength but greatly reduces those properties related to failure, such as tensile strength and ductility. Hughes attributes this loss of fracture resistance to the sharp notches at the root of flakes. However, this widely held belief presupposes that the flakes act as cracks. This is probably only true if the flakes are formed on bifilms; the bifilm providing the crack. It is the presence of the crack that has to be viewed as the principal cause of failure.

In terms of the bifilm hypothesis, graphite would be expected to grow on oxide bifilms. Thus, if Mg treatment were carried out to eliminate silica bifilms, spheroids would be created in suspension, floating upwards only very slowly as a result of their small size. However, many Mg addition techniques are extremely turbulent, so that, unfortunately, large quantities of MgO and Mg silicates are expected to be created by the mechanical entrainment of the liquid surface. These new bifilms will be permanent defects formed from highly stable magnesia (MgO) or magnesium silicate ($MgSiO_3$). Before pouring into the mould, some of these will fortunately float out in the ladle, adding to the Mg-rich slag. Thus, given a reasonable time for separation (an interesting and clearly important process variable that seems not well researched), not all of these Mg-rich bifilms find their way into the castings to impair the structure and properties.

Unfortunately again, on pouring into the mould, additional large quantities of Mg-rich oxide bifilms are likely to be re-introduced, particularly if the mould filling system is a poor design.

On contact with an oxide bifilm, a spheroid would tend to attach firmly to the film, thereby reducing the overall energy of the system. Further growth of the spheroid would then necessarily spread over the plane of the film. The symmetrical spherical constraint previously provided by the surrounding austenite is now also destroyed, aiding the non-spherical development. Thus the spheroid will grow to become significantly mis-shapen.

This effect can be seen in Figures 6.31 and 6.32(b). In Figure 6.31, several oxide bifilms have been straightened by the growth of dendrites to lie along 100 planes. The nodules attached to the bifilms are clearly poorly shaped. Additional mis-shapen nodules elsewhere in the structure are expected to be lying on random areas of bifilm not straightened by dendrite growth. In Figure 6.32(b), it seems likely that the bifilm lies along the grain boundary, decorated with mis-shaped nodules, ferrite and (probably) some scattering of porosity or cracks.

The phenomenon of mis-shapen nodules has up to now appeared a mystery. The effect is therefore predicted to be associated with either (1) insufficient dwell time for the damage introduced during Mg addition to float out or (2) poor casting practice, in which an otherwise nicely inoculated and spherodised melt is re-contaminated with bifilms.

It is interesting to predict that more time after the spherodising treatment to allow the melt to clear, together with a properly designed filling system, or counter-gravity filling system, should completely solve this problem. A step in the right direction was taken by Takita and colleagues (1999) who observe that nodules are converted from mis-shapes to spheroids by the use of a filter to take out the 'inclusions produced by inoculation'. This positive step contrasts with that taken by Liu and co-workers (1992) who, after the Mg addition, added 'postinoculants'. The postinoculants were highly successful to increase the nodule count, but led to a disastrous fall in nodularity. This was almost certainly a result of adding the inoculants through the melt surface. The inoculants would, of course, have their own oxide skins, which would have doubled up with the entrained surface oxide of the melt to create an asymmetrical bifilm with an alloy oxide on one side and a Mg-rich silicate on the other. On contact with growing graphite particles, this major source of fresh bifilm contamination would lead to the growth of non-spheroids.

Furthermore, of course, the significant reduction of properties associate with malformed spheroids cannot be the direct result of the shape of the spheroids because they occupy such a small volume fraction of the alloy. The loss of properties is predicted to be the result of the presence of the bifilms in the melt, occupying a vastly greater cross-sectional area than the spheroids. These extensive bifilms will act as cracks in the casting, significantly reducing properties.

Exploded nodular graphite

'Exploded' spheroids (Figure 6.52) are commonly seen in irons subject to graphite flotation, and especially if the composition of the iron is sufficiently hyper-eutectic (Sun and Loper, 1983; Druschiz and Chaput, 1993).

This undesirable morphology is not easily explained at this stage as a result of relatively little experimental work to clarify the problem. Cole (1972) suggested they had suffered re-melting as a result of being carried by convection in and out of hot zones of the liquid. This seems most unlikely, however, because if the nodule had grown uniformly in a compact morphology the uniform graphite would be expected to have a substantially uniform rate of dissolution; solidification and re-melting would be expected to be reversible.

Because exploded nodules appear exclusively in the flotation region of hyper-eutectic irons two far more likely factors are.

1. Nodules growing in a sufficiently hyper-eutectic melt (Sun and Loper, 1983) will experience an enhanced driving force for growth because of the carbon supersaturation that develops as the melt cools. This will encourage growth instabilities leading to 'dendritic' rather than 'planar' growth, leading to exploded rather than smooth spheroid surfaces.
2. The nodules may have nucleated early in the liquid phase and grown without the benefit of the mechanical constraint of the austenite. When not mechanically pressurised to remain spherical, the nodule will be free to grow more like a dendrite, developing instabilities that grow into projections to its growth front, finally developing the characteristic exploded forms. Evidence for mechanical restraint as a powerful effect is presented in the section on nodular graphite previous.

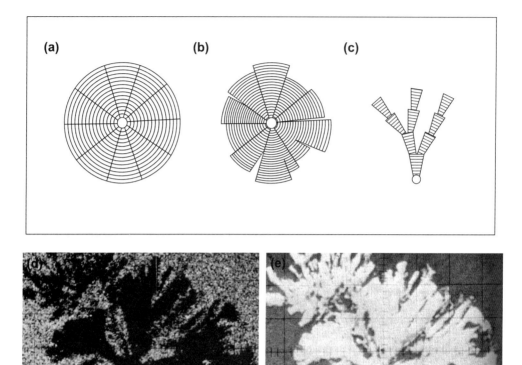

FIGURE 6.52

(a) Spheroid and (b) malformed spheroid; (c) chunky graphite (after Liu et al., 1983; (d) SEM iron image of an exploded spheroid; (e) electron image (Cole, 1972).

6.5.8 COMPACTED GRAPHITE IRON

If the addition of magnesium is more carefully controlled to some level intermediate between spheroidal and flake iron, compacted ('vermicular' based on the Italian for 'worm-like') graphite is the result (Figure 6.53).

In our bifilm model, it is clear that most of the oxide bifilms will be quickly dissolved by the addition of Mg. However, small patches may remain if the Mg addition is not too high; the tiny patches on which the original nuclei sat will be resistant to dissolution because they will be stabilised by their attachment to the nuclei. (Naturally, it will have been energetically favourable for the nuclei to attach to the bifilm, so that the combination of nucleus and film will enjoy a reduction in overall energy, stabilising the combination.) Only half of the bifilm will be retained in this way, its distant 'twin' half not enjoying the protective influence of the nucleus will dissolve and disappear because it is separated by an

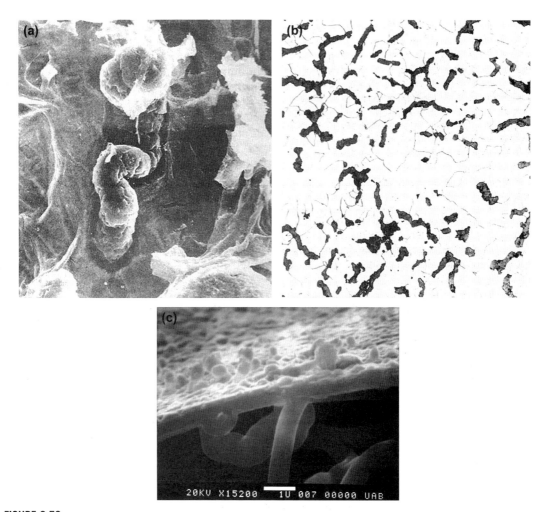

FIGURE 6.53

CGI viewed by (a) SEM deep etching, (b) optical metallography (Stefanescu et al., 1988) and (c) carbon wool fibre (growing off a lustrous carbon surface film on a grey iron) as a possible vapour phase–equivalent of compacted graphite (Campbell and Naro, 2010).

air gap. Only the small part of one half of the bifilm together with its unbonded interface, the remnants of the layer of air, will remain (Figure 6.54(b)).

The subsequent nucleation of graphite on the nucleus will result in rapid spreading of growth around the nucleus. On arrival at the non-wetted interface belonging to the residual patch of bifilm this spreading will be arrested (Figure 6.54(c)). The graphite has now grown to reach the residual layer of air on the remaining patch of half of the bifilm. The further growth of graphite is forced to occur not radially, but in general unidirectionally away from the bifilm residue (Figure 6.54(d)). Clearly, the growth cannot now be spherical, but its exact form is not possible to predict. Cole (1972) had observed a fine, unidirectional spiral structure similar to the worm-like growth mode clearly seen in Figure 6.53. Liu and

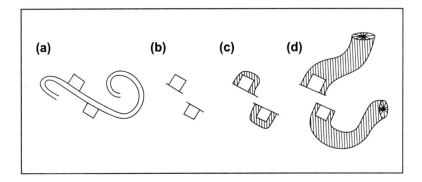

FIGURE 6.54

Formation of CGI by addition to (a) nuclei on bifilms of just sufficient Mg to (b) eliminate most of the silica-rich bifilms except for the remnant attached to the nuclei. (c) Inoculation promotes graphite initiation on the nuclei; (d) growth cannot occur by wrapping completely around the nucleus, so that initial growth cannot be spherically symmetrical, possibly favouring unidirectional growth.

colleagues (1980) found that the growth direction is along the C axis (0001 direction perpendicular to the basal planes) and appears to develop by a spiral dislocation mechanism as witnessed by the coarse and irregular spirals that they observe.

The great sensitivity of the compacted graphite morphology to magnesium concentration is corroborated by the proposed bifilm mechanism. If the Mg level is too low, residual bifilms will encourage flake graphite, whereas if the Mg level is too high, the limited stability enjoyed by the residual bifilm patches will be overcome, and the last remaining patches of bifilm will be dissolved, allowing the growth of totally spherical grains.

After nucleation of the compacted graphite form, a number of workers (for instance, Su, 1982 and Mampaey, 2000) found that the continued growth of the graphite appears to occur solely in the liquid. Because the graphite stays in contact with the liquid, it transfers its expansion directly to the liquid, reducing feeding requirements (Altstetter and Nowicki, 1982). This welcome and valuable good behaviour is in contrast to ductile iron in which the graphite transfers its expansion to its surrounding solidified shell, expanding the casting, and the mould, thereby increasing feeding requirements. Alonso et al. (2014) appeared to find this behaviour in their study of graphite expansion of flake and nodular irons.

As a fascinating final thought concerning compacted graphite iron (CGI), the fibrous graphite seen growing from the vapour phase in Figure 6.35, and shown in close up in Figure 6.53(c) appears to be identical in form to the graphite grown from the melt as CGI or 'worm-like' graphite, and might therefore be viewed as a three-dimensional model, giving an insight into the internal structure of CGI. These free-growing vapour growth phase and liquid growth phase forms of graphite appear to be identical. It is possible that the carbon concentration in each of these environments is similar, resulting in similar kinetics of deposition, and similar growth modes.

6.5.9 CHUNKY GRAPHITE

Chunky graphite is often observed concentrated in the centers of heavy sections of nodular iron castings. 'Chunky' is not a particularly helpful descriptive adjective for this variety of graphite. Its 'chunkiness' is only apparent under the microscope at high magnification; otherwise, it simply appears to be fine, irregular, branched and interconnected fragments (Figures 6.52(c) and 6.55). Once again, the properties of nodular iron are reduced. However, it seems the loss of properties may, once again, be at least partly associated with the short diffusion distances between branches of the graphite filaments, promoting the development of ferrite instead of the stronger pearlite eutectoid phase (Liu et al., 1983).

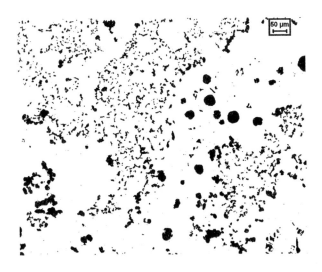

FIGURE 6.55

Graphite nodules and areas of fine, chunky graphite in the thermal centre of a 200 mm cube casting (Kallbom, 2006).

Liu and co-workers (1980, 1983) found evidence that chunky graphite grows along the C-axis direction, as does both nodular and compacted graphite. Furthermore, they reported observations on spheroids that exhibit gradual degeneration, gradually taking on the growth forms of chunky graphite. Thus they concluded that chunky graphite is a degenerate form of spheroidal graphite, and their work implied that chunky graphite grows out from spheroids. Itofugi and Uchikawa (1990) confirmed the identical growth modes of spheroidal, compacted and chunky graphites as illustrated schematically in Figure 6.52.

All these workers observed the characteristic form of chunky graphite, as an apparently 'stop/start' growth in the C-direction consisting of nearly separate pyramidal 'chunks' linked by a narrow neck, like beads on a string (Figure 6.52(c)). The individual chunk sections comprise layers parallel to the basal plane, but only nanometres thick. This characteristically lumpy growth may be the result of a pulsating or irregular advance of the growth front, with the austenite advancing to nearly grow over the top of the graphite, forming the nearly pinched-off neck of the graphite, only to be overtaken again because the carbon in solution will now buildup in the liquid ahead of the front, accelerating the next phase of growth of the graphite until the local carbon concentration is depleted once again.

Observations by Kallbom and co-workers (2006) are consistent with an origin associated with bifilms. They observe the chunky graphite to be concentrated in the center of heavy sections, explained by the growth of the freezing front pushing bifilms ahead, and explaining their observations of 'stringers' of graphite nodules. These features are almost certainly sheets of oxides decorated with graphite nodules that have been nucleated on the oxide (analogously to those seen in Figure 6.32). The earlier paper by these authors (Kallbom, 2005) described how the outer several centimetres of the casting can be perfect, with good nodularity, good strength and ductility, but the structure can change abruptly, over a short distance equivalent to only a nodule diameter, to chunky graphite. Thus the central volume of the casting is weak and brittle. Because the problem is buried in the centre of thick section castings it can be difficult to find by non-destructive methods.

All previous evidence that suggests chunky graphite requires both the presence of bifilms combined with an absence of nuclei. It is possible to imagine a mechanism in which following a turbulent pour, many bifilms will be pushed ahead of the forest of growing dendrites, forming, at times, a distinct and abrupt separation of the outer dendritic region from the inner residual liquid. The presence of the concentration of bifilms in the centre will suppress the normal pattern of free circulation that would ensure a good supply of nodules from the outer, cooler regions into the hot central region. Furthermore, because so much time is available, particles such as nuclei and nodules already in suspension in the centre will have time to float out, or existing nodules to dissolve (because they will be unstable at these higher temperatures) depleting the centres of nuclei so that spheroids cannot form.

For those nuclei now floated out to the edge of this region of higher temperature and enhanced segregation, any nodules formed on the nuclei will not enjoy the benefit of a surrounding austenite shell, so that their growth mechanism will more nearly resemble that of an exploded spheroid. Because these will be at the boundary of the central region, their growth is most likely to be an extension of the nodules along the C-axis (Liu et al., 1980) in the direction of the gradually advancing solidification front. In the absence of nuclei, the whole region would be expected to fill with this continuous growth form. The extended size of chunky graphite regions, much larger than cells of other types of graphite (Itofugi and Uchikawa, 1990), corroborate the absence of nuclei in these regions.

The presence of bifilm cracks concentrated in the chunky graphite regions would explain the poor properties that are observed; it is difficult to see how otherwise a continuous graphite phase could reduce properties, particularly because in other irons (such as CGI) a continuous graphite phase is associated with excellent and reproducible tensile properties.

It is hoped that in the near future the explanation for the origin of chunky graphite might be clarified and confirmed by further careful experiments. The key word here is 'careful'. For instance, the experiment by Asenjo and co-workers (2009) involving the placement of inoculants in different branches of a runner system to simulate the casting of separate moulds from one melt. In this way, it should have been possible to compare the effects of different mould inoculation in separate heavy castings. The idea was clever, but regrettably experimentally flawed because, in common with most iron casting, the runners were not designed to be pressurised and fill on a single pass. Thus a reverse flow is likely to have contaminated the mould cavities, and all the cast material would have suffered from turbulence and air entrainment. All the cavities would therefore have been contaminated with varying amounts of inoculants from neighbouring cavities, and all would have contained unknown quantities of oxide bifilms. Clearly, in the future, much greater sophistication of melting and casting will be required for experiments designed to clarify the solidification mechanisms for cast irons.

6.5.10 WHITE IRON (IRON CARBIDE)

Work by Rashid and Campbell (2004) demonstrated the nucleation and growth of carbides on oxide bifilms in vacuum-cast Ni-base superalloys (seen later in Figures 6.62 and 6.63). It would be expected therefore that an analogous reaction would occur in Fe-C alloys, because the austenite forming during solidification also possesses a closely similar fcc structure.

Carbides in irons appear to form preferentially at grain boundaries, and appear often to be associated both with residual graphite (sometimes as nodules, malformed nodules or flakes aligned with the boundary) and pores all forming on the same boundaries (Figure 6.32(b)). These are all clues to the possible presence of an otherwise invisible bifilm.

Faubert et al. (1990) have studied carbides in heavy section austenitic ductile iron. Towards the top of their castings they find degradation of properties more serious than they would have expected from the carbides themselves. They suspected that the real impairment was caused by the presence of films that had floated into this region. The 'films' would, of course, have been 'bifilms', in line with their profound effect on properties. Bifilms would segregate to grain boundaries, and possibly actually constitute the boundary. The presence of the bifilm is not only inferred from (1) the cracked carbides but also (2) from the linear rows of nodules seen in micrographs from this work, (3) from the pores as the residues of air bubbles trapped between the films, (4) from the graphite flakes sitting in the boundary (called by the authors, unflatteringly, 'degenerate' graphite). Both graphite and carbides are expected to form on the wetted, outer surfaces of the bifilms. The presence of the central unbonded region (including the pores) between the films, constituting the crack through the interiors of the carbides, explains the *apparent* brittleness of the carbides. These intermetallic compounds would otherwise be expected to be strong, resistant to failure by cracking at the modest stresses that can be induced by solidification and cooling. The plasticity and crack resistance of intermetallic phases has been discussed in Section 6.3.7.

Stefanescu (1988) quoted the work of Hillert and Steinhauser (1960) in which the growth of iron carbide eutectic (ledeburite) occurs by the spreading of carbide (cementite) across a plane, followed by the development of a rod type of eutectic at right angles. It is tempting to consider that the original planar expansion would have been facilitated by growth across the surface of a bifilm. The bifilm would originally have been randomly crumpled, but would have been straightened by the progress of the carbide across its face, thus creating an essentially planar crack that would constitute a serious defect in the carbide. Associated branching cracks would have arisen from irregular folds in the bifilm.

Although carbides in grey irons are feared for their disastrous effect on machinability, they are of course desirable in those components required to withstand wear and abrasion. Furthermore, a complete white iron structure is required before subjecting the iron to malleabilising treatments. Although the role of malleable iron has been greatly eroded by the rise of ductile iron, for some products and some foundries the economics are not so different, so that at this time a respectable tonnage of malleable iron is still produced. It will be interesting to see the eventual outcome of this competition.

6.5.11 GENERAL

Several workers have noted that the initial stages of formation and growth of the graphite particles start with a spheroidal shape, regardless of whether these are destined to grow into spheroids or flakes, or any other type of graphite (for instance, Jianzhong, 1989; Itofugi et al., 1983). This observation is consistent with the scenarios portrayed in Figures 6.39, 6.40 and 6.47(d). The spread of the graphite around the originating nucleus is always nearly spherical, or partly spherical. Even the partly spherical forms will appear spherical in some sections.

Overall, it seems clear that both nucleation and growth mechanisms influence graphite morphology in both flake and spheroidal forms, although these mechanisms dominate to different extents in different circumstances. For compacted, undercooled coral and chunky graphites nucleation occurs once to initiate each cell, after which continuous growth dominates the development of the structure, leading to continuous branching morphologies. Indeed, the finer, fairly continuous forms, coral, chunky and compacted graphites, are so similar that they often seem to be confused in the casting literature (see Figure 6.56). They do seem similar in the sense that they all appear to be more or less coupled growth forms, advancing together with the austenite.

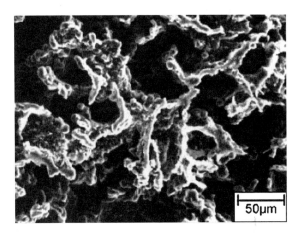

FIGURE 6.56

SEM image of deeply etched CGI closely resembling coupled eutectic coral structure, illustrating the overall similarity of these structures.

Courtesy T Prucha, AFS 2009.

If, as proposed here, the various forms of graphite are significantly influenced by the presence or absence of bifilms, it would explain the historical reluctance of cast irons to give up their secrets and allow an understanding. This seems typical of bifilm phenomena.

Furthermore, it seems likely that the principle cause of reduced mechanical properties in all cases of non-spheroidal forms of cast irons is the presence of various kinds of oxide bifilms acting as cracks. Ductile iron owes its ductility to graphite that forms in the absence of oxide bifilms, but when oxide bifilms (mainly magnesia-rich) are entrained by poor casting technology, even ductile iron can be seriously reduced in ductility, possibly failing brittlely by the 'plate fracture' mechanism as we have seen.

6.5.12 SUMMARY OF STRUCTURE HYPOTHESIS

1. Cast iron melts normally contain oxide double films (mainly SiO_2 bifilms) in suspension.
2. Inoculation produces oxy-sulfide particles which nucleate on silica-rich oxide bifilms (and unintentionally introduces entrained oxide bifilms to further aid the growth of flake graphite).
3. Graphite nucleates on the oxy-sulfide particles, and grows primary graphite, spreading over the bifilms, straightening the bifilms and forming flakes of crystallographically near-perfect graphite; the presence of the bifilms, as cracks, trapped inside or alongside graphite flakes accounts for the relatively poor tensile properties of flake irons.
4. Compacted graphite nucleates on oxy-sulfide nuclei occupying bifilm residues. Growth occurs by fibrous filaments extending freely into the liquid.
5. Spheroidal graphite nucleates on oxy-sulphide nuclei free from bifilms. The spherical growth morphology is initiated by the freedom that the graphite has to wrap completely around the nucleus, but is later importantly encouraged by mechanical constraint of the austenite matrix; the absence of bifilms, eliminated by the Mg addition, explains the high mechanical properties.
6. Eutectic coral graphite nucleates on (currently unknown) nuclei at low temperatures, expanding to form cells of coupled growth with austenite, consisting of highly faulted continuous branching filaments of graphite in the austenite matrix; bifilms play no part in this growth mode.
7. Mis-shaped spheroids appear to be spheroids that have encountered an oxide (probably Mg-rich) bifilm, subsequently growing along the bifilm and losing sphericity. The presence of the bifilm also destroys the symmetrical mechanical constraint of the austenite that favours sphericity.
8. Chunky graphite occurs in heavy section ductile iron in regions. The central regions are isolated by, and probably full of, bifilms created by the poor casting technique. Thus convection and the redistribution of nuclei and nodules are suppressed, and existing nuclei can float out, so that the central region becomes devoid of nuclei. Graphite therefore develops akin to coral morphology. As such, it may be a coupled eutectic form, and its 'beads on a string' morphology may result from an unstably advancing growth front.
9. Exploded spheroids may be the result of growth in the liquid, without the benefit of the mechanical constraint of an austenite shell. They are favoured by (1) high CEV and (2) the presence of oxide bifilm which interfere with the sphericity of growth.
10. Carbides form at very low temperatures on oxide bifilms. The presence of bifilms in the carbides explains the apparently brittle behaviour of these strong, crack-resistant intermetallics.

6.6 STEELS

There are such a wide variety of steels of widely differing properties that it is possible only to generalise with extreme caution. In general, we shall consider the simplest of steels; the carbon steels (often more accurately referred to as carbon/manganese steels, because Mn is such a common additional alloying element) and stainless steels. Stainless steels fall into main groups: ferritic, austenitic and duplex (i.e. mixed ferrite and austenite).

Other steels are commonly known from their microstructures, including ferritic, austenitic, pearlitic, martensitic and bainitic. Tool steels (another wide category) encompass forming and cutting tools often with high Mo or W additions to form hard carbides.

Although astonishing strengths available in the final worked and heat-treated products, sometimes approaching or even in the GPa range, from the point of view of the casting technologist, the key differences between the steels and the light alloys are as follows.

1. The high melting and casting temperatures encouraging more severe and faster reaction with the environment.
2. The possibility of benefiting from a partly or completely melted surface oxide film.
3. The higher strengths of the steels result in higher internal stress during freezing, leading to higher driving forces for the initiation of such defects as shrinkage porosity, hot tearing and cracking.
4. The higher density of steels requires stronger, reinforced moulds and cores to resist flotation and heavy weights to prevent the lifting of copes.
5. The greater difference in density between the metal and its oxides encourages the faster flotation of entrained defects after pouring events.

6.6.1 CARBON STEELS

Steelmaking practice for the production of carbon steels traditionally starts from pig iron produced from the blast furnace. The high carbon in the iron, in the region of 3–4%, is the result of the liquid iron percolating down through the coke in the furnace stack. (A similar situation exists in the cupola furnace used in the melting of cast iron used by iron foundries.) Oxygen is added to oxidise the carbon down to levels more normally in the range of a few tenths of a percent. The bonus from the burning of the carbon is the huge and valuable increase of temperature that is needed to keep steels molten. Oxygen to initiate the CO reaction is added in various forms, traditionally as shovelfuls of granular FeO thrown onto the slag, but in modern steelmaking practice by spectacular jets of supersonic oxygen. The stage of the process in which the CO is evolved as millions of bubbles is so vigorous that it is aptly called a 'carbon boil'.

After the carbon is brought down into specification, the excess oxygen that remains in the steel must be lowered by deoxidising additions to prevent a 'carbon boil' as a result of the positive segregation of carbon and oxygen during freezing. Common deoxidisers have been manganese, silicon and/or aluminium. In modern practice, a more complex cocktail of deoxidising elements is added as an alternative or in addition. These often contain small percentages of rare earths to control the shape of the non-metallic inclusions in the steel. It seems likely that this control of shape is the result of reducing the melting point of the inclusions so that they become at least partially liquid, adopting a more rounded form that is less damaging to the properties of the steel. Also, of course, such compact spherical inclusions float out more rapidly than solid film-type deoxidation products, so that the steel is much cleaner, and properties are higher and more uniform as a result of the absence of bifilm-type cracks.

In most steel foundries, only *steel melting* is carried out from scrap steel (not made from pig iron, as in *steelmaking*). Because the carbon is therefore already low, there is often no requirement for a carbon boil.

The potentially significant problem that remains in the absence of a carbon boil is that hydrogen may remain in the melt. In contrast to oxygen in the melt that can quickly be reduced by the use of a deoxidiser, there is no quick chemical fix for hydrogen. Hydrogen can only be encouraged to leave the metal by providing a dry and hydrogen-free environment. Hydrogen then will gradually evaporate off from the melt, tending to equilibrate with its surroundings. If a carbon boil can be induced, the loss of hydrogen will be rapid. However, if this cannot be induced, possibly in some artificial way, and if environmental control is insufficiently good, or is too slow, then the comparatively expensive last resort is vacuum de-gassing, although the use of argon-oxygen de-gassing treatment is now more common which can efficiently flush hydrogen down to very low levels sometimes required for large castings whose dimensions (measured in large fractions of metres) exceed the distance that hydrogen can diffuse out during heat treatments.

6.6.2 STAINLESS STEELS

The percentage ferrite in cast stainless steels at room temperature depends mainly on the composition of the alloy. This is summarised in the Schoefer diagram presented in Figure 6.58. The equivalent Cr and Ni contributions are estimated from the weight percentage of alloying elements:

$$Cr_{equiv} = \%Cr + 1.5(\%Si) + 1.4(\%Mo) + \%Nb - 4.99$$
$$Ni_{equiv} = \%Ni + 30(\%C) + 0.5(\%Mn) + 26(\%N - 0.02) + 2.77$$

Fully austenitic and fully ferritic stainless steels suffer a number of cracking and intergranular corrosion problems that seem to me to be the result mainly of oxide bifilm problems, most probably as a result of dendrite straightening, leading to a population of cracks throughout the alloy.

However, most cast austenitic steels do not have the single phase structure that their name implies; their structure is usually duplex. Ferrite in the range 5–25 volume percent is found to be valuable for (1) improved strength; (2) improved weldability and castability (reducing cracking problems); and (3) improved resistance to corrosion, stress corrosion cracking and intergranular corrosion attack in certain corrosive environments. It seems possible that the formation of the austenite during the growth of the primary ferrite dendrites might interrupt and hamper bifilm straightening in some way.

The advantages of the duplex structure are maximised at about 50/50 volume percent mixture (actually encompassing the small range of approximately 40–60 volume percent). It is *only* these steels that are known in the industry as '*duplex stainless steels*'.

An additional method of classifying stainless steels is according to their stainlessness, or ability to resist corrosion as quantified by their pitting resistance equivalent. A commonly used equation for ranking stainless compositions is given by Davidson (1990):

$$PRE = \%Cr + 3.3(\%Mo) + 16(\%N)$$

Four categories of improving levels of pitting resistance equivalent can be listed (together with typical examples):

	Pitting Resistance Equivalent	Example
1.	Approximately 25	(23Cr steels with zero Mo)
2.	30–36	(22Cr steels)
3.	32–40	(High-alloy 25Cr steels)
4.	>40	(Super duplex steels 25Cr-7Ni-4Mo)

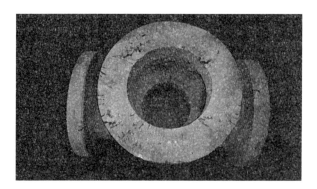

FIGURE 6.57

Bifilm cracks revealed by red dye in a 20Cr-20Ni-6Mo steel casting. The alignment of the cracks by the growth of columnar grains is evident. The deep red indications are most probably bubble trails.

Courtesy S Scholes.

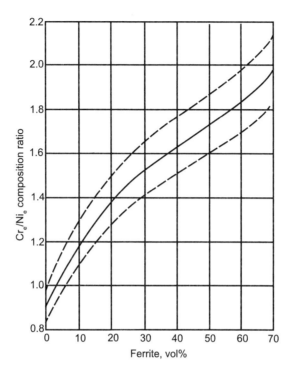

FIGURE 6.58

Schoefer diagram for estimating the ferrite content of steel castings in the composition range 16-26Cr, 6-14Ni, 0-4Mo, 0-2Mn, 0-2Si, 0-1Nb, 0-0.2C, 0-0.19N. Broken lines denote limits of uncertainty.

One of the standard problems with many cast stainless steels is the formation of the infamous sigma phase (σ) usually at grain boundaries. As with most intermetallics, the phase appears to be brittle, exhibiting cracks and leading to failure by cracking. Once again, this maligned intermetallic has all the familiar signs of having nucleated and grown on an oxide bifilm, thereby naturally exhibiting cracks in a material that would normally be expected to be strong and crack-free (see Section 6.3.7). Once again, the avoidance of this problem lies not necessarily in metallurgical control but in the use of appropriate casting technology. Contact pouring and a naturally pressurised filling system is predicted to largely eliminate bifilms, removing these favoured substrates, and thereby suppressing the formation of sigma phase. A large benefit to tensile properties would result from the elimination of the bifilm cracks, and a perhaps smaller benefit from the presence of additional solutes in solution (because the sigma phase may not precipitate in the absence of an attractive substrate) that would enhance strength. The pitting resistance equivalent might also improve, and possibly the resistance of some of the stainless steels to such other imponderables such as stress corrosion cracking.

6.6.3 INCLUSIONS IN CARBON AND LOW-ALLOY STEELS: GENERAL BACKGROUND

Svoboda and colleagues (1987) reported on a large programme carried out in the United States, in which more than 500 macro-inclusions were analysed from 14 steel foundries. This valuable piece of work has given a definitive description of the types of inclusions to be found in cast steel, and the ways in which they can be identified. A summary of the findings is presented in Figure 6.59 and is discussed later.

Each inclusion type can be identified by (1) its appearance under the microscope and (2) its composition.

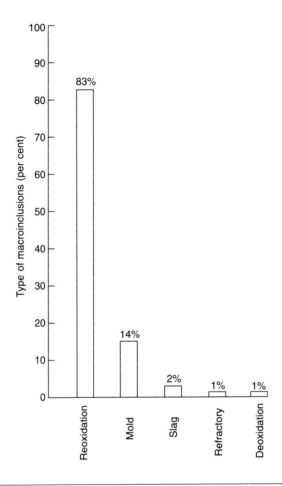

FIGURE 6.59

Distribution of types of macro-inclusions in carbon and low-alloy steel castings, from a sample of 500 inclusions in castings from 14 foundries (Svoboda et al., 1987).

1. Acid slags can be identified by their high FeO content (typically 10–25%), and glass-like microstructure.
2. Basic slags and furnace slags from high-alloy melts can be traced by the calcia (lime), alumina, and/or magnesia that they contain.
3. Refractories from furnace walls and/or ladles have characteristic layering, flow lines and a pressed and sintered appearance including sintered microporosity. Their compositions are reminiscent of those of the refractories from which they originated (e.g. pure alumina, pure magnesia, phosphate bonded aggregates etc.).
4. Moulding sand is identified from the shape of residual sand grains and from its composition high in silica.
5. Mould coat material is normally easily distinguished because of its composition (e.g. alumina, zircon).
6. Deoxidation products are always extremely small in size (typically less than 15 μm) and are composed of the strongest deoxidisers. These inclusions are likely to have formed at two distinct stages. (1) During the initial addition of strong deoxidiser to the liquid steel, when small inclusions will be nucleated in large numbers as a result of the high supersaturation of reactive elements in that locality of the melt. Any larger inclusions will have some opportunity to float out at this time. (2) During solidification and cooling. These stages will be discussed further later.

7. Re-oxidation products are large in size, usually 2–10 mm in diameter, and consist of a complex mixture of weak and strong deoxidisers. In carbon steels, the mixture contains aluminium, manganese and silicon oxides. In high-alloy steels the mixture often contains a dark silica-rich phase, and a lighter coloured Mn + Cr oxide-rich phase. Entrapped metal shot is found inside most of these inclusions. The shot is probably incorporated by turbulence (rather than by chemical reduction of the FeO by the strong deoxidisers). These larger inclusions have been previously known as *ceroxide* defects, not as a result of their content of cerium and other powerful rare earth deoxidisers as I had originally expected; the name was simply made up from '*ceramic oxides*'.

Most research on the formation of inclusions in steels is directed, disappointingly, only at the diffusion reactions between the raw metallic deoxidiser, such as aluminium and liquid steel. The inclusions generated in the supersaturation surrounding the deoxidants during its first seconds of melting and dispersing are interesting, including mixes of dendritic, star-like and spherical forms (Tiekink, 2010). Van Ende and co-workers (2010) did not comment on the presence of the crack that separated the liquid Al sitting upon their sample of liquid steel. This is clearly present in all their work. Even when bifilms are clear they are nearly always ignored. The inter-diffusion between deoxidisers and liquid steel elucidate the processes occurring during *deoxidation* but the much more important processes involved in *re-oxidation*, including entrainment during pouring, have mainly been overlooked so far.

It seems clear that the re-oxidation products are bifilms of various sorts, somewhat scrambled, and often unable to re-open as a result of their formation from partially molten oxide mixtures. They confirm their identity by their content of droplets of metal and bubbles of residual air.

The re-oxidation products are formed during the tapping of the furnace into the ladle. In general they will largely have floated out of the metal, forming the slag layer on the top of the melt in the ladle during the several minutes that it takes to lift the ladle out of the pit and travel to the casting station. Only the pour into the mould is expected to create the re-oxidation products that are observed. There is excellent evidence for this expectation; Melendez, Carlson and Beckermann (2010) developed a computer model that predicts the locations of re-oxidation inclusions in the mould. They achieve good descriptions of the awful mess castings become when they adopt a filling system design which contains a conical funnel basin, uniform internal diameter refractory tubes for the sprue, runners and ingates, and finally gates into parts of the casting that are not at the lowest points of the mould cavity. This filling system was clearly designed to break all the rules, and so be as bad as possible, to illustrate the accuracy of the computer model in predicting castings destined for the scrap heap.

These efforts to predict the distribution of re-oxidation defects seem curiously misplaced when it is reasonably widely known (Kang, 2005; Puhakka, 2010), and certainly predictable, that if a properly designed filling system is applied to the casting the re-oxidation inclusions disappear! Clearly, the so-called re-oxidation problems during casting are entrainment problems.

6.6.4 ENTRAINED INCLUSIONS

Previously, most inclusions introduced from outside sources have been called exogenous inclusions, but this name, besides being ugly, is unhelpful because it is not descriptive. 'Entrained' indicates the mechanism of incorporation as an impact of opposed surfaces, as occurs in the incorporation of a droplet, or the folding-in of the surface. Also, the word entrained draws attention to the fact that as a necessary consequence of their introduction to the melt, such inclusions have passed through its surface, and so will be wrapped in a film of its surface oxide. Depending on the dry or sticky qualities of the oxide, and the rate at which the wrapping may react with the particle, the entrained inclusion and its wrapping can act later as an initiation site for porosity or cracks. Metal too, may become entrapped in the entraining action, and thus form the observed shot-like particles.

Svoboda has determined the distribution of types of macro-inclusion in carbon and low alloy cast steels from the survey (Figure 6.59). The results are surprising. He finds that re-oxidation defects comprise nearly 83% of the total macro-inclusions. These are our familiar bifilms created by the surface turbulence during the transfer of the melt from the furnace into the ladle, and from the ladle, through the filling system and into the mould. In addition, he found nearly 14% of macro-inclusions were found to be mould materials. Because we know that mould materials are also introduced to the melt as part of an entrainment process, it follows that approximately 96% of all inclusions in this exercise were entrainment defects resulting from damaging pouring actions.

Only approximately 4% of inclusions were due to truly extraneous sources; the carryover of slag, refractory particles and deoxidation products.

This sobering result underlines the importance of the reaction of the metal with its environment after it leaves the furnace or ladle. The pouring, and the journey through the running system and into the mould, are all opportunities for reaction of those elements that were added to reduce the original oxygen content of the steel in the furnace. The unreacted, residual deoxidiser remains to react with the air and mould gases. Such observations confirm the overwhelming influence of reactions during pouring or in the running system as a result of surface turbulence; these effects are capable of ruining the quality of the casting.

However, good running systems are not usually a problem for small steel castings. I used to think that large steel castings were a separate matter because of the high velocities that the melt necessarily suffers. This is partly a consequence of the use of bottom-pour ladles and partly the result of the long fall of the melt down tall sprues. However, even these challenges are now known to be solvable.

The historical use of rather poor filling system designs has given steel the reputation for a high rate of attack on the mould refractories. Unfortunately, the attempt at a solution to this problem has resulted in the use of pre-formed refractory tubes for the running system. The joining of these standard pipe shapes means that nicely tapered sprues cannot easily be provided, with the result that much air goes through the running system with the metal. The chaos of surface turbulence in the runner, and the splashing, foaming and bursting of bubbles rising through the metal in the mould cavity, will mean that re-oxidation product problems are an automatic, unavoidable penalty.

It follows that a common feature of steel foundries is that the foundry often employs more welders in the 'upgrading' department than people in the foundry making castings. Common black humour among steel foundry workers includes 'More weld metal goes out of the foundry door than cast metal' and 'the drag is made in the foundry and the cope in the welding shop'. These regrettable jibes follow from the surface turbulence caused during pouring. The difficulties are addressed in 'casting manufacture', in which pouring basins are recommended to be eliminated and the filling system is recommended to be a naturally pressurised design moulded in sand. In the meantime, we shall examine the problems caused by the current poor filling systems.

Solid oxide surface films

Some liquid steels have strong, solid oxide films covering their surface. The high melting point of these oxides ensures that the surfaces of these liquid metals appear to be perfectly dry. They occur on many stainless steels rich in chromium and molybdenum, especially the super duplex stainless steels. In casting above about 250 kg in weight, the filling systems are sufficiently large to pass bifilms up to 100 mm across or more. Entrained air bubbles and surface turbulence in the mould cavity create even more films in situ in the forming casting. Clusters of bifilms are identified on radiographs as resembling faint, dispersed microshrinkage porosity. When attempting to grind away such regions, hollowing out deep cavities in the casting in an effort to eliminate such apparently porous metal, it is common to check periodically with the red penetrant dye to ascertain whether the region is yet free from defects. The bifilms in these regions appear as an irregular red-coloured spider's web. Figure 6.57 shows a typical valve casting in a 20Cr-20Ni-6Mo steel which is notorious for its problems. The bifilm cracks are mainly at grain boundaries, of course, and are noticeably aligned by the growth of the columnar grains at the outer edge of the casting. The casting has received some attempt at cleaning up and grinding out of defects. The deep red indications are almost certainly bubble trails because these tend to hold a lot of dye.

The grain boundary bifilms are often somewhat opened up by cooling strains. When viewed under the optical microscope they have given rise to the description the 'loose grain effect' in some stainless steel foundries because the grains appear to be separated from each other by deep cracks at the grain boundaries, as though they might rattle if shaken. This is, of course, nearly true. The bifilms are segregated by the growing grains, being pushed ahead, and thus usually come to rest in the boundaries between grains. Thermal strains during cooling open the bifilms to give the crazed appearance of the microstructure. This maze of thin, deep cracks often has to be excavated completely through walls sometimes 100 mm in thickness and sometimes greater before these regions can be re-built by welding.

However, in small castings of these particular steels, such as many lost wax castings, there now seems to be evidence that the ingates can be sufficiently narrow (measured in millimetres) so that strong, rigid plates of oxide cannot pass through (Cox et al., 1999). Thus, paradoxically, alloy steels which are notoriously difficult for the pouring of large castings can be used to make small castings that are relatively free from defects.

Low carbon/manganese and low alloy steels are typically deoxidised with Si, Mn and Al in that order. They can suffer from a stable alumina film on the liquid if the final deoxidation with Al has been carried out too enthusiastically. This causes similar problems to those described previous. In addition, I have known ingots of high alloy steels and Ni-base alloys to break up on the first stroke of a forging press after high levels of Al deoxidation have been used before casting. Those ingots that survived the forge usually cracked later during rolling or extrusion. When Al was reduced and Ca added (the effect of the Ca addition is explained later) the problem was solved for some alloys; the ingot forged and rolled like butter. For other alloys, the solution has not worked. We need to do far more research into the achievement of liquid oxidation production on molten steels.

Partly liquid surface oxide films

When adding the usual level of final deoxidation with Al, approximately only 1 kg or less Al per 1000 kg steel, most low carbon/manganese and low alloy steels do not usually suffer severe internal defects. Because of the high melting temperatures of such steels, the surface oxides contain a mix of SiO_2, MnO and Al_2O_3, amongst other oxide components. The mix is usually partially molten. On being entrained during pouring, the internal turbulence in the melt tumbles the films into sticky agglomerates. Because of the presence of the liquid phases that act as an adhesive, the bifilms cannot re-open, and grow by agglomeration. The matrix becomes therefore relatively free from defects in this way. Also, the oxide conglomerates are now rather compact. Their compactness and their low density cause them to float out rapidly, gluing themselves to the surface of the cope as a 'ceroxide' defect. Cope defects are common *surface defects* in these steels. In castings weighing a 10,000 kg or more the defects can sometimes grow to the size of a fist. They are, of course, labour-intensive to dig out and repair by welding. However, their compact form makes this job somewhat easier and not quite in the league of the extensive webs of bifilms bridging the walls of some stainless castings.

Liquid surface oxide films

Over recent years, it has become popular to give a final deoxidation treatment with calcium in the form of calcium silicide or ferro-silicon-calcium because the steel has been found to be much cleaner. This is quickly understood from the following. Alumina and calcia both have melting points in excess of 2000°C. The dry Al_2O_3 surface oxide (if deoxidising with only Al, plus possibly Si and Mn) is converted by addition of Ca into a liquid oxide of approximate composition $Al_2O_3 \cdot CaO$ that has a melting point at or lower than 1400°C. Any folding-in of the liquid film that may now occur will quickly be followed by agglomeration of the impacting liquid films into droplets. The compact form and low density of the droplets will ensure that they float out quickly and will be assimilated into the original liquid eutectic film at the surface, leaving the steel without defects. Any inclusions that remain will be small and spherical, and thus having a minimal impact on properties.

The previous example of the substantial benefits of targeting a liquid condition for the surface film on the steel is a critical factor which has been largely overlooked in the focus for the development of deoxidation reactions. Over recent years, I have revolutionised my approach to quality issues in steel castings by emphasising the importance of attaining a liquid surface film on the steel before and during pouring.

With a liquid surface film, entrainment of the surface no longer results in the creation of bifilms. Thus, Lind and Holappa (2010) in their review of the effects of calcium additions to alumina in steels, record the benefits to ductility and toughness of high-strength low-alloy steels and high quality structural steels by Al + Ca deoxidation.

The reviewers go on to describe how MnS inclusions in flat rolled plate and sheet grades of steels elongate to form stringers or platelets, lowering the through-thickness ductility and toughness. This behaviour seems most likely to be the result of alumina bifilms forming elongated stringers or platelets, onto which manganese sulphide precipitates. The central alumina bifilms are too thin to see easily, but form the cracks that reduce the through-thickness properties.

The boron containing steels might be an important class of steels with liquid surface oxides because of the role of the very low melting point oxide BO_2. The high strength and toughness of these steels may owe much to bifilm reduction, or possibly bifilm adhesion as a result of the BO_2 forming a liquid sandwich inside higher melting point and stable oxide films such as Al_2O_3 or Cr_2O_3. The glue formed by the boron oxide will not be strong, but it will be enormously better than nothing, and would be expected to be particularly effective in preventing the bifilm from shearing. The filled bifilm, or possibly absence of bifilms, may also explain the extraordinary effect which tiny boron additions make to the hardenability of steels. The cleaner metal (or metal with filled bifilms) would have significantly higher thermal conductivity, conferring a much improved penetration of the action of a quench, thus aiding the depth of hardening heat treatments.

Hadfield manganese steels appear to be another example of an alloy which benefits from the oxidation of its major constituents, creating a liquid MnO_2-rich surface film which is probably something like a manganese silicate MnO_2SiO_2 or $MnSiO_4$. The absence of bifilms in this cast steel explains its formidable toughness, despite poor filling system designs that would ensure poor properties in any ordinary steel. It is used for such punishing applications as rail track crossings and points, and the only (rare) failures tend not to be related to cracking but to improper bolting and support problems as a result of mechanical maintenance issues.

Aluminium nitride

In passing, it seems worth mentioning a class of defect that has been the subject of huge amounts of research, but which has never been satisfactorily explained. A tentative explanation is presented here. The phenomenon is the so-called '*rock candy fracture*'. This type of defect was seen when the ductility of the casting was especially low, despite the metal appearing to have precisely the correct chemistry and heat treatment. The fracture surface was characterised by intergranular facets that on examination in the scanning electron microscope were found to contain aluminium nitride. Naturally, the aluminium nitride was concluded to be brittle.

This defect seems most likely to be an entrained surface film. The film would probably originally consist of alumina, but would also contain some enfolded air. The nitrogen in the entrained air would be gradually consumed to form aluminium nitride as a facing to the crack. The defect would, of course, be pushed by the growing dendrites into the interdendritic spaces, particularly to grain boundaries. The central crack in the bifilm would give the appearance of the nitride being brittle. On examination, only the nitrogen is likely to be detected, constituting four-fifths of the air, and the oxygen being in any case not easily analysed. The defect is analogous to the plate fracture defect in ductile irons, and the planar fracture seen in Al alloys and other alloy systems (Figures 2.44, 2.46 and 6.30).

Thus in this case, despite the chemistry of the steel being maintained perfectly within specification, the defect would come and go depending on chance entrainment effects. Such chance effects could arise because of slight changes in the running system, or the state of fullness of the bottom-pour ladle, or the skill of the caster, etc. It is not surprising that the defect remained baffling to metallurgists and casters for so long.

6.6.5 PRIMARY INCLUSIONS

When the liquid alloy is cooling, new phases may appear in the liquid that precede the appearance of the bulk alloy. We shall deal with the formation of the primary metallic matrix phase in Section 6.6.7. Whether any newly forming dense phase gets called a phase or an inclusion largely depends on whether it is wanted or not: keen gardeners will appreciate the similar distinction between 'plants' and 'weeds'!

New phases that precede the appearance of the bulk alloy are especially likely following the additions to the melt of such materials as deoxidisers or grain refiners, but may also occur because of the presence of other impurities or dilute alloying elements.

For instance, in the case of steel that has a sufficiently high content of vanadium and nitrogen, vanadium nitride, VN, may be precipitated according to the simple equation:

$$V + N \leftrightarrow VN$$

Whether the VN phase will be able to exist or not depends on whether the concentrations of V and N exceed the solubility product for the formation of VN. To a reasonable approximation the solubility product is defined as:

$$K = [\%V] \cdot [\%N]$$

where the concentrations of V and N are written as their wt%. More accurately, a general relation is given by using, instead of wt%, the activities a_V and a_N, in the form of a product of activities:

$$K' = a_V \cdot a_N$$

It is clear then that VN may be precipitated when V and N are present, where sometimes V is high and N low, and vice versa, providing that the product %V × %N (or more accurately, $a_V \times a_N$) exceeds the critical value K (or K'). It is interesting to speculate that [N] may be high very close to the surface where the melt may be dissolving air. Thus the formation of a surface film of VN may be more likely. It is also necessary to bear in mind that even though all thermodynamic conditions for the formation of VN are met (meaning the solubility product is exceeded), the compound may still not form. This is because there is often a substantial nucleation problem that has to be overcome. This is either achieved by extremely high supersaturations of either V or N or both, or by the provision of a favourable nucleus. As we have seen repeatedly, oxide films are common substrates for the formation of many precipitating intermetallics. The film itself may not be the nucleant, but the nucleant will almost certainly have already nucleated on the film. Thus the film is the starting point for growth, starting at the site of a suitable incumbent nucleus, but spreading out over the film as its favoured growth substrate.

In the case of deoxidation of steel with aluminium, the reaction is somewhat more complicated:

$$2Al + 3O = Al_2O_3$$

and the solubility product now takes the form:

$$K'' = [a_{Al}]^2 \cdot [ao]^3$$

where the value of K'' increases with temperature. Again, the surface conditions are likely to be different from those in the bulk, with the result that a surface film of AlN or Al_2O_3 is to be expected, even if concentration for precipitation in the bulk are not met.

We have considered examples of nucleation at various points in the book, especially in Section 5.2.2. At this stage, we shall simply note that any primary inclusions form before the arrival of the matrix primary phase. Thus they appear in a sea of liquid. During this 'free-swimming' phase, primary inclusions have been thought to grow by collision and agglomeration (Iyengar and Philbrook, 1972). Whether this is true or not probably depends on the nature of the inclusions.

For instance, it is not clear whether all liquid inclusions will coalesce even if they impinge. It seems probable that coalescence may be hindered by the presence of a surface film. However, if coalescence does occur, droplets would be expected to result in large spherical inclusions whose compact shape will enable them to float rapidly to the surface and become incorporated into a slag or dross layer which can be removed by mechanically raking off or can be diverted from incorporation into the casting by the use of bottom-pouring ladles.

For solid inclusions, any agglomeration process that may occur might take the form of loosely adhering aggregates or clouds. However, agglomerations apply to particles. Entrained solid alumina films or other entrained solid films will not, in general, be particles but will be messy crumpled masses of double films that on a polished section might easily appear as a cloud of particles, particularly if parts of the alumina bifilm are so thin as to avoid detection, so that only thicker fragments are visible. Furthermore, at steel-casting temperatures, an extremely thin entrained and ravelled alumina film may condense into arrays of compact particles, analogous to the way in which a sheet of liquid metal breaks up into droplets (a spectacular example is given in Figure 2.13), an effect driven by the reduction of surface energy. Subsequent working by rolling or extrusion will elongate films or arrays of separated particles thus explaining the observed alumina 'stringers' (an extremely poor name because these are clearly not one-dimensional linear features but two-dimensional arrays of planar features) often seen in wrought steels. The occasional cracks and pores associated with alumina inclusions will almost certainly be the residue of the central unbonded interface of the original alumina bifilm (it will not be expected to be the result of so-called brittleness of the alumina phase, nor its effect to initiate cracks in the matrix). Work to clarify these suppositions would be welcome.

Hutchinson and Sutherland (1965) have studied the formation of open-structured solids. They found that flocs can form by the random addition of particles. If these particles are spherical and adhere precisely at the point at which they first happen to encounter the floc, then the floc builds up as a roughly spherical assembly, with maximum radius R, and about half the number of spheres within a region $R/2$ from the centroid. The central core has an almost constant density of 64% by volume of spheres. Occasional added spheres will penetrate right into the heart of the floc. Graphite nodules in ductile iron appear to be a good example of this kind of flocculation; melts of hyper-eutectic ductile irons suffer a loss of graphite by the floating out of loose flocs of spherulites (Rauch et al., 1959).

We have only touched on examples of oxides and nitrides as inclusions in cast metals. Other inclusions are expected to follow similar rules and include borides, carbides, sulphides and many complex mixtures of many of these materials. For instance, carbo-nitrides are common, as are oxy-sulphides. In C-Mn steels, the oxide inclusions are typically mixtures of MnO, SiO_2, and Al_2O_3 (Franklin et al., 1969) and in more complex steels deoxidised with ever more complex deoxidisers the inclusions similarly grow more complex (Kiessling, 1978).

In his substantial review of inclusions in steels, Kiessling points out that steel that contains only as little as 1 ppm oxygen and sulphur will contain more than 1000 inclusions/g. Thus it is necessary to keep in mind that steel is a composite product, and probably better named 'steel with inclusions'. Even so, steels are often much cleaner than light alloy castings that might contain 10 or 100 times more inclusions, partly helping to explain the relatively poor ductility and absence of a fatigue limit exhibited by Al-based casting alloys compared with steels.

Finally, of course, not all inclusions will be formed during the liquid phase of the metal alloy. Many, if not most, will be formed later as the metal freezes. These are termed secondary inclusions, or second phases, and are dealt with in the following section.

6.6.6 SECONDARY INCLUSIONS AND SECOND PHASES

After the primary alloy phase has started to freeze, usually in the form of an array of dendrites, the remaining liquid trapped between the dendrite arms progressively concentrates in various solutes as these are rejected by the advancing solid. Because the concentration ahead of the front is increased by a factor $1/k$, where k, the partition coefficient, can often be a rather small number, greatly enhancing the segregation effect, the number of inclusions can be greatly increased compared to those that occurred in the free-floating stage in the liquid. However, the size population is usually different, being somewhat finer and more uniform as a result of the smoothing action of diffusion in the tiny volumes between the dendrites.

The secondary inclusions or second phases form at or close to the freezing front. One of the most common and important second phases is a eutectic. We have already seen how microsegregation can lead to the formation of eutectic at bulk compositions that are much below those expected from the equilibrium phase diagram.

However, the nucleation of a first phase is likely to prevent the subsequent nucleation of any other phase that might also require one of the same elements for its composition. The availability of solute is clearly limited by a naturally occurring 'first come first served' principle.

In the subsequent observation of inclusions in cast steels, those that have formed in the melt before any solidification are, in general, rather larger than those formed on solidification within the dendrite mesh. The possible exceptions to this pattern are those inclusions that have formed in channel segregates, where their growth has been fed by the flow of solute-enriched liquid. Similarly, in the cone of negative segregation in the base of ingots, the flow of liquid through the mesh of crystals would be expected to feed the growth of inclusions trapped in the mesh, like sponges growing on a coral reef feeding on material carried by in the current. In Figure 5.44, the peak in inclusions in the zone of negative segregation is composed of macro-inclusions that may have grown by such a mechanism. Elsewhere, particularly in the region of dendritic segregation around the edge of the ingot, there are only fine alumina inclusions.

Sulphides

It would not be right to leave the subject of inclusions without mentioning the special importance of the role of sulphide inclusions in cast steel. The ductility of plain carbon steel castings is sensitive to the type of sulphide inclusions that form.

Type 1 sulphides have a globular form. They are produced by deoxidation with silicon.

Type 2 sulphides take the form of thin grain boundary films that seriously embrittle the steel. They usually form when deoxidising with aluminium, zirconium or titanium.

Type 3 sulphides have a compact form and do not seriously impair the properties of the steel. They form when an excess of aluminium or zirconium (but not apparently titanium!) is used for deoxidation.

Mohla and Beech (1968) investigated the relation between these sulphide types and concluded that the change from type 1 to type 2 is brought about by a lowering of the oxygen content. Additionally, it seems that the new mixed sulphide/oxide phase has a low interfacial energy with the solid, allowing it to spread along the grain boundaries. Also, it might constitute a eutectic phase. Type 3 sulphides were thought by Mohla and Beech to be a primary phase. In common with all other investigators of sulphide embrittlement of steels, it is clear that these authors were less than happy with their tentative findings because doubts and confusion still existed.

However, type 2 inclusions have all the hallmarks of an entrained bifilm. It is significant that this type of inclusion forms only when the melt is deoxidised with Al or other powerful deoxidisers that are known to create solid films on the melt. The surface film might originally have been enriched with the other highly surface-active element, sulphur. The entrainment of an oxide film would in any case be expected to form a favourable substrate for the precipitation of sulphides. The film would naturally be pushed into the interdendritic regions by the growing dendrites, so that it would automatically sit at grain boundaries.

Even so, an explanation of type 3 sulphides remains elusive. These results illustrate the complexity of the form of inclusions, and the problems to understand their formation. Much additional research is required to elucidate the mechanism of formation of these defects.

A final question we should ask is 'How do inclusions in the liquid become incorporated into the freezing solid?'

It seems that for small inclusions, especially those that are in the relatively quiet region of the dendrite mesh, the particles are pushed ahead of the front, concentrating in interdendritic spaces.

For larger inclusions, generally larger about 10 μm in diameter, trapping between dendrite arms is only likely if the inclusion is carried directly into the mesh by an inward-flowing current. This may be the mechanism by which inclusions are originally trapped within the cone of negative segregation, where they subsequently grow to large size, feeding on solute-rich liquid percolating through the dendrite mesh (Figures 5.44 and 5.45).

Where the front is relatively planar and strong currents stir the melt, the larger inclusions are not frozen in to the advancing solid as a consequence of the velocity gradient at the front. Delamore (1971) found that those particles that do approach the interface cannot be totally contained within the boundary layer, and as a result spin or roll along it because of the torque produced by the velocity gradient. In this way the larger particles finally come to rest in the centre of castings. Rimming steels benefited for the same reason from an absence of large inclusions in their pure rim.

Steel inclusion summary

The liquid metal in the melting furnace is probably fairly clean at a late stage of melting because of the large density difference and plenty of time for flotation. However, before casting, deoxidation by Si, Mn and Al etc. creates a large population of fine inclusions in situ in the melt. A proportion of the inclusions generated from this action separate quickly by gravity. There has been much research on these in situ–generated inclusions (see, for instance, Tiekink et al., 2010).

Much less research has been carried out on the entrainment inclusions created during the two major pouring events. These are as follows.

1. The transfer of the melt from the furnace into the transfer ladle creates a fresh, dense crop of inclusions, the air reacting with the exposed melt surface, oxidising the residual deoxidiser additions that remain in solution in the steel. These surface inclusions, often in the form of a surface film, become entrained into the bulk melt by the action of the pour. They are known as re-oxidation inclusions. Fortunately, a large proportion of this contribution towards the loss of quality of the steel will float out in the time taken to lift the ladle out of the pit and transfer it to the casting station. This huge volume of defects constitutes most of the layer of slag on the surface of the ladle by the time the pouring of castings begins.
2. A second crop of re-oxidation inclusions are now entrained by the pouring action into the mould. Depending on the freezing time and geometry of the casting, this late crop of inclusions may little or no time to float out.

Huge numbers of inclusions, generally known in the trade as re-oxidation products, result from these turbulent pouring processes, although, clearly, the second pour is far more damaging to the product than the first.

However, in addition to these sources, steels are also noted for the number of very fine additional inclusions that form later, during solidification. The remaining unreacted deoxidising elements are concentrated in the interdendritic liquid, where more inclusions pop into being by a nucleation and growth process. The interdendritic regions are small, limiting the size to which such inclusions can grow. Svoboda et al. (1987) observe that these inclusions are often also associated with small amounts of MnS. This is to be expected because both manganese and sulphur will also be concentrated in the interdendritic spaces. It is possible that, because of their small size, some of the deoxidation products may be pushed ahead of the growing dendrites and so become the nuclei for precipitates that arrive later.

As solidification proceeds, other inclusions may form by this concentration of segregated solutes in the interdendritic spaces. These may include nitrides such as TiN, carbides such as TiC, sulphides such as MnS and oxy-sulphides etc. In general, they will be most concentrated and largest in size in the regions between grains, and in regions of the cast structure where segregation is highest, such as in channel segregates and the tops of feeders. They may also enjoy late excursions into the casting as this remaining enriched metal from the feeder is sucked into the casting during the very last moments of feeding. The region under the feeder is known for its segregation problems resulting from the concentration of light elements, particularly carbon, but it is probable that the residual liquid will also carry plenty of additional unwanted solutes and solid debris.

Later still, further precipitation of inclusions will occur in the solid state. In general, these are up to 10 times finer as a result of the much lower coefficient of diffusion in the solid. The driving force for their appearance is the decreasing solubility of the elements as the temperature falls. Many hardening reactions are driven in this way; for instance, the formation of aluminium and vanadium carbides and nitrides in steels and the precipitation of $CuAl_2$ phase in Al-4.5Cu alloy. The hardening is the consequence of the very fine size and spacing of the precipitate, making the inclusions effective as impediments to the movement of dislocations.

The precipitation of inclusions in the liquid and solid states creates populations of phases that are entirely different from entrained inclusions. Such intrinsically formed (in situ) inclusions grow atom by atom from the matrix and so are in perfect atomic contact with the matrix. They would *never* be expected therefore to initiate volume defects such as pores and cracks.

In contrast, the entrained inclusion is characterised by its unbonded wrapping of oxide, so it can easily nucleate pores and cracks from this lightly adherent coat; it easily peels away during subsequent plastic working or creep. Alumina-rich inclusions in rolled steel are often seen to be associated with cavities. These are usually assumed to be the result of the brittle breakup of the inclusion during working, or the tearing away of the matrix from sharp corners acting as stress raisers. It seems likely to me that neither of these explanations is correct because alumina is strong, and volume defects are difficult if not impossible to initiate in the solid state. The real explanation is most probably the fact that cracks and pores that are evident in such strings of inclusions are the fragments of the original alumina bifilm. This will never completely close despite much plastic working, partly as a result of the inert and stable nature of the ceramic phase, alumina and partly as a result of the residual argon gas in the entrained layer of air. The residual argon is highly insoluble in metals.

6.6.7 NUCLEATION AND GROWTH OF THE SOLID

During the cooling of the liquid steel, a number of particles may pre-exist in suspension, or may precipitate as primary inclusions. The primary iron-rich dendrites will nucleate in turn on some of these particles. The work by Bramfitt (1970) illustrated how only specific inclusions act as nuclei for iron.

Bramfitt carried out a series of elegant experiments to investigate the effect of a variety of nitrides and carbides on the nucleation of solid pure iron from the liquid state. In this case, of course, the solid phase is the body centred cubic delta-iron (δFe). In his work he found that his particular sample of iron froze at approximately 39°C undercooling (i.e. 39°C below the equilibrium freezing point). Of the 20 carbides and nitrides that were investigated, 14 had no effect and the remaining six had varying degrees of success in reducing the undercooling required for nucleation.

The results are shown in Figure 6.60. They give clear evidence that the best nuclei are those with a lattice plane giving a good atomic match with a lattice plane in the nucleating solid. Extrapolating Bramfitt's theoretical curve to the value for the supercooling of his pure liquid iron indicates that any disregistry between the lattices beyond approximately 23% means that the foreign material is of no help in nucleating solid iron from liquid iron.

Another interesting detail of Bramfitt's work was that a number of additions were ineffective because they either melted or dissolved in the liquid iron as it was cooled to promote freezing. This consequent lack of effectiveness was despite, in some cases, quite low values of disregistry. This underlines the perhaps self-evident point (but often forgotten) that any addition has to be present in solid form for it to nucleate another solid. In addition, all of Bramfitt's work was concerned with the nucleation of δFe, the body-centred cubic form of iron.

6.6 STEELS

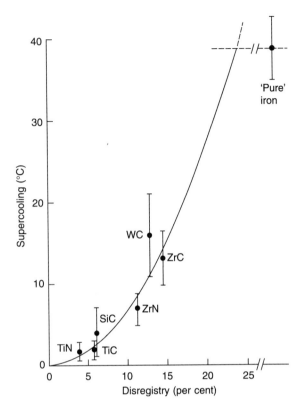

FIGURE 6.60

Supercooling of liquid iron in the presence of various nucleating agents.

Data from Bramfitt (1970).

The fcc form of iron, γFe, or austenite, has been considerably more resistant to past attempts to nucleate it. Until recently no one had succeeded to identify any previously existing solid that acted as a nucleus for the γFe phase. In 2013, Moumeni and colleagues found a mixed sulphide of Mn and Ti (Mn,Ti)S, in the centres of austenite dendrites in cast irons, indicating that they may have acted as nuclei. This has yet to be confirmed. Japanese workers led by Mizumoto (2008) claimed to have successfully grain refined an austenitic stainless steel (12Cr 18Ni with $Cr_{eq}/Ni_{eq} = 0.83$; the values of this ratio are explained later). Even here, the work was carried out only on a laboratory scale (melt weight 90 g) and had to be performed in seconds otherwise the niobium carbide contained in their master alloy dissolved so quickly as to become ineffective. Again, no mechanical testing was carried out to ascertain whether there appeared to be any benefit to this procedure.

Jackson (1972) listed a large number of additives that were unsuccessful in attempts to refine austenitic steels. His first success was the addition of FeCr powder together with floor sweepings! This impressively economic but hardly commendable formula caused him to persevere, searching for more scientifically chosen additions. He later found that calcium cyanamide was quite useful, but required the nitrogen content in the alloy to be raised to 0.3% to be successful in 18:8 stainless steels. At this level of gas content, severe nitrogen porosity is the unwelcome product. Jackson was able to define satisfactory conditions for 18/10/3Mo stainless steel, again providing that the nitrogen content was above 0.08 wt%. The modest improvement that he reports in mechanical properties he attributes not so much to the reduction in grain size as to the increase in the alloying effect of nitrogen! This work is further complicated by the expectation that the compound calcium cyanamide will probably have decomposed at steel casting temperatures.

The grain refinement of austenitic stainless steels therefore remains a challenge to future metallurgists. It seems likely that only those cast austenitic stainless steels that solidify first to δFe, before their subsequent transformation in the solid state to γFe, are definitely able to benefit from grain refinement resulting from a nucleation process.

Suutala (1983) proposed a factor that allows the prediction of whether the steel will solidify to austenite or primary ferrite (delta iron); this is the ratio of the chromium equivalent to the nickel equivalent, Cr_{eq}/Ni_{eq}, where the chromium and nickel equivalents are calculated from (elements in wt%):

$$Cr_{eq} = \%Cr + 1.5Si + 1.37Mo + 2Nb + 3Ti$$

$$Ni_{eq} = \%Ni + 22C + 0.31Mn + 14.2N + Cu$$

(Readers will note a close similarity with the definition of the Schoefer equations given earlier that define the structures of stainless steels at room temperatures.) Suutala proposes the ratio $Cr_{eq}/Ni_{eq} = 1.55$ is the critical value; at values higher than this, the solidification changes from primary austenite to ferrite. (This value applies for shaped castings and ingots. The equivalent value for welds because of their much faster rates of freezing is 1.43.) However, careful work by Ray and co-workers (1996) suggest that the division between austenite and ferrite is not sharp, but is better defined by a parameter the ferrite potential (FP) that assesses the fraction of primary ferrite in austenitic stainless steels

$$FP = 5.26(0.74 - Cr_{eq}/Ni_{eq})$$

Barbe and co-workers (2002) found that an FP value below 3.5 is useful for mainly ferritic steels, corresponding to a maximum $Cr_{eq}/Ni_{eq} = 1.40$. The retention of some austenite limits the grain growth of ferrite, and is suggested to reduce the susceptibility that ferritic steels have to 'clinking', i.e. cracking during continuous casting. When slabs are cooled to room temperature cracks are found on the slab edges. Sometimes the slabs crack into two. This problem is reminiscent of problems in continuously cast direct chill aluminium alloys, particularly the strong 7000 series alloys. These appear to suffer from oxide bifilms entrained during the first few seconds of casting in which the metal falls and splashes turbulently during the first seconds of the filling of the mould. The bifilm cracks created in this moment float randomly into higher regions of the ingot during the casting operation, causing it to crack catastrophically, sometimes weeks after being cast. It is dangerous to be near such an event. The high Cr ferritic steels might be expected to behave analogously as a result of entrained Cr-rich films.

Roberts et al. (1979) confirmed that only ferritic material was refinable with titanium additions, and confirm that TiC and TiN have lattices that are good fits with ferrite, but poor fits with austenite. Baliktay and Nickel (1988) reported that titanium additions can also refine the grain size of the widely used high-strength stainless steel 17-4-PH. The ratio $Cr_{eq}/Ni_{eq} = 2$ approximately for this material, confirming that it solidifies to ferrite, in agreement with the finding by Roberts. More recent work (Wang et al., 2010) has demonstrated good refinement under laboratory conditions in a 12Cr steel ($Cr_{eq}/Ni_{eq} = 25$) using an addition of Fe-Ti-N.

Now, let us stop this account of the grain refinement of cast steels and stand back a little to take stock of the situation in which we find ourselves. Whereas it is common knowledge that the grain refinement of wrought steels is a valuable feature, the grain refinement of cast steels does not seem necessarily beneficial. Having recorded these struggles to achieve significant grain refinement in cast steels, the question arises, 'Is it worth it?' The answer is not clear as the following reports indicate.

Doubt is inferred from the work by Campbell and Bannister (1975) on the ferritic alloy Fe-3Si. They showed that the best refinement was obtained by the addition of TiB_2 to the melt. However, on metallographic examination the grain boundaries were found to be surrounded by a phase that appeared to be iron boride which would have done little for the mechanical properties. These were not tested, but were likely to have been impaired. It would be valuable to explore further whether conditions could be found in which grain refinement would improve properties.

Encouragement that useful results might be gained is given by Church et al. (1966), but even these results are not all good news. Their work on a high-strength steel, 0.33C-0.7Mn-0.3Si-0.8Cr-1.8Ni-0.25Mo-0.040S-0.040P, revealed that although grain refinement was successfully accomplished with 0.60Ti, the benefit was negated by the presence of interdendritic films of titanium sulphide, causing severe embrittlement. However, toughness and ductility could be improved by smaller additions of titanium in the range 0.1–0.2%, which was still successful in achieving grain refinement. The doubt remains that much of this research was undermined by poor casting technique, introducing quantities of deleterious bifilm cracks, particularly at grain boundaries. Sulphur would be expected to precipitate preferentially on such substrates, giving the impression of brittle sulphide films at these locations.

In contrast to the limited success from attempts to nucleate grains, other approaches involving 'seeding', using granular metal additions of the same composition as the melt, have been repeatedly confirmed by different workers over the past half-century as potentially successful (Jackson's addition of FeCr powder is an instance, although the floor sweepings are probably not; other genuine instances include Campbell and Caton, 1977).

Part of the reason for the overall disappointing progress on the grain refinement of cast steels, that might appear to be a potentially important metallurgical advance, lies in the special properties of steels. First, many steels transform and recrystallize more than once to different crystal structures as they cool, thereby automatically providing a fine grain size at room temperature (even though, of course, the original segregated boundaries may still be in place). Secondly, unlike Al alloys with long-lived bifilm populations, they do not benefit from being 'cleaned up' by sedimentation of the bifilms by the grain refining addition, because the high density of steels ensures they are already generally clean by flotation at the time of the grain refining addition. No significant additional cleaning is possible, thus the major advantage of a cleaning action plus grain refinement enjoyed by Al alloys does not happen and is not necessary for steels.

6.6.8 STRUCTURE DEVELOPMENT IN THE SOLID

The grain structure of the steel that forms on solidification may turn out to be the same as that seen in the finished casting. However, this would be somewhat unusual. It happens only in those cases in which the metal is a single phase from the freezing point down to room temperature. Examples include some austenitic stainless, and some ferritic stainless. However, even the ferritic stainless may undergo a transformation to martensite or bainite depending mainly on its carbon content. The transformer steel, Fe-3.25%Si, is a common ferritic steel that, on a polished and etched section, clearly displays at room temperature a structure that has changed little from that originating during solidification. For that reason, it is a useful model alloy for research.

Even in these single-phase materials, there is opportunity for grain boundary migration, possibly grain growth, and possibly recrystallisation. Complete recrystallisation would be expected in those parts of castings that had been subject to considerable plastic deformation during cooling. This would be expected, for instance, at junctions of flanges that restrain the contraction of the casting.

In materials that change phase during cooling to room temperature, the situation can be very much more complicated. Low-carbon and low-alloy steels are a good example, illustrating the problems of understanding a structure that, after freezing, has undergone at least two further phase changes during cooling to room temperature. Figure 6.61 lists the changes.

(a) The liquid solidifies to delta iron dendrites.
(b) When solidification is complete, the principal grain boundaries (shown as full lines) have their positions delineated and to some extent fixed in position by segregates, particulate inclusions and bifilms. The slight misalignments between parts of the dendrite raft result in a network of less important subgrain boundaries (shown as broken lines).
(c) During cooling and differential contraction of the casting, the plastic strains will create dislocations that will migrate to form an additional network of new subgrain boundaries. These are, of course, all low angle boundaries and may not be readily visible.
(d) On reaching the temperature for the formation of the gamma-iron phase, austenite grains will nucleate on the original grain boundaries or other discontinuities, particularly bifilms because the residual 'air gap' between the films will reduce the strain energy required for the nucleation of a new phase which requires to change both its volume and shape. Their growth into the delta-grains will sweep away most traces of the subgrain network.
(e) When the conversion to austenite is complete the original delta-grain boundaries (shown as broken lines) will still usually be discernible as ghost boundaries because of the fragmentary lines of segregates, particularly second phases decorating bifilms.
(f) Further cooling strains will generate a new subgrain structure.
(g) Austenite will start to convert to ferrite, usually nucleating at grain corners and boundaries, once again likely to be associated with bifilms as a result of the reduction in strain energy, sweeping away the substructure once again.

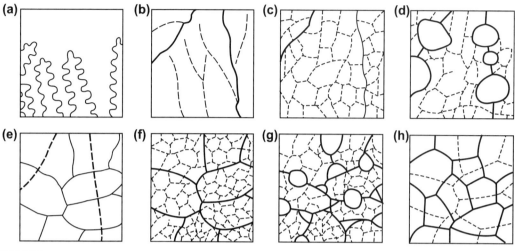

FIGURE 6.61

Successive stages of grain evolution in a low carbon steel, from its freezing point to room temperature (see text for full explanation).

(h) The final ferrite grains will again show the ghost boundaries of the previous austenite grains because these will have experienced sufficient time at temperature to have gathered some segregates by diffusion to the boundary.

(i) Subsequently, a further series of subgrains may be created, although by now the temperature is sufficiently low that any strains will generate fewer dislocations, and that such dislocations will not be sufficiently mobile at lower temperature to migrate into low-energy positions, forming low-angle boundaries; thus, the alloy will have become sufficiently strong to retain any further strain as elastic strain. The structure of the alloy will no longer be affected during further cooling, but elastic stresses will build up.

The final structure on a polished section will be a grain size that has been refined by two successive phase changes (but possibly coarsened a little by intervening grain growth) and may still retain ghost boundaries of delta iron and austenite. The underlying structure of the original delta iron dendrites will probably still be present, as can be revealed by etching to highlight the differences in chemical composition.

This sequence of events neglects the other many phase changes that can occur in some steels. Thus transformation of austenite to pearlite is usual for carbon/manganese steels, or transformations to martensite or bainite is also possible at higher cooling rates.

For a formal review of the development of structure in castings, see Rappaz (1989). Further detailed work on cast structures has been carried out during extensive work on the structures of welds in steels. For a review of this work, see Sugden and Bhadeshia (1987). This work draws attention to the complicating effects of the formation of Widmanstatten and acicular ferritic structures, and the presence of martensite, bainite, pearlite and retained austenite. The solidification morphology of the steel in this review of welding seems to be principally cellular or possibly cellular/dendritic (i.e. dendrites without side branches). Also, of course, in successive weld deposits there are the additional effects of the subsequent heat treatment of the previous runs in the laying down of the subsequent deposits.

6.7 NICKEL-BASE ALLOYS

The reader does not have to spend much time in a foundry casting steels and Ni-base alloys to realise that even though many steels are not easy, there are special difficulties with casting good Ni-base castings. This has always been a mystery to those involved with casting these high-temperature metals; a mystery we shall explore and seek to solve during the course of this chapter.

There are of course a wide range of Ni-base alloys, together with their cousins, the Co-base alloys. When alloyed with Cr, the alloys can achieve great resistance to heat and oxidation. For a special class of highly creep-resistant alloys, strong and oxidation resistant at high temperatures, the name '*high-temperature alloy*' is often used.

The class of alloys known as '*superalloys*' has, additionally, several wt% of Al as the major strengthening element, precipitating Ni_3Al (the so-called gamma prime phase, written as the Greek letter γ'). The '*gamma prime*' phase is an extremely fine and stable second phase. Ti similarly contributes to some extent as Ti_3Al. Many superalloys appear to contain most of the elements of the periodic table, making them probably more complex than any other metallic alloys. Recent developments have seen the incorporation of many rare and expensive elements into high temperature alloys. In particular, hafnium (Hf) has been used and noted for its action to reduce porosity. This somehow has been attributed to its lower coefficient of expansion (Chen and Knowles, 2003), but seems much more likely to be the result of a stronger oxide on the surface of the liquid, thus holding up on the lip of the crucible or in the mouth of the mould, and so avoiding incorporation as bifilms into the casting.

The high quantity of highly oxidisable elements, Al, Cr, Ti, Hf etc. in the high-temperature Ni-base alloys is the reason for the problems suffered by these alloys during casting. These elements all react rapidly with oxygen (and possibly with nitrogen) in the environment, creating solid films on the liquid surface. The entrainment of the solid surface film leads to bifilm creation and hence pore and crack defects. In their common wall-to-wall form, bifilms can also be expected to be a barrier to the growth of single crystals. As for steels, the conversion of the solid alumina-based surface film into a liquid surface film with the aid of Ca addition is a great help, avoiding bifilm formation, and thus raising properties, but does not seem to be effective for all alloys. For those alloys for which this phenomenon is effective the technique is important and is mentioned repeatedly later. Even here, it is possible that for some alloys where this would be helpful, Ca may be considered a harmful impurity for other reasons. The superalloys are complicated!

6.7.1 AIR MELTING AND CASTING

Many Ni-base alloys are melted and cast in air, both into ingots for subsequent working into plate, sheet, bar, tube and wire etc. and into shaped castings. Such castings behave similarly to steels, so that deoxidisers such as Al require to be added to avoid a carbon boil reaction during freezing. Alternatively, Niu and colleagues (2003) described how carbon is useful to deoxidise, eliminating more than 50% of the oxygen during the melting stage, even though Al is useful later to reduce the oxygen content further. Once again, as for steels, if Al would normally be used for deoxidation, a 50:50 mixture of Al and Ca is recommended to be used instead. For some Ni-base alloys, this technique is valuable to avoid the embrittling effects of alumina bifilms, replacing these with liquid oxide eutectic that, even if entrained by poor casting technique, still has the chance to spherodise and float out of the casting. At this time, it is not clear which alloys respond successfully to this deoxidation technique. More research would be so welcome.

For a high-temperature alloy, its alloy content of Al means that no additional Al will be required for deoxidation because the alloy will naturally have a low oxygen content (although sometimes not so low to prevent a carbon boil occurring during freezing). However, the alloy will definitely form a strong aluminium oxide bifilm on pouring; although there is only a few percentage of Al in the alloy, the high temperature encourages its extremely rapid growth. Thus a small addition of Ca immediately before pouring is valuable to liquefy the surface oxide, and thereby reduces cracking on solidification or during subsequent working.

Two factors currently work against success in the Ni casting industry.

1. For some reason lost in the mists of the early history of Ni-base alloy casting, it seems that most if not all melts are treated with Mg. A more inappropriate alloy addition can hardly be imagined: any residual Mg will ensure the development of a strong, solid MgO film on the liquid which would be highly damaging to the liquid metal after a turbulent pour. For confirmation, a study of Inconel 718 by Chen and colleagues (2012) found all the oxides were MgO or spinel $MgAl_2O_4$. Work by Ren (2014) shows that Mg deoxidation of steels can form globular

liquid oxides if Al is present together with Ti in sufficient concentration: if no Ti is present the oxides are solid at 1600°C but if 0.2Ti is present the melting point lowers to 1450°C. Thus the highly dangerous addition of Mg might be made safe by the presence of other alloy additions. The once-sacrosanct Mg addition requires to be critically reassessed immediately; it may be permissible for some alloys, whereas others will be made even more difficult to cast.

2. Ni-base alloys have a melting point sometimes 100°C below that of steels, so that Ni-base casters generally pour at temperatures well below those used for steels. In addition, of course, as is usual in foundries, casters understandably work with concepts such as minimising superheat, providing just sufficient to keep the alloy liquid during casting. These circumstances cooperate to ensure that the surface oxide on the liquid metal is solid, making Ni-base alloys difficult to cast and forge without cracking. This cracking behaviour has always been a mystery for such a malleable metal. These alloys may require much higher casting temperatures to help liquefy their surface oxide if turbulent filling systems are retained. Alternatively, *of course*, non-turbulent filling could also eliminate cracking.

6.7.2 VACUUM MELTING AND CASTING

One of the principal castings made from Ni-base (and some Co-base) alloys are turbine blades mainly for aero engines, but also nowadays for power generating turbines. As usual, these alloys contain aluminium and titanium as the principal hardening elements together with high levels of chromium for high temperature oxidation resistance. All of these elements can assist to create oxide films. Because such castings are produced by investment (lost wax) techniques, the running systems have been traditionally poor. It is usual for such castings to be top poured, introducing severe surface turbulence, and creating high scrap levels. Even most bottom-gated filling systems, still filled by pouring from a highly placed furnace, designed to fill the mould cavity in an uphill direction, are still poor; the defects introduced in the metal during the turbulence of the pour are simply carried over into the mould cavity to spoil the casting, or interfere with the progress of directional solidification or single crystal growth. When studying faults in single crystals, Carney and Beech (1997) found oxides at the root of most stray grains.

Turbine blades that have failed from major bifilm defects have caused planes to crash and have cost lives. Clearly, the problem is not trivial and the consequences are tragic. It is not satisfactory that we continue to live with gravity-poured blades fitted in aero engines. Such gravity-poured castings necessarily contain bifilm defects, many of which are not detectable by either X-ray radiography or by dye penetrant inspection (DPI). In contrast, blades produced by counter-gravity filling could probably be guaranteed free from filling defects. Furthermore, perhaps we would neither have to check them by radiography nor by DPI. If we *have* to live with a defective production technique for parts for aircraft engines, more discriminating testing by multi-resonance techniques would ensure greater safety.

When melting Ni-base superalloys in vacuum, a slag collar usually builds up from the accretion of oxides in suspension in the alloy. Some of these oxides are from the skins of the charge materials, but many are from the population of bifilms generated during the prealloying production process. Here the alloy maker blends his alloy in a large melting furnace. The melt is poured into launders and poured again through drops of several metres, falling into long steel tubes in which it solidifies into cast rods (again this occurs in vacuum, but the vacuum is not sufficiently good to prevent the creation of oxide bifilms of course). It is known that the larger diameter tubes (for the preparation of larger logs to be melted for the production of larger castings) have higher oxide content than the slimmer tubes in which turbulence would have been better suppressed.

Casting in vacuum is probably essential for such products as turbine blades. The wall thickness of these products is so small, becoming measured in micrometres at the trailing edges of blades, that any backpressure of gas will inhibit filling.

It is quite clear, however, that casting in vacuum is not a complete solution from the point of view of eliminating solid surface films. A good industrial vacuum is around 10^{-4} torr (approximately 10^{-2} Pa). This is not good enough to prevent the formation of bifilm defects (from entrainment of the surface oxide or nitride film) during pouring.

Equilibrium theory predicts that not even the vacuum of 10^{-18} torr (10^{-20} Pa) that exists in the space of near-Earth orbit is good enough to prevent the formation of Al_2O_3 (the most likely and most usual oxide) because 10^{-40} torr

(10^{-42} Pa) is required to dissociate this particular oxide. However, recent research on pure liquid aluminium has revealed that the equilibrium prediction is hugely inaccurate as a result of the alumina dissociating to a sub-oxide Al_2O. This occurs at around 10^{-8} Pa, reducing to 10^{-6} Pa at approximately 1000°C, causing a 'wind' of Al_2O to flow away from any Al_2O_3 on the surface of liquid Al, decomposing the Al_2O_3 and conveying away its oxygen in the form of the sub-oxide. The wind of sub-oxide vapour prevents oxygen from arriving at the surface of the liquid metal (Giuranno, 2006; Molina, 2007). The aluminium surface can then slowly become perfectly clean, free from oxide (Aguilar-Santillan, 2009). Even without the action of the evaporation of the sub-oxide, Zemcik (2015) reminds us that alumina can be reduced by carbon in only modest vacuum conditions at the temperatures used for casting. Unfortunately, he omits to check whether other oxides can be similarly reduced in casting conditions.

Despite this beneficial evaporation of the aluminium sub-oxide, and possible reduction by carbon, it remains a fact that films are found entrained in current vacuum cast alloys. Perhaps they are not oxides, but nitrides? Perhaps they are oxides of Cr, Ti, Nb, Hf or any one of the long list of elements sometimes present in the alloy which carbon is unable to reduce at these pressures and temperatures.

We need to give some further consideration to the fact that the surface films on high-temperature Ni-based alloys might in some cases be AlN. The Ni-base superalloys are well known for their susceptibility to react with nitrogen from the air and so become permanently contaminated, especially when the casting is cooled by opening the furnace door to air immediately after pouring. It seems more likely that the contamination is actually a nitride film problem rather than a problem caused by nitrogen in solution. In any case, in air, the reaction to the nitride may be favoured even if the rates of formation of the oxide and nitride are equal, simply because air is 80% nitrogen. Niu et al. (2002) reported in many superalloys that there are an order of magnitude more nitride than oxide inclusions. These authors reported higher porosity and loss of rupture life with higher nitrogen, lending some support to the concern that the nitrogen is in the form of nitride bifilms.

These complications and uncertainties about the films on the liquid metal emphasise, if emphasis is needed, that the real solution to entrainment problems is not to attempt to prevent the formation of the surface film by, for instance the quest for better vacuums, but simply to avoid *entrainment* of films.

Gravity pouring using a well-designed bottom-gated filling system might therefore be a significant improvement on most current filling system designs. However, the ultimate answer would be the complete avoidance of any type of pouring, using a counter-gravity system of filling. It would make beautiful castings that would probably not require expensive testing (the multi-resonance technique is fast and low cost). Furthermore, the easy automation of counter-gravity would greatly lower costs and enhance quality.

Whatever the films are, oxides or nitrides, an example is seen in Figure 6.62 that happens in this case to be an oxide. Other examples of film-like defects in vacuum cast superalloys can be seen in the work of Ocampo (1999) and Malzer (2007). Ocampo described a Co-base alloy with continuous grain boundary films decorated with porosity and carbides which are the typical signatures of the presence of bifilms at the boundaries. Figure 6.63 also shows a bifilm crack surrounded by carbides.

The precipitation of topologically close-packed (TCP) phases in some single crystal alloys seems to be associated with porosity and with preferred fracture planes. Work by Graverend (2012) and Shi et al. (2014) clearly showed TCP phases on {111} planes thought to be the habit plane of the precipitate, but more likely the preferred close packed plane favouring the straightening of a bifilm, on which the TCP has formed (Figure 6.64). This interpretation is strongly confirmed by the observation of decohesion only on one side of the TCP phase, and transverse cracks in the TCP phase on the opposing side of the crack, exactly as is typical of bifilm structure. It is interesting to compare the exactly analogous structure in an Al alloy (Figure 6.20(b)).

There is an interesting situation with the Ni-base alloys containing Hf. It seems that when these alloys are cast, a curious unexplained glassy surface defect on the blades that appears to run down the casting surface like a congealed river, but if a filter is placed in the filling system the defect does not occur. It seems likely that the defect represents an attack of the ceramic mould by HfO surface films generated by turbulence. The HfO will most likely form a low-melting-point oxide mixture with the components of the mould, often mainly Al_2O_3. If the filter is present, any HfO already present will be reduced by filtering out, and the reduced turbulence after the filter will not generate more HfO.

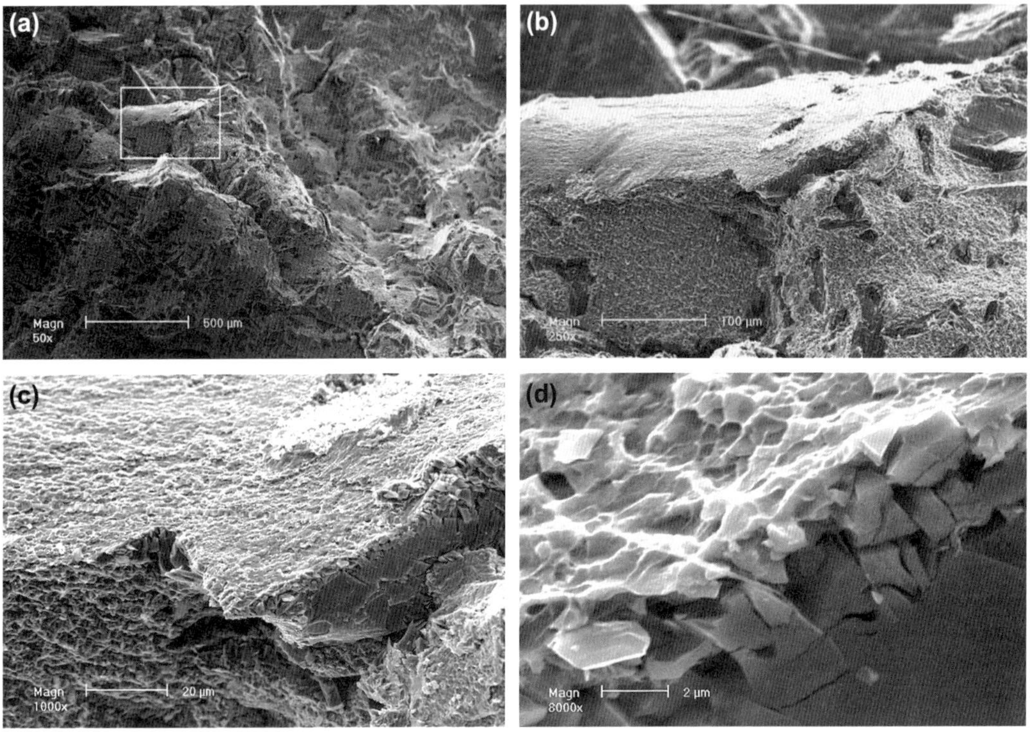

334 CHAPTER 6 CASTING ALLOYS

FIGURE 6.62

(a) to (d) show successively closer SEM images of the fracture surface of a Ni-based superalloy melted and cast in vacuum. The light-grey oxide clearly occupies more than 10 mm^2 of surface. The selected region, like a glacier, is only about 1 μm thick, but has a 20 μm deep underlay of carbides. At its high temperature of formation, the oxide film has recrystallised (Rashid and Campbell, 2004).

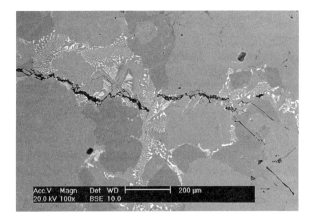

FIGURE 6.63

SEM image of a Ni-base alloy casting polished through a freckle containing segregated carbides precipitated on a bifilm crack (Rashid and Campbell, 2004).

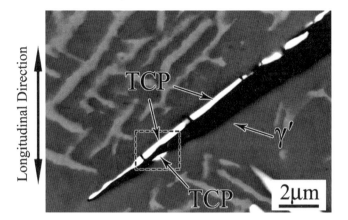

FIGURE 6.64

A bifilm crack in a Ni-based superalloy, revealed by its precipitate of a topologically close packed (TCP) phase (only a fragment of the phase is on one side of the bifilm crack) with characteristic short transverse cracks as a result of folds in one of the constituent oxide films of the bifilm (Figure 6.20(b) shows an analogous defect in an Al-Si alloy. Courtesy Professor Qiang Feng (Shi et al., 2014).

The larger bifilm defects, clearly seen on fracture surfaces, may be less common than those that give rise to the 'stars at night' symptoms created by clouds of bifilm fragments that intersect the surface of vacuum investment superalloy castings (for instance, reported by Wauby, 1986) when viewed in ultraviolet light after treatment by fluorescent dye penetrant. However, it is difficult to be sure. The fluorescent dye can penetrate the folded bifilms, and so exude again during inspection under ultraviolet light. Some indications are large and bright, indicating large internal cavities. Others can appear on both front and back of blades, possibly indicating that they completely penetrate the wall. A casting made with a good filling system (probably of lower cost than a conventional filling system, for those who assert that quality has to be paid for) can display zero indications. In the presence of a group of investment foundry personnel, I recall casting a blade that was put through the DPI process. It was then presented to an experienced inspector who, after a long and puzzled inspection, concluded it had not undergone immersion in the dye.

Further important aspects of Ni-base alloys, particularly for turbine blades, are discussed in Chapter 17 (Controlled Solidification Techniques).

6.8 TITANIUM

Titanium is a curious and special metal. In its liquid state, it is practically impossible to handle without it reacting with everything. However, in its solid state, it can be wonderfully resistant to oxidation and corrosion, whilst being impressively strong and having a density occupying an interesting 'halfway house', partway between the Al alloys and steels.

Its chemically inert behaviour makes it favoured for use for valves and fittings in the chemical industry, marine applications and in prostheses for the human body, including replacement hip and knee joints. Its unique combination of lightness and strength make it ideal for aircraft, particularly undercarriage structures, and blades for some of the cooler parts of turbines.

Despite these glowing advantages, the whole technology of melting and casting of Ti and its alloys is, in my view, deeply disappointing as a result of poor melting and casting technology. Parallels with the casting of superalloy turbine blades are clear. For this reason, I am risking unpopularity to state my views, because I believe that the industry has the capability to develop what I believe to be its true, very much greater potential which has so far not been attained.

6.8.1 Ti ALLOYS

Naturally, the metallurgy of Ti alloys is an extensive subject that cannot be covered in any detail here. The reader is recommended to Polmear's book *Light Alloys* (2006). However, for a quick overview in a few sentences, the metallurgy of Ti is divided into (1) the low-temperature α-alloys generally stabilised by O, N, C and slightly by Sn and Zr and (2) the higher temperature β-alloys, stabilised at room temperatures by Mo, V, Cr and Fe. Common α-alloys include commercially pure Ti (actually a Ti-O alloy) and Ti-5Al-2.5Sn. Probably the most common of all Ti alloys is the mixed αβ-alloy Ti-6Al-4V (often known simply as 6-4 alloy).

Other important alloys include the titanium aluminides, TiAl and Ti_3Al both of which exist over a range of compositions and usually contain significant additional alloys. They have excellent specific stiffness, oxidation resistance and good creep resistance up to 700°C. Disadvantages include significant microsegregation as a result of the 'double cascade' effect of the two significant peritectic reactions that occur in series during its solidification. The alloy also suffers from all the typical symptoms of high bifilm content, including brittleness, low thermal conductivity and difficulty to machine, creating chipping, cracking and grain pullout.

Despite these drawbacks, much development work is in progress with the aluminides because of their huge potential for weight reduction in gas turbines for aircraft.

6.8.2 MELTING AND CASTING Ti ALLOYS

There is a fundamental challenge with the melting and casting of Ti and its alloys. Ti reacts with and dissolves practically everything, so there are few materials in which liquid Ti can be contained. In addition, the high melting temperature creates other technical problems, particularly rapid loss of temperature during casting.

Much Ti is therefore melted in a water-cooled copper crucible. Roberts (1996) describes the early history of the process in which melting was carried out in an inert atmosphere or vacuum by electric arc, electron beam or by plasma in a hemispherical copper crucible. After the liquid metal had been poured out, a hemispherical shell of frozen metal remained. When this was lifted out of the crucible, its shape was remarked to be like a skull. Hence early workers referred to the process as 'melting in a skull' which became shortened to 'skull melting'.

One of the serious issues that arises as a result of melting in a cold crucible is that relatively little average superheat can be achieved in a melt. With modern induction heating typical average superheats in vacuum are around 15°C, whereas an argon atmosphere seems to 'keep the melt warm' achieving around 35°C (Rishel, 1999). These very limited values have promoted workers to adopt (erroneously, in my view) techniques for extremely rapid filling of moulds, such as centrifugal casting. The melt is poured down a central downsprue and radially out to the moulds along spoke-like runners. Moulds are arranged around the outside of a rotating platform, carefully balanced before spinning. All this happens in a large chamber providing an environment of vacuum or inert gas.

The disastrous turbulence generated by this approach ensures that castings are full of defects (Cotton, 2006). Nevertheless, this problem seems to be accepted as normal in the Ti casting industry. Castings are repaired by grinding out defects and refilling by gas-tungsten arc welding. Finally, all Ti alloy castings are subjected to hot isostatic pressing (hipping) as standard procedure. Hipping appears to be especially effective for Ti alloys because pores are not only closed, but also appear to be effectively welded, because any oxide or nitride film on the original pore is taken into solution. Thus the casting defects are mainly repaired by these additional (expensive) procedures.

Harding (2006) has demonstrated the casting of TiAl by tilt pouring. Compared with centrifugal casting, this is a vast improvement in terms of reduced damage to the product. Even so, tilt is not an easy process at the low superheats available from a cold crucible. In addition, many castings have a geometry for which a tilt-filling solution is impossible without a free fall at some point in the mould and consequent damage to the liquid.

My approach would be to use counter-gravity to fill the mould. This is a powerful filling technique that can apply to practically all geometries of castings. The melt is transferred vertically into the mould by dipping a riser tube into the centre of the melt in the crucible and applying a differential pressure. This is a simple technique. In addition, a gentle uphill fill would make good use of the limited superheat because most moulds even up to a metre or so high could be filled within a very few seconds without exceeding 0.5 or perhaps 1.0 m/s. Such a technique would provide an essentially defect-free casting. Furthermore, of course, the castings would not require hipping, or, if hipping were insisted upon (as would be likely with many quality specifications apparently formulated without regard to logic) would not benefit further. Counter-gravity would ensure that the most frequent inclusions, solidified droplets of tungsten metal (W) from the splatter of weld electrodes in scrap and returned material, would remain sitting firmly on the bottom of the crucible, and thus not appear in the castings. The other major inclusion type, hard alpha particles, again mainly from foundry returns, are less easily dealt with, and largely invisible to X-rays. They can be detected by sophisticated ultrasound techniques.

It is doubtful if a low melt temperature would be a problem when filling moulds by counter-gravity. If temperature were a problem, I would opt to trial a CaO crucible. This material (lime) is one of the most available and low-cost materials, in contrast to the more usual ceramic material for melting Ti, cerium oxide, which is a potentially highly suitable material, but rare and costly. A calcia crucible allows significantly greater temperatures, probably 100°C or more, to be reached in the melt. Perhaps more importantly, these temperatures are reached at lower power densities so the melt is subjected to much less turbulence. In a cold crucible, the melt jumps about in the crucible and may be a problem to prevent it jumping out of the crucible when using very high power in an effort to achieve a tolerable casting temperature. Thus much entrainment of the liquid surface will occur with all this violent activity of the surface. Although researchers are concerned for oxygen that will be picked up from the CaO by the melt, the real problem may be the pickup not of oxygen itself, but of oxide films entrained by the turbulence in the cold crucible. Thus the action of CaO to introduce oxygen into solution in the alloy may be harmless, or relatively harmless, whereas oxide films introduced by turbulent melting will act as cracks and will therefore seriously reduce properties, particularly ductility. As usual, there is no shortage of avenues to explore.

A further potential benefit of a ceramic crucible is that there is some chance of its accretion of sundry debris in suspension in the melt (the 'fly paper' principle). Thus a melt would be expected to gradually become clean. This action occurs with practically all other ceramic and refractory containment of liquid metals; in fact the buildup of a collar of 'slag' is a common complaint with a melting process, but in reality is probably an unappreciated benefit. Such a scavenging action might be enhanced by the use of a very small amount of flux or slag. Such a liquid second phase afloat the main melt, occupying the meniscus valley at the crucible wall, is a common technique to encourage suspended material to stick to and be absorbed by the second liquid.

One of the most exciting developments in the history of titanium has to be the work of Schwandt and Fray (2014) who demonstrate the melting of a titanium alloy in an alumina or magnesia crucible, in air. The melt is protected against the absorption of oxygen by melting under a CaF_2-rich flux which transfers cathodic protection to the liquid metal from an iridium anode and titanium cathode. The technique might be most useful in the use of ceramic crucibles in vacuum, thereby gaining the benefit of adequate casting temperatures without turbulence-inducing energy inputs needed for the cold crucible technique.

Chandley (2000) described a quite different approach to reducing the contamination problems during melting. Titanium is heated under vacuum in a graphite-lined, induction-heated crucible. The vacuum is then replaced by argon and liquid aluminium is poured into the heated solid titanium and the power turned on. The exothermic reaction between the metals forms the TiAl molten alloy which is then transferred by suction counter-gravity into the investment shell mould.

Mould materials for Ti and its alloys are also problematic because of the high reactivity of Ti. Machined graphite moulds were the first to be used for Ti in 1954 (Cotton, 2006), soon to be followed by rammed graphite (Antes, 1958). More recently, oxide-based materials that can be investment cast are becoming the norm. Cheng et al. (2014) found some alumina is allowable to dilute the expense of yttria face coats. In all graphitic and oxide-based moulds, reaction with the mould stabilises the Ti alpha phase, thus converting the surface of the casting to this hard, brittle phase, known as the 'alpha skin'. The alpha skin has to be removed from the casting by machining or chemical dissolution. Unfortunately, this high-cost operation adds to the high cost of the alloy, the high cost of processing and the high cost of shipping. The selection of Ti and its alloys for a casting requires a dedicated and determined buyer, usually one who has no choice but to opt for a titanium product.

Permanent metal moulds for simple shapes have the great advantage of avoiding the creation of an alpha skin, but such moulds are often themselves expensive, consisting of alloys based on either Mo, W or Nb (Choudry, 1999).

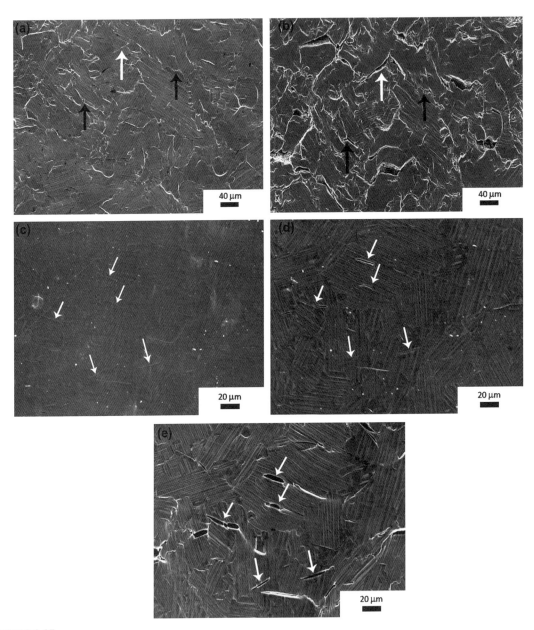

FIGURE 6.65

Cracks opening with time during creep at 700°C in Ti-45Al-TiB$_2$ alloy (a, b) at 300MPa showing development of cracks at B phase (white arrow) and colony boundaries (black arrows); (c,d,e) evolution of cracks with time at 400MPa only at B phase.

Courtesy Munoz-Moreno (2013)

6.8.3 SURFACE FILMS ON Ti ALLOYS

Mi (2002) has produced evidence that films do occur and can occasionally be seen by careful examination under the scanning electron microscope in castings.

In general, however, titanium alloys may not be troubled by a surface film. Certainly during the hot isostatic pressing (hipping) of these alloys, any oxide seems to go into solution. Careful studies have indicated that a cut (and, at room temperature, presumably oxidised) surface of the intermetallic alloy, TiAl, can be diffusion bonded to full strength across the joint, and with no detectable discontinuity when observed by transmission electron microscopy (Hu and Loretto, 2000). It seems likely, however, that the liquid alloy may exhibit a *transient film* in some conditions, such as the oxide on copper and silver and the graphite film on cast iron. Transient films are to be expected where the film-forming element is arriving from the environment faster than it can diffuse away into the bulk. This is expected to be a relatively common phenomenon because the rates of arrival, rates of surface reaction and rates of dissolution are not likely to be matched in most situations.

In conditions for the formation of a transient film, if the surface happens to be entrained by folding over, although the bifilm formed in this way is continuously dissolving, it may survive sufficiently long to create a legacy of permanent problems. These could include the initiation of porosity, tearing or cracking, prior to the complete disappearance of the bifilm. In this case, the culprit responsible for the problem would have vanished without trace.

Other consequences of the transient presence of bifilms in Ti alloys may be the morphology of the borides in some alloys of titanium aluminide. These are of extremely variable lengths in castings, short from some melts, but long from others, their lengths appearing to be bafflingly independent of casting parameters, and wildly out of the control of the metallurgist. This is typical of bifilm problems. In addition, on close examination, the boride particles are often seen to be associated with crack-like porosity along their lengths (too thin to be easily seen in Figure 6.64), typical of that which would occur if a bifilm had started to separate a little during the growth of the boride over its surfaces. Munoz-Moreno and colleagues (2013) studied centrifugally cast and hipped Ti-45Al alloy, finding, significantly, cracks associated with both boride particles and colony boundaries in creep tests at 700°C (Figure 6.65). It is expected that centrifugally cast material will have a high density of bifilms from this turbulent casting process, some of which will be favoured substrates

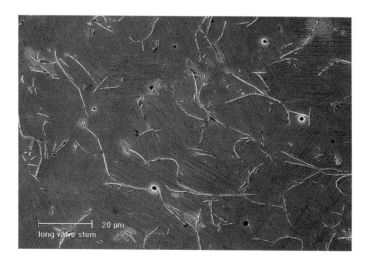

FIGURE 6.66

A titanium aluminide casting poured in argon atmosphere. Round pores may be argon bubbles, but all fine, crack-like pores are associated with platelet growth of borides (white linear features in image) and may indicate the bifilm substrates for the borides.

for the precipitation of borides, whose growth will then straighten the bifilms, and some will be pushed into grain (colony) boundaries. Chandravanshi (2013) studied creep fracture surfaces of an Alpha Ti alloy, finding clear decohesion from borides. Similarly, Cowen and Boehlert (2007) observed borides on the polished surfaces of specimens of Ti-22Al-26Nb alloy tested in creep. Immediately after the stress was applied, they saw the borides opening like cracks. As is often assumed for graphite in cast iron, the borides may also be weak, but their strengths are not known and may actually be high across transverse planes, in which case tensile failure would only be expected if the borides were formed on bifilms. Research is needed to clarify this (Figure 6.66).

Appendix I

THE 1.5 FACTOR

Experimental results for side gated 99.8 Al plate castings plotted in Figure A1 show that casting time t_c may be estimated for the plates and other castings from an equation:

$$t_c = 1.5 \times Casting\ volume/Initial\ casting\ rate \tag{1}$$

This is equivalent to

$$Initial\ fill\ rate/Average\ fill\ rate = 1.5 \tag{2}$$

These experimental results give support to the value of 1.5 chosen by previous authors, particularly those of the British Non-Ferrous Research Association (now no longer with us) researching for the UK Admiralty (Ship Department 1975).

Exploring the 1.5 factor further by a theoretical approach is not quite so straightforward, but an attempt is outlined here.

Considering Figure A2, the velocity at the base of the sprue is given by

$$V_2 = (2gH)^{0.5}$$

If the areas of the base of the sprue is A_2 and the mould cavity is of uniform area A_C the initial velocity of rise in the mould will be given by

$$V_i = (A_2/A_C) \cdot (2gH)^{0.5}$$

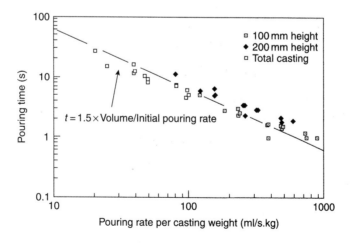

FIGURE A1

Experimental demonstration of the relation between initial and average filling rates.

Data from Runyoro and Campbell (1992).

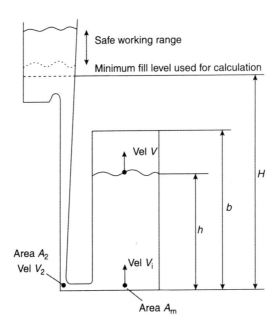

FIGURE A2

Schematic view of the filling of a uniform section casting.

Similarly, at some later instant, when the melt has reached height h, the net head driving the filling is now reduced to (H−h) so that the rate of rise is now

$$V = (A_2/A_C) \cdot (2g(H-h))^{0.5}$$

Substituting dh/dt for the rate of rise V, rearranging and integrating between the limits of time $t = 0$ at $h = 0$, and $t = t_c$ at $h = b$, we find the casting time (the time to fill the mould) t_c is given by

$$t_c = (A_C/A_2) \cdot (2/g)^{0.5} \left[H^{0.5} - (H-b)^{0.5}\right]$$

Now writing simple definitions of the initial rate of castings Q_i and the average rate of casting Q_{av} in such units as volume of liquid per second, defined by the appropriate velocity times the area, we have

$$Q_i = A_2 \cdot (2gH)^{0.5}$$
$$Q_{av} = A_C \cdot b/t_c$$

It follows that

$$Q_i/Q_{av} = 2H^{0.5}\left[H^{0.5} - (H-b)^{0.5}\right]/b$$

This solution to the filling problem is interesting. There are various combinations of H and b that can fulfil the conditions defined by the equation. For instance, if $H = b$, then $Q_i/Q_{av} = 2$, which is actually an obvious result, meaning simply that the average is half of the start and finishing rates.

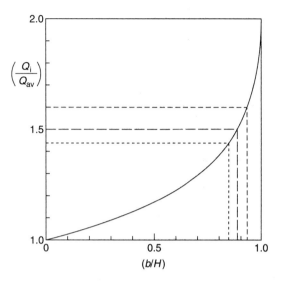

FIGURE A3

The relation between the initial and average fill rates for a uniform section casting as a function of the relative heights of the casting and the pouring basin.

On the other hand, $Q_i/Q_{av} = 1.5$ only when $b = 0.89H$. This represents an intriguing result, indicating that for most castings the top of the pouring basin is on average only about 10% higher than the height of the casting. Thus it seems the factor 1.5 is quite fortuitous, and results simply from the geometry we happen to select for most of the castings we make. If, in general, we were to raise (or lower) our pouring basins in relation to the tops of our castings, the factor would have to be revised.

However, all is not as bad as it seems. Figure A3 shows that the factor 1.5 does not change rapidly with changes in relative height of basin, varying over reasonable changes in basin height of b/H from 85% to 95% from roughly 1.45 to 1.60. These changes are of the same order as errors arising from other factors such as frictional losses etc. and so can be neglected for most practical purposes.

Appendix II

THE BERNOULLI EQUATION

Daniel Bernoulli represents the revered name in flow. He published his equation in 1738 in one of the first books on fluid flow. This magnificent result is the one used for all descriptions of flow in pipes and channels. Whole books are devoted to its application.

There are, of course, excellent examples of the power of Bernoulli's equation. Sutton (2002) has made good use of the equation to describe the pressures along a long runner, explaining the early partial filling of gates at different positions along the runner, and thus resulting in part-filled castings. He used the equation in its simplest form, derived from a statement of conservation of energy along a flow tube as illustrated schematically in Figure A4:

$$p_1/\rho g + v_1^2/2g + z_1 = p_2/\rho g + v_2^2/2g + z_2 = \text{constant}$$

Where all the component terms of this equation have units of length, conveniently metres. For this reason, each term can be regarded as a 'head'. Thus $p/\rho g$ = pressure head, $v^2/2g$ = kinetic or velocity head, and z = potential or elevation head.

In application to the running system used by Sutton (Figure A5) at location 1, the height above the centreline of the runner is 0.5 m, the kinetic head at this point is zero because the melt has zero downwards velocity, and the elevation head z is considered zero because the runner is horizontal. At point 2, the pressure head requires to be known because this is the pressure raising the melt level in the vertical ingate. The elevation is zero once again, and the velocity head is close to 0.25 m, easily deduced from the total fall height and allowing for a small loss factor of 0.70 (probably overestimated because I think

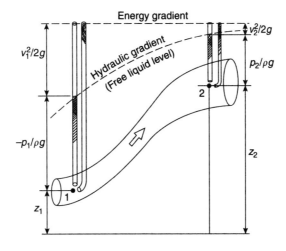

FIGURE A4

A pictorial representation of the factors in the Bernoulli equation.

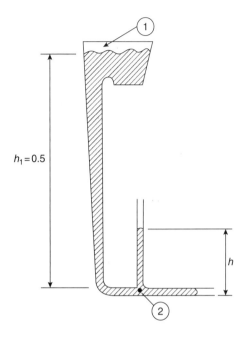

FIGURE A5

An example of the use of the Bernoulli equation by Sutton (2002) to calculate the rise of metal in a vertical gate.

this should be more like 0.80 or even 0.85) as a result of the turn at the base of the sprue. Thus the Bernoulli equation becomes

$$0.5 + 0 + 0 = p_2/\rho g + 0.25 + 0$$

Thus

$$p_2/\rho g = 0.25 \text{ metre}$$

It is not necessary to find p_2 alone; the whole term is the height distance. Thus, this answer would be the same for aluminium or iron.

Sutton found that because of this kinetic head, ingates were filling before the runner was fully filled. The first impression in his multi-impression mould was only about 200 mm above the runner so that metal entered the mould cavity under only about 50-mm net head. The result was a premature dribble into the cavity that quickly froze. The arrival of melt at the intended full flow rate a few seconds later was too late to remelt and thus assimilate the frozen droplets. An apparently mis-run casting was the result.

In general, however, the application of the Bernoulli equation to filling systems is not quite so straightforward as has sometimes been assumed. There are various reasons for this.

1. In general, Bernoulli's equation relates to steady state flow. However, of course, in filling systems, most of the interest necessarily lies in the priming of the flow channels. In this situation, the surface tension of the advancing meniscus can be important, as enshrined in the Weber number. If the priming is not carried out well, the casting is likely to suffer severely.

2. The surface tension of liquid metals is more than 10 times higher than that of water, and even higher still compared with most organic liquids. Thus pressures resulting from surface tension have been neglected and are neglectable for such common room temperature liquids on which most flow research has been conducted. The additional pressure generated because of the curvature of the meniscus at the flow front, and the curvature at the sides of a flow stream affect the behaviour of metals in many examples involved in the filling of moulds. For instance, at the critical velocity that is targeted in mould filling, the effects of surface tension and flow forces are equal. At velocities lower than this, surface tension dominates.
3. The presence of the oxide (or other thin, solid film) on the surface of an advancing liquid is a further complication, and is not easily allowed for. The flow adopts a stick-slip motion as the film breaks and re-forms. The advance of the unzipping wave is a classic instance that could not be predicted by a purely liquid model such as that described by Bernoulli.
4. The frictional losses during flow, which can be explicitly cited in Bernoulli's equation, are known to be important. However, in general, although they are assumed to be known, they have been little researched in the case of the flow of liquid metals. Furthermore, it is unfortunate that most of the research to date in this field has used such poor designs of filling systems that the existing figures are almost certainly misleading. The losses need to be confirmed by new, careful, accurate studies, supplemented by accurate computer simulation together with video X-ray radiography of real flows.
5. The presence of oxide films floating about in suspension is another uncertainty that can cause problems. The density of such defects can easily reach levels at which the effective viscosity of the mixture can be very much increased (although it is to be noted that viscosity does not appear explicitly in the Bernoulli equation). The suppression of convection in such contaminated liquids is common. Flow out of thick sections and into very thin sections can be prevented completely by blockage of the entrance into the thin section.

From this list, it is clear that the application of Bernoulli is more accurate for thicker section flows where surface effects and internal defects in the liquid are less dominant. As filling systems are progressively slimmed, and casting sections are thinned, Bernoulli's equation has to be used with greater caution.

As a result of the problems of the application of Bernoulli to the priming of the filling system, it has been relatively little used in this book because the concentration of effort has focussed on the control of the priming of the system. The subsequent flow of the system when completely filled, as nicely described by Bernoulli, is, with the greatest respect to the great man and his magnificent equation, much less important.

Appendix III

The recording of the lists of choices when drafting up a new methoding design for a casting must be carefully recorded. If there is found to be any problem with the first design, a second iteration can be worked out in the light of what appears to be needed (for instance, the casting may be required to fill a little faster). Table A1 can be used as a hard copy, allowing up to four iterations for a new filling system design. Alternatively, of course, the table can form the basis of a computer spread sheet, so that iterations can be performed rapidly and recorded digitally.

Table A1 Running System Record

			Design 1 Signature Date	Design 2 Signature Date	Design 3 Signature Date	Design 4 Signature Date
Casting Name						
Part Number						
Customer						
Alloy						
Casting weight	M_C	kg				
Rigging weight	M_R	kg				
Total pour weight	$M_C + M_R = M$	kg				
Fill time selected	t	s				
Average flow rate	M/t	kg/s				
Design flow rate	2M/t	kg/s				
Volume flow rate	$Q = 2M/\rho t$	m^3/s				
Height in basin	h	m				
Working basin depth	h to 2h	m				
Velocity into sprue	$V_1 = (2gh)^{1/2}$	m/s				
Area sprue top	$A_1 = Q/V_1$	m^2				
Velocity exit sprue	$V_2 = (2gH)^{1/2}$	m/s				
Area sprue exit	$A_2 = Q/V_2$	m^2				
Radius to runners	$R = A_2^{1/2}$	m				
Number of runners	n					
Area of each runner	A_2/n	m^2				
Select critical velocity	$V_C = 0.5-1.0$	m/s				
Total gate area	$A3 = A_2 \cdot V_2/V_C$	m^2				
Basin volume (1 s)	Q	m^3				

Appendix IV

RATE OF DELIVERY OF STEEL FROM A BOTTOM-POUR LADLE

The following is an example of how the nomogram is used.

A ladle contains 5000 kg of steel, from which we wish to pour a casting of total weight 1250 kg. Thus, in Figure A6, we follow the arrows from the start point to junction A. From here, a horizontal line connects to the next figure, where we select a pouring nozzle for the ladle of 60-mm diameter. At this junction B, we drop a vertical line down to intersect with the line denoting that our ladle is about 1.5 m internal diameter. From

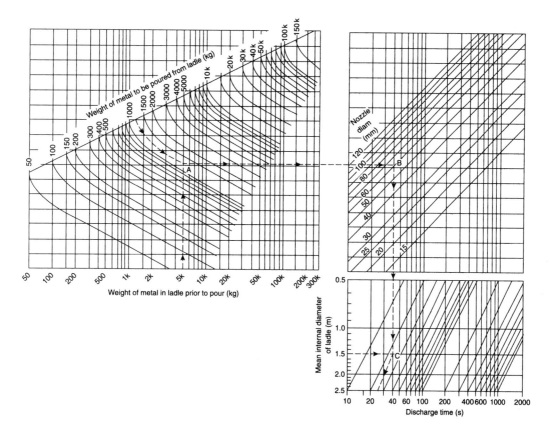

FIGURE A6

Rate of delivery of metal from a bottom-pour ladle.

957

this junction C, we continue with a parallel line to the family of sloping lines, to find that our casting will pour in approximately 23 s.

Interestingly, the reader can check that the next 1250 kg casting in line (now starting with a ladle of 5000 − 1250 = 3750 kg) will be found to pour in about 29 s, and the next in 34 s, and the next in 77 s, as the ladle progressively empties.

References

Abbas, M., Ray, N.K., Deb, P., Mallick, J.P., 1993. Indian Foundry J. 39 (9), 12–16.
Adams, C.M., 1958. Thermal considerations in freezing (Chapter). In: Liquid Metals and Solidification. ASM, Cleveland, Ohio, pp. 187–217.
Adams, A., 2001. Mod. Cast. 91 (3), 34–36.
Adams, C.M., Taylor, H.F., 1953. Am. Found. Soc. Trans. 61, 686–696.
Adefuye Segun, 1997. The Fluidity of Al-Si Alloys (Ph.D. thesis). University of Birmingham, UK (Supervisor N R Green).
Afseth, A., Nordlien, J.H., Scamans, G.M., Nisancioglu, K., 2000. In: ASST 2000 (2nd Internat. Symp. Al Surface Science Technology) UMIST Manchester, UK. Alcan + Dera, pp. 53–58.
Agema, K.S., Fray, D.J., 1988. Agema Ph.D. thesis, Dept Materials, Cambridge, UK.
Aguilar-Santillan, J., 2009. Metall. Mater. Trans. B 40B (3), 376–387.
Ahamed, A.K.M.A., Kato, H., 2008. Int. J. Cast Met. Res. 21 (1–4), 162–167.
Ahn, J.-H., Berghezan, 1991. Mater. Sci. Technol. 7 (7), 643–648.
Ainsworth, M.J., Griffiths, W.D., 2006. TAFS 114, 965–977.
Akita, K.K., UK Patent 1397821 filed 10 April 1972.
Alexopoulos, N.D., Tiryakioglu, M., 2009. Mater. Sci. Eng. A. http://dx.doi.org/10.1016/j.msea.2008.12.026.
Ali, S., Mutharasan, R., Apelian, D., 1985. Met. Trans. 16B, 725–742.
Aliravci, C.A., Gruzleski, J.E., Dimayuga, F.C., 1992. TAFS 100, 353–362.
Allen, D.I., Hunt, J.D., 1979. In: Solidification and Casting of Metals. Sheffield Conference, 1977, Metals. Soc., pp. 39–43.
Allen, A.G., Howard, J.C., Howard, J.F., Neenan, P.A. UK Patent Specification 1013851. Application date 31 January 1963.
Allen, N.P., 1932. J. Inst. Met. 49, 317–346.
Allen, N.P., 1933. J. Inst. Met. 51, 233–308.
Allen, N.P., 1934. J. Inst. Met. 52, 192–220.
Allen, A.G., 25 January 1968. Foundry Trade J. 159–161.
Almarez, G.M.D., Aburto, A.D., Gomez, E.C., January 2014. Metall. Mater. Trans. A 45A, 280–286.
Alonso, G., Stefanescu, D.M., Suarez, R., Loizaga, A., Zarrabeitia, G., 2014. Int. J. Cast Met. Res. 27 (2), 87–100.
Alsem, W.H.M., van Wiggen, P.C., Vader, M., 1992. Light metals. In: Cutshall, E.R. (Ed.), The Minerals. Metals and Materials Soc., USA, pp. 821–829.
Altstetter, J.D., Nowicki, R.M., 1982. TAFS 90, 959–970.
American Foundrymen's Society, 1987. Green Sand Additives. AFS, Detroit, USA.
Anderson, J.V., Karsay, S.I., 1985. Br. Found. 492–498.
Anderson, S.H., Foss, J.W., Nagan, R.M., Jhala, B.S., 1989. TAFS 97, 709–722.
Anderson, G.R., 1985. In: Patt-tech 85 Conference. Oxford, UK.
Anderson, G.R., 1987. TAFS 95, 203–210.
Andresen, P.L., Chou, P.H., Morra, M.M., Nelson, J.L., Rebak, R.B., 2009. Metall. Mater. Trans. A 40A (12), 2824–2836.
Andrew, J.H., Percival, R.T., Bottomley, G.T.C., 1936. Iron Steel Spec. Rep. 15, 43–64.
Angelov, G., June 1969. Russ. Cast. Prod. 282–284.
Angus, H.T., 1976. Cast Iron. Butterworths, London.
Anson, J.P., Gruzleski, J.E., 1999. Trans. Am. Found. Soc. 107, 135–142.
Anson, J.P., Stucky, M., Gruzleski, J.E., 2000. Trans. Am. Found. Soc. 108, 419–426.
Antes, H.W., Norton, J.T., Edelman, R.E., 1958. TAFS 66, 135–142.
Appa Rao, G., Srinivas, M., Sarma, D.S., 2004. Mater. Sci. Technol. 20, 1161–1170.
Archibald, J.J., Smith, R.L., 1988. "Casting" from Metals Handbook, vol. 15. ASM, 214–221.
Armbruster, D.R., Dodd, S.F., 1993. TAFS 101, 853–856.
Armstrong, G.L., Martin, W., 1974. TAFS 82, 253–255.
Arnold, F.L., Prestley, J.S., 1961. Trans. Am. Found. Soc. 69, 129–136.
Arnold, F.L., Jorstad, J.L., Stein, G.E., 1963. Curr. Eng. Pract. 6, 10–15.
Aryafar, M., Raiszadeh, R., Shalbafzadeh, A., 2010b. J. Mater. Sci. 45, 3041–3051.
Arzt, E., Ashby, M.F., Easterling, K.E., February 1983. Metall. Trans. A 14, 211–221.

Asbjornsonn, 2001. Ph.D. thesis, Department of Materials, University of Nottingham.
Ashton, M.C., Buhr, R.C., 1974. Phys. Met. Div. Internal Report PM-1-74-22. Canada Dept Energy Mines and Resources.
Ashton, M.C., 1990. In: SCRATA 34th Conference. Sutton Coldfield, Paper 2. Steel Castings Research and Trade Assoc, Sheffield, UK.
Ashton, M.C., January 1991. Met. Mater. 12–17.
Askeland, D., Holt, M.L., 1975. TAFS 83, 99–106.
Atwood, R.C., Lee, P.D., 2000. In: Sahm, P., Hansen, P. (Eds.), Modeling of Casting Welding and Advanced Solidification Processing Conf. Aachen.
Avey, M.A., Jensen, K.H., Weiss, D.J., 1989. TAFS 97, 207–212.
Aylen, P., Foulard, J., Galey, J., 1965. TAFS 73, 311–316.
Bachelot, F. 1997. M.Phil. thesis, University of Birmingham, UK.
Bachmann, P.K., Messier, R., 1984. Chem. Eng. News 67 (20), 24–39.
Backerud, L., Chai, G., Tamminen, J., 1990. Solidification Characteristics of Aluminum Alloys, Volume 2, Foundry Alloys. AFS/Skanaluminium (printed in USA).
Badia, F.A., Rohatgi, P., 1969. TAFS 77, 402–406.
Badia, F.A., 1971. TAFS 79, 347–350.
Bahreinian, F., Boutorabi, S.M.A., Campbell, J., 2005. In: Tiryakioglu, M., Crepeau, P.N. (Eds.), Shape Casting: The John Campbell Symposium. TMS, pp. 463–472.
Bailey, A, Davenport, A.J., April 2002. Final Year Project. University of Birmingham, Dept. Metall, UK.
Bak, C., Degois, M., Schissler, J.M., 1980. TAFS 88, 301–312.
Baker, W.A., 1945. J. Inst. Met. 71, 165–204.
Baker, W.F., 1986. TAFS 94, 215–218.
Bakhtiarani, F.N., Raiszadeh, R., 2011. Met. Mat. Trans. B 42 (2), 331–340.
Balewsky, A.T., Dimov, T., 1965. Br. Found. 78 (7), 280–283.
Baliktay, S., Nickel, E.G., 1988. In: Seventh World Conf. Invest. Casting, Munich paper 10.
Ball, R., May 1998. Foundryman 157–160, 175.
Ball, R., Hardcastle, P., January 1991. Foundry Trade J. 11 (25), 52–53.
Barbe, L., Bultink, I., Duprez, L., De Cooman, B.C., 2002. Mater. Sci. Technol. 18, 664–672.
Barkhudarov, M.R., Hirt, C.W., January–February 1999. In: Die Casting Engineer, pp. 44–47.
Barlow, G., 1970. Ph.D. thesis, University of Leeds.
Barrett, L.G., 1967. TAFS 75, 326–329.
Barton, R., 28 February 1985. Foundry Trade J. 117–126.
Basdogan, M.F., Bennett, G.H.J., Kondic, V., 1983. Solidification technology in the foundry and cast house. In: Warwick Univ. Conf. 1980. Metals Soc. Publication, pp. 240–247.
Bastien, P., Armbruster, J.C., Azou, P., 1962. In: 29 Internat. Found. Congress Detroit, pp. 400–409.
Bates, C.E., Monroe, R.W., 1981. TAFS 89, 671–686.
Bates, C.E., Scott, W.D., 1977. TAFS 85, 209–226.
Bates, C., Wallace, J.F., 1966. TAFS 74, 174–185.
Batty, O., 1935. TAFS 43, 75–106.
Bean, X., Marsh, L.E., 1969. Metal. Prog. 95, 131–134.
Beck, A., Schmidt, W., Schreiber, O., 1928. US Patent 1788185.
Beckermann, C., 2002. Int. Mater. Rev. 47 (5), 243–261.
Beech, J., April 1974. The Metallurgist and Materials Technolo-gist, pp. 129–232.
Beeley, P.R., Smart, R.F., 2009. Investment Casting. Maney Materials Science.
Beeley, P.R., 1972. Foundry Technology. Butterworths, London.
Beeley, P.R., 2001. Foundry Technology, second ed. Butterworth-Heinemann, London.
Benaily, N., 1998. Inoculation of Flake Graphite Iron (M.Phil. thesis), Univ. Birmingham, UK.
Bennett, S., Moody, T., Vrieze, A., Jackson, M., Askeland, D.R., Ransay, C.W., 2000. TAFS 108, 795–803.
Benson, L.E., June 1938a. Foundry Trade J. 527–528.
Benson, L.E., June 1938b. Foundry Trade J. 543–544.
Benson, L.E., 1946. J. Inst. Met. 72, 501–510.

Beretta, S., Murakami, Y., 2001. Metall. Mater. Trans. B 32B, 517–523.
Berger, R., 1932. Fonderie Belge 17.
Berry, J.T., Taylor, R.P., 1999. TAFS 107, 203–206.
Berry, J.T., Watmough, T., 1961. TAFS 69, 11–22.
Berry, J.T., Kondic, V., Martin, G., 1959. TAFS 67, 449–476.
Berry, J.T., Watmough, T., 1961. TAFS 69, 11–22.
Berry, J.T., Luck, R., Taylor, R.P., 2005. In: Tiryakioglu, M., Crepeau, P.N. (Eds.), Shape Casting: The John Campbell Symposium. TMS, pp. 113–122.
Berthelot, M., 1850. Ann. Chim. 30, 232.
Bertolino, M.F., Wallace, J.F., 1968. TAFS 76, 589–628.
Betts, B.P., Kondic, V., 1961. Br. Found. 54, 1–4.
Bex, T., November 1991. Mod. Cast. 56.
Bhaumik, S., Moles, V., Gottstein, G., Heering, C., Hirt, G., 2010. Adv. Eng. Mater. 12 (3), 127–130.
Biel, J., Smalinskas, K., Petro, A., Flinn, R.A., 1980. TAFS 88, 683–694.
Bindernagel, I., Kolorz, A., Orths, K., 1975. TAFS 83, 557–560.
Bindernagel, I., Kolorz, A., Orths, K., 1976. AFS Int. J. Cast Met. 1 (4), 42–45.
Birch, J., March 2000a. Diecast. World 18–19.
Birch, J., September 2000b. Diecast. World 174 (3570), 28.
Birch, J., March 2000c. Diecast. World 174, 35.
Bird, P., Savage, W., 1996. TAFS 104, 321–324.
Bird, P.G., 1 March 1989. Foseco Technical Service Report MMP1.89. Examination of the Factors Controlling the Flow Rate of Aluminium through DYPUR Units.
Birol, Y., 2012. Mater. Sci. Technol. 28 (8), 924–927.
Bishop, H.F., Ackerlind, C.G., Pellini, W.S., 1952. TAFS 60, 818–833.
Bishop, H.F., Myskowski, E.T., Pellini, W.S., 1955. TAFS 63, 271–281.
Bisuola, V.B., Martorano, M.A., 2008. Metall. Mater. Trans. A 39A (12), 2885–2895.
Biswas, P.K., Rohatgi, P.K., Dwarakadasa, E.S., 1985. Br. Found. 78, 511–516.
Biswas, P.K., Pillai, R.M., Rohatgi, P.K., Dwarakadasa, E.S., 1994. Cast. Met. 7 (2), 65–83.
Boenisch, D., Patterson, W., 1966. TAFS 74, 470–484.
Boenisch, D., 1967. TAFS 75, 33–37.
Boenisch, D., March 1988. In: Conference "Ensuring Quality Castings". BCIRA, Univ. of Warwick, pp. 29–31, paper 18.
Bonsak, W., 1962. TAFS 70, 374–382.
Boom, R., Dankert, O., Veen van, A., Kamperman, A.A., October 2000. Metall. Mater. Trans. B 31B (5), 913–919.
Bossing, E., 1982. TAFS 90, 33–38.
Bounds, S.M., Moran, G.J., Pericleous, K.A., Cross, M., 1998. In: Thomas, B.G., Beckerman, C. (Eds.), Modelling of Casting, Welding and Solidification Processing Conf. VIII, San Diego. TMS, pp. 857–864.
Boutorabi, S.M.A., Din, T., Campbell, J., 1992. Univ. Bham, unpublished work.
Bower, T.F., Brody, H.D., Flemings, M.C., 1966. Trans. AIME 236, 624.
Bracale, G., 1962. TAFS 70, 228–252.
Bradley, F.J., Hooper, J.A., Kannan, S., Balakrishna, J.V., Heinemann, S., 1992. TAFS 100, 917–923.
Bramfitt, B.L., 1970. Met. Trans. 1, 1987–1995.
Brandes, E.A., Brook, G.B. (Eds.), 1992. Smithells Metals Reference Book, seventh ed. Butterworths.
Brauer, H.E., Pierce, W.M., 1923. Trans. Am. Inst. Min. Metall. 68, 796–832.
Bridge, M.R., Stephenson, M.P., Beech, J., 1982. Met. Technol. 9, 429–433.
Bridges, D., 1999. Wheels and Axles. I. Mech. E. Seminar, London.
Briggs, C.W., Gezelius, R.A., 1934. TAFA 42, 449–476.
Briggs, L.J., 1950. J. Appl. Phys. 21, 721–722.
Brimacombe, J.K., Sorimachi, K., 1977. Met. Trans. 8B, 489–505.
Britney, D.J., Neailey, K., 2003. J. Mater. Process. Technol. 138, 306–310.
Bromfield, G., July 1991. Foundryman 261–265.
Brondyke, K.J., Hess, P.D., 1964. Trans. Met. Soc. AIME 230 (7), 1542–1546.

Brookes, B.E., Berckermann, C., Richards, V.L., 2007. Int. J Cast Met. Res. 20 (4), 177–190.
Brown, N., Rastall, D., 1986. European Patent Application No 87111549.9 filed 10 August 1987. p. 9.
Brown, J.R., 1970. Br. Found. 63, 273–279.
Brown, J.R., 1992. Met. Mater. 8, 550–555.
Bryant, M.D., Moore, A., 1971. Br. Found. 64, 215–229, 306–307.
Buhrig-Polackzek, A., Santos de, A., 2008. In: Metals Handbook, Volume 15, Casting. ASM, Ohio, USA, pp. 317–329.
Burchell, V.H., 1969. Br. Found. 62, 138–146.
Burton, B., Greenwood, G.W., 1970. Met. Sci. J. 4, 215–218.
Busby, A.D., 1996. TAFS 104, 957–968.
Butakov, D.K., Mel'nikov, L.M., Rudakov, I.P., Maslova Yu, N., 1968. Lit. Proizv. 4, 33–35.
Butler, T., Lund, J.N., February 2003. Mod. Cast. 40–42.
Butler, C.J., 1980. UK Patent GB 2020714.
Byczynski, G.E., Cusinato, D.A., 4–5 April 2001. In: First International Conf. Filling and Feeding of Castings. University of Birmingham, UK and Int. J. Cast Met. Res. 2002, 14 (5), 315–324.
Caceres, C.H., 2000. TAFS 108, 709–712.
Caceres, C.H., 2004. In: Nie, J.F., et al. (Eds.), Proc. 9th Internat. Conf. Al Alloys. Inst Materials Engineering Australia Ltd, pp. 1216–1221.
Caine, J.B., Toepke, R.E., 1966. TAFS 74, 19–22.
Caine, J.B., Toepke, R.F., 1967. TAFS 75, 10–16.
Campbell, J., Bannister, J.W., 1975. Met. Technol. 2 (9), 409–415.
Campbell, J., Caton, P.D., 1977. In: Institute of Metals Conference on Solidification, Sheffield, UK, pp. 208–217.
Campbell, J., Clyne, T.W., 1991. Cast. Met. 3 (4), 224–226.
Campbell, J., Isawa, T., 1994. UK Patent GB 2284168 B (Filed 4 February 1994).
Campbell, J., Naro, R.L., 2010. Lustrous carbon in grey iron. TAFS 114, 6 paper 10–36.
Campbell, J., Olliff, I.D., June 1971. AES Cast Met. Res. J. 55–61.
Campbell, J., Tiryakioglu, M., 2010. Effect of Sr on Porosity. Met. Mat. Trans. 26 (3), 262–268.
Campbell, H.L., 1950. Foundry 78, 86, 87, 210, 212, 213.
Campbell, J., 1967. Trans. Met. Soc. AIME 239, 138–142.
Campbell, J., 1968a. The Solidification of Metals. ISI Publication 110, pp. 19–26.
Campbell, J., 1968b. Trans. Met. Soc. AIME 242, 264–268.
Campbell, J., 1968c. Trans. Met. Soc. AIME 242, 268–271.
Campbell, J., 1968d. Trans. Met. Soc. AIME 242, 1464–1465.
Campbell, J., 1969a. Feeding Mechanisms in Castings. Cast. Met. Res. J. 5 (1), 1–8.
Campbell, J., 1969b. The non-equilibrium freezing range and its relation to hydrostatic tension and pore formation in solidifying binary alloys. Trans. AIME 245, 2325–2334.
Campbell, J., 1971. Metallography 4, 269–278.
Campbell, J., 1980. Solidification technology in the foundry and casthouse. In: Warwick Conference, Metals Soc. Publication 1981, vol. 273, pp. 61–64.
Campbell, J., 1981. Int. Met. Rev. 26 (2), 71–108.
Campbell, J., 1988. Mater. Sci. Technol. 4, 194–204.
Campbell, J., 1991a. Cast. Met. 4 (2), 101–102.
Campbell, J., 1991b. Castings. Published by Butterworth Heinemann (now Elsevier).
Campbell, J., September 1991c. Metals and Materials, p. 575.
Campbell, J., 1994. Cast. Met. 7 (4), 227–237.
Campbell, J., 2000a. Ingenia 1 (4), 35–39.
Campbell, J., 2000b. The concept of net shape for castings. Mater. Des. 21, 373–380.
Campbell, J., 2003. Castings. Elsevier, Oxford, UK pp (a) 178–181; (b) 161–162; (c) 158–160.
Campbell, J., 2006a. Mater. Sci Technol. 22 (2), 127–145 and (8), 999–1008.
Campbell, J., 2006b. Modeling of Entrainment Defects during Casting. In: San Antonio, USA. TMS Annual Congress.
Campbell, J., 2007. AFS Int. J. Metalcast. 1 (1), 7–20.
Campbell, J., 2008. Mater. Sci. Technol. 24 (7), 875–881.

Campbell, J., 2009a. Incipient grain boundary melting. Mater. Sci. Technol. 25 (1), 125–126.
Campbell, J., 2009b. A hypothesis for graphite formation in cast irons. Metall. Mater. Trans. B 40B (6), 786–801.
Campbell, J., 2009c. Stress corrosion cracking of Mg alloys. Metall. Mater. Trans. A 40A (7), 1510–1511.
Campbell, J., May 2009d. Discussion of "Effect of Sr and P on Eutectic Al-Si Nucleation and Formation of beta-Al$_5$FeSi in Hypoeutectic Al-Si Foundry Alloys". Metall. Mater. Trans. A 40A, 1009–1010.
Campbell, J., 2010a. Stress corrosion cracking of stainless.... Metall. Mater. Trans. A 41A (5), 1101.
Campbell, J., 2011a. In: Tiryakioglu, M., Crepeau, P., Campbell, J. (Eds.), The Origin of Griffiths Cracks. Shape Casting Symposium TMS Annual Congress, San Diego, CA.
Campbell, J., 2011b. The origin of Griffith cracks. Metall. Mater. Trans. B 42B, 1091–1097.
Campbell, J., September 2014. Metall. Mater. Trans. A 45A, 4193.
Campbell, J., 2015. Quality Castings: A Personal History of the Cosworth Casting Process. AFS.
Cao, X., Campbell, J., 2000. Am. Found. Soc. Trans. 108, 391–400.
Cao, X., Campbell, J., July 2003. Metall. Mater. Trans. A 34A, 1409–1420.
Cao, W., Fathallah, R., Castex, L., 1995. Mater. Sci. Technol. 11 (9), 967–973.
Cao, P., Qian, M., StJohn, D.H., Frost, T.M., 2004. Mater. Sci. Technol. 20 (5), 585–592.
Cao, X., 2001. Ph.D. thesis. Department of Metallurgy, University of Birmingham, UK.
Cao, X., 2009. Personal communication.
Capello, G.P., Carosso, M., May 1989. AGARD Report, No. 762.
Cappy, M., Draper, A., Scholl, G.W., 1974. TAFS 82, 355–360.
Carlberg, T., Fredriksson, H., 1979. In: Solidification and Casting of Metals Conf., Univ. Sheffield, UK, 1977. The Metals Society, pp. 115–124.
Carlson, C., Beckermann, C., 2009. Metall. Mater. Trans. A 40A, 163–175 and 3054–3055.
Carlson, K.D., Beckermann, C., 2008. Proc 62nd SFSA Conf. Paper No 5.6, Steel Founders' Soc of America.
Carlson, B.E., Pehlke, 1989. TAFS 97, 903–914.
Carlson, G.A., 1975. J. Appl. Phys. 46 (9), 4069–4070.
Carne, C.A., Beech, J., 1997. In: Beech, J., Jones, H. (Eds.), Solidification Processing. Sheffield University, UK, pp. 33–36.
Carte, A.E., 1960. Proc. Phys. Soc. 77, 757–769.
Carter, S.F., Evans, W.J., Harkness, 1. C., Wallace, J.F., 1979. TAFS 87, 245–268.
Celik, M.C., Bennett, G.H.J., April 1979. Met. Technol. 138–144.
Chadwick, H., Campbell, J., 1997. University of Birmingham, unpublished research.
Chadwick, G.A., Yue, T.M., January 1989. Met. Mater. 6–12.
Chadwick, G.A., Yue, T.M., 1991. Hi-Tech Metals R&D Ltd, Southampton, personal communication. (See Yue, T.M.).
Chadwick, P., 1963. Int. J. Mech. Sci. 5, 165–182.
Chadwick, H., 1991. Cast. Met. 4 (1), 43–49.
Chalmers, 1953. Quoted by Flemings, M.C., 1974
Chakrabarti, I., Campbell, J., 2000. University of Birmingham, unpublished research.
Chamberlain, B., Zabek, V.J., 1973. TAFS 81, 322–327.
Chambers, L.W., 1980 (April 24th). Foundry Trade J. 802.
Chandley, G.D., 1976. TAFS 84, 37–42.
Chandley, G.D., 1983. TAFS 91, 199–204.
Chandley, G.D., 1989. Cast. Met. 2 (1), 2–10.
Chandley, D., 2000. Metall. Sci. Technol. 18 (1), 8–11.
Chandravanshi, V., Sarkar, R., Kamat, S.V., Nandy, T.K., January 2013. Metall. Mater. Trans. A 44A, 201–211.
Chang, J.-K.(B.), Taleff, E.M., Krajewski, P.E., 2009. Metall. Mater. Trans. A 40A (13), 3128–3137.
Changyun, L., Shiping, W., Jingjie, G., Hengzhi, F., 2006. Int. J. Cast Met. Res. 19 (4), 237–240.
Charbonnier, J., Perrier, J.J., January 1983. Giesserei 70 (2), 50–55.
Charles, J.A., Uchiyama, I., July 1969. J. Iron Steel Inst. 207, 979–983.
Chaudhury, S.K., Apelian, D., 2006. Int. J. Cast Met. Res. 19 (6), 361–369.
Chechulin, V.A., 1965. In: Gulyaev, B.B. (Ed.), Gases in Cast Metals. Consultants Bureau Translation, pp. 214–218.
Chegini, S., Raiszadeh, R., 2014. Int. J. Cast Met. Res. 27 (6), 349–356.
Chen, X.G., Engler, S., 1994. TAFS 102, 673–682.

Chen, X.-G., Fortier, M., 2010. J. Mater. Process. Technol. 210, 1780–1786.
Chen, Q.Z., Knowles, D.M., 2003. Mater. Sci. Technol. 19 (4), 447–455.
Chen, C.O., Ramberg, F., Evensen, J.D., 1984. Metal. Sci. 18, 1–5.
Chen, G.J., Liu, S.H., Ren, B.L.T., 1989. TAFS 97, 335–338.
Chen, X.G., Klinkenberg, F.J., Ellerbrok, R., Engler, S., 1994. TAFS 102, 191–197.
Chen, T.J., Ma, Y., Li, Y.D., Lu, G.X., Hao, Y., 2010. Mater. Sci. Technol. 26 (10), 1197–1206.
Chen, X.-C., Shi, C.-B., Guo, H.-J., Wang, F., Ren, H., Feng, D., December 2012. Metall. Mater. Trans. B 43B, 1593–1607.
Chen, Z., Mo, Y., Nie, Z., August 2013. Metall. Mater. Trans. 44A, 3910–3920.
Cheng, X., Yuan, C., Blackburn, S., Withey, P.A., 2014. Mater. Sci. Technol. 30 (14), 1758–1764.
Chernov, D.K., December 1878. Reports of the Imperial Russian Metallurgical Society (see Russkoe Metalurgicheskol Obshchestro 1, 1915).
Chiesa, F., 1990. TAFS 98, 193–200.
Chisamera, M., Riposan, I., Barstow, M., 1996. TAFS 104, 581–588.
Cho, J.-I., Loper, C.R., 2000. Trans. Am. Found. Soc. 108, 359–367.
Cho, Y.H., Lee, H.-C., Oh, K.H., Dahle, A.K., 2008. Metall. Mater. Trans. 39A (10), 2435–2448.
Choudhury, A., Blum, M., Scholz, H., Jarczyk, G., Busse, P., February 1999. In: Proc. 1999 Internat. Symp. Liquid Metal Processing and Casting; Santa Fe, NM, pp. 244–255.
Chu, M.G., 2002. Light Met. 899–907.
Chung, Y., Cramb, A.W., 2000. Metall. Mater. Trans. B 31B, 957–971.
Church, N., Wieser, P., Wallace, J.F., 1966. TAFS 74, 113–128. Also in Br. Found. 1966 59, 349–363.
Churches, D.M., Rundman, K.B., 1995. TAFS 103, 587–594.
Chvorinov, N., 1940. Giesserei 27, 177–186, 201–208, 222–225.
Chvorinov, N., 1940. Giesserei 10, 177–186, 201, 222 and 27 (31 May) 201–208.
Cibula, A., 1955. Proc. IBF 45, A73–A90. Also in Foundry Trade J. 1955 98, 713–726.
Clark, J.C., April 1989. BCIRA Report Number 1769, 181–193.
Clausen, C., August 1992. Mod. Cast. 39.
Claxton, K.T., 1967. The Influence of radiation on the inception of boiling in liquid sodium. In: UK Atomic Energy Authority Research Group Report AERE-R5308. Also in Proc. Internat. Conf. on the Safety of Fast Reactors (CAE). Aix-en-Provence, September 1967, p. II-B-8-1.
Claxton, K.T., 1969. Private communication. UKAEA, Harwell, UK.
Clegg, A.L., Das, A.A., 1987. Br. Found. 80, 137–144.
Clyne, T.W., Davies, G.J., 1975. Br. Found. 68, 238–244.
Clyne, I.W., Davies, G.J., 1979. In: Solidification and Casting of Metals. Metals Soc. Conference, Sheffield, 1977, pp. 275–278.
Clyne, T.W., Davies, G.J., 1981. Br. Found. 74, 65–73.
Clyne, T.W., Kurz, W., 1981. Met. Trans. 12A, 965–971.
Clyne, T.W., Wolf, M., Kurz, W., 1982. Met. Trans. 13B, 259–266.
Clyne, T.W., 1977. Ph.D. thesis, University of Cambridge.
Coble, R.L., Flemings, M.C., 2 February 1971. Met. Trans. 409–415.
Cochran, C.N., Belitskus, D.L., Kinosz, D.L., 1977. Metall. Trans. 8B, 323–332.
Cole, G.S., Cisse, J., Kerr, H.W., Bolling, G.F., 1972. TAFS 80, 211–218.
Cole, G.S., 1972. TAFS 80, 335–348.
Colwell, D.L., 1963. TAFS 71, 172–176.
Cook, R., Kearns, M.A., Cooper, P.S., 1997. In: Huglen, R. (Ed.), Light Metals. TMS, pp. 809–814.
Cotton, J.D., Clark, L.P., Phelps, H.R., June 2006. J. Met. 13–16.
Cottrell, A.H., 1964. The Mechanical Properties of Matter. Wiley, p. 82.
Couture, A., Edwards, J.O., 1966. TAFS 74, 709–721, 792–793.
Couture, A., Edwards, J.O., 1967. AFS Cast Met. Res. J. 3 (2), 57–69.
Couture, A., Edwards, J.O., 1973. TAFS 81, 453–461.
Cowen, C.J., Boehlert, C.J., 2007. Metall. Mater. Trans. A 38A, 26–34.
Cox, M., Harding, R.A., 2007. Mater. Sci. Technol. 23 (2), 214–224.

Cox, M., Wickins, M., Kuang, J.P., Harding, R.A., Campbell, J., 2000. Mater. Sci. Technol. 16, 1445–1452.
Cox, M., Harding, R.A., Green, N.R., Scholl, G.W., 2007. Mater. Sci. Technol. 23 (9), 1075–1084.
Creese, R.C., Sarfaraz, A., 1987. TAFS 95, 689–692.
Creese, R.C., Sarfaraz, A., 1988. TAFS 96, 705–714.
Creese, R.C., Xia, Y., 1991. TAFS 99, 717–727.
Crossley, F.A., Mondolfo, L.F., 1951. J. Met. 3, 1143–1154.
Cunliffe, E.L., 1996. The minimal gating of aluminium alloy castings (Ph.D. thesis), Metals and Materials Department, University of Birmingham, UK.
Cunningham, M., 1988. Stahl Speciality Co, Kingsville, MO, USA. Private communication.
Cupini, N.L., Prates de Campos Filho, M., 1977. In: Sheffield Conf. "Solidification and Casting of Metals" Metals Soc. 1979, pp. 193–197.
Cupini, N.L., de Galiza, J.A., Robert, M.H., Pontes, P.S., 1980. Solidification technology in the foundry and cast house. In: Metals Soc. Conf, pp. 65–69.
Czerwinski, F., November 2008. J. Met. 82–86.
D'Errico, F., Rivolta, B., Gerosa, R., Perricone, G., November 2008. J. Met. 70–75.
Dai, X., Yang, X., Campbell, J., Wood, J., 2003. Mater. Sci. Eng. A354 (1–2), 315–325.
Dantzig, J.A., Rappaz, M., 2009. Solidification. EPFL Press, Lausanne, Switzerland, pp. 486–490.
Das, A.A., Chatterjee, S., 1981. Metall. Mater. Technol. 13 (3), 137–142.
Das, C.R., Albert, S.K., Bhaduri, A.K., Murty, B.S., May 2013. Metall. Mater. Trans. A 44A, 2171–2186.
Dasch, J.M., Ang, C.C., Wong, C.A., Waldo, R.A., 2009. J. Mater. Process. Technol. 209, 4638–4644.
Dasgupta, S., Parmenter, L., Apelian, D., 1998. In: AFS 5th Internat. Conf. Molten Aluminum Processing, pp. 285–300.
Datta, N., Sandford, P., 1995. In: 3rd AFS Internat. Permanent Mold Casting of Aluminum Conference Paper 3, p. 19.
Davidson, R.M.R., 1990. Mater. Perform. 29 (1), 57–62.
Davies, G.J., Shin, Y.K., 1980. Paper 78. In: Solidification Technology in the Foundry and Casthouse. Metals Soc. Conference, Warwick, pp. 517–523 (Published Metal Soc. 1983).
Davies, I.G., Dennis, J.M., Hellawell, A., 1970. Metall. Trans. 1, 275–280.
Davies, V. de L., 1963. J. Inst. Met. 92, 127.
Davies, V. de L., 1964–1965. J. Inst. Met. 93, 10.
Davies, V. de L., 1970. Br. Found. 63, 93–101.
Davis, K.G., Magny, J.-G., 1977. TAFS 85, 227–236.
Davis, K.G., Internat, A.F.S., March 1977. Cast Met. J. 23–27.
Dawson, J.V., Kilshaw, J.A., Morgan, A.D., 1965. TAFS 73, 224–240.
Dawson, J.V., 1962. BCIRA J. 10 (4), 433–437.
Daybell, E., 1953. Proc. Inst. Br. Found. 46, B46–B54.
De Sy, A., 1967. TAFS 75, 161–172.
Delamore, G.W., Smith, R.W., 1971. Met. Trans. 2, 1733–1743.
Delamore, G.W., Smith, R.W., Mackay, W.B.F., 1971. TAFS 79, 560–564.
Denisov, V.A., Manakin, A.M., 1965. Russ. Cast. Prod. 217–219.
Dennis, K., Drew, R.A.L., Gruzleski, J.E., 2000. Alum. Trans. 3 (1), 31–39.
Devaux, H., 1987. In: Moreau, R.J. (Ed.), Measurement and Control in Liquid Metals Processing. Nijhoff, The Netherlands, pp. 107–115.
DeYoung, D.H., Dunlay, M.J., 2002. US Patent 6334978.
Dickhaus, C.H., Ohm, L., Engler, S., 1993. TAFS 101, 677–684.
Dietert, H.W., Fairfield, H.H., Brewster, F.S., 1948. TAFS 56, 528–535.
Dietert, H.W., Doelman, R.L., Bennett, R.W., 1970. TAFS 78, 145–156.
Dimayuga, F.C., Handiak, N., Gruzleski, I.E., 1988. TAFS 96, 83–88.
Dimmick, T., 2001. Mod. Cast. 91 (3), 31–33.
Din, T., Campbell, J., 1994. University of Birmingham, UK, unpublished work.
Din, T., Rashid, A.K.M.B., Campbell, J., 1996. Mater. Sci. Technol. 12 (3), 269–273.
Din, T., Kendrick, R., Campbell, J., 2003. TAFS 111, 91–100 (paper 03-017).
Dinayuga, F.C., Handiak, N., Gruzleski, J.E., 1988. TAFS 96, 83–99.

Dion, J.-L., Couture, A., Edwards, J.O., 1978. TAFS 86, 309–314 and AFS Int. Cast Met. J. 1979 4 (2), 7–13.
Dion, J.-L., Fasoyinu, F.A., Cousineau, D., Bibby, C., Sahoo, M., 1995. TAFS 103, 367–377.
Dionne, P., Dickson, J.I., Bailon, J.P., 1984. TAFS 92, 693–701.
Disa Industries, July 2002. Foundry Trade J. 27.
Dispinar, D., Akhtar, S., Nordmark, A., Di Sabatino, M., Arnberg, L., 2010. Mater. Sci. Eng. A 527 (16–17), 3719–3725.
Dispinar, D., Campbell, J., 2004. Int. J. Cast Met. Res. 17 (5), 280–286 and 278–294.
Dispinar, D., Campbell, J., April 2005. University of Birmingham, UK, unpublished work.
Dispinar, D., Campbell, J., 2006. Int. J. Cast Met. Res. 19 (1), 5–17.
Dispinar, D., Campbell, J., 2007. J. Mater. Sci. 42, 10296–10298.
Dispinar, D., Campbell, J., 2011. Mater. Sci. Eng. A 528 (10), 3860–3865.
DiSylvestro, G., Faist, C.A., 1977. TAFS 85, 627–642.
Divandari, M., Campbell, J., 11–13 October 1999. In: AFS 1st Internat. Conf. on Gating Filling and Feeding of Aluminum Castings, pp. 49–63.
Divandari, M., Campbell, J., 2000. Alum. Trans. 2 (2), 233–238.
Divandari, M., 1998. University of Birmingham, UK, unpublished work.
Dixon, B., 1988. In: Solidification Processing Conference 1987. Inst. Metals, pp. 381–383.
Dodd, R.A., Pollard, W.A., Meier, J.W., 1957. TAFS 65, 100–117.
Dodd, R.A., 1950. Ph.D. thesis, Dept. Industrial Metallurgy, Univ. Birmingham, UK.
Dodd, R.A., 1955a. Hot Tearing of Casting: A Review of the Literature. Dept. Mines Ontario, Canada. Research Report PM184. Also in Foundry Trade J. 1956, 101, 321–331.
Dodd, R.A., 1955b. Hot Tearing of Binary Mg-Al and Mg-Zn Alloys. Dept Mines Ontario, Canada. Research Report PM191.
Dolan, G.P., Flynn, R.J., Tanner, D.A., Robinson, J.S., 2005. Mater. Sci. Technol. 21 (6), 687–692.
Donahue, R., Anderson, K., 2008. ASM Handbook "Casting", vol. 15, pp. 640–645.
Donahue, J., June 2006. INCAST 11.
Dong, S., Niyama, E., Anzai, K., 1995. ISIJ Int. 35, 730–736.
Donoho, C.K., 1944. Trans. Am. Found. Assoc. 52, 313–332.
Doremus, G.B., Loper, C.R., 1970. TAFS 78, 338–342.
Dorward, R.C., 2001. Oxid. Met. 55 (1/2), 69–74.
Double, D.D., Hellawell, A., 1974. Acta Metall. 22, 481–487.
Doutre, D.A., Hay, G., Wales, P., European Patent 20000951137 filed 26 July 2000.
Doutre, D., 1998. In: Wilkinson, D.S., Poole, W.J., Alpes, A. (Eds.), Advances in Industrial Materials. The Metallurgical Soc. of CIM.
Draper, A.L., Gaindhar, J.L., 1975. TAFS 83, 593–616.
Draper, A.B., 1976. TAFS 84, 749–764.
Drezet, J.M., Commet, B., Fjaer, H.G., Magnin, B., 2000. In: Sahm, P.R., Hansen, P.N., Conley, J.G. (Eds.), Modeling of Casting, Welding and Advanced Solidification Processes IX, pp. 33–40.
Drouzy, M., Mascre, C., 1969. Metall. Rev. 14, 25–46.
Drouzy, M., Jacob, S., Richard, M., 1980. Int. Cast Met. J. 5 (2), 43–45.
Druschitz, A.P., Chaput, W.W., 1993. TAFS 101, 447–458.
Durrans, I., 1981. Thesis. University of Oxford.
Durville, P.H.G., 1913. Br. Patent 23719.
Eastwood, L.W., 1951. In: AFS Symposium on Principles of Gating, pp. 25–30.
Edelman, R.E., Saia, A., 1968. TAFS 76, 222–224.
Edelson, B.J., Baldwin, W.M., 1962. Trans. ASM 55, 230.
Eigenfeld, K., 1988. 8th Europe Kolloquium der ne. Metallgiesserei im CAEF Paris.
Einstein, A., 1906. Ann. Phys. 19, 289 and 1911, 34, 591.
Elliott, H.E., Mezoff, J.G., 1947. TAFS 55, 241–253.
Elliott, H.E., Mezoff, J.G., 1948. TAFS 56, 223–245, 279–285.
Ellison, W., Wechselblatt, P.M., 1966. TAFS 74, 350–356.
El-Mahallawi, S., Beeley, P.R., 1965. Br. Found. 58, 241–248.
El-Sayed, M.A., Salem, H.A.G., Kandeil, A.Y., Griffiths, W.D., August 2014. Metall. Mater. Trans. B 45B, 1398–1406.

REFERENCES

Emadi, D., Whiting, M., Djurdjevic, M., Kierkus, W.T., Sokolowski, J., 2004. MJoM 10, 91–106.
Emami, S., Sohn, H.Y., Kim, H.G., August 2014. Metall. Mater. B 45B, 1370–1379.
Emamy, G.M., Campbell, J., 1995a. Int. J. Cast Met. Res. 8, 13–20.
Emamy, G.M., Campbell, J., 1995b. Int. J. Cast Met. Res. 8, 115–122.
Emamy, G.M., Campbell, J., 1997. Trans. AFS 105, 655–663.
Emamy, G.M., Taghiabadi, R., Mahmudi, M., Campbell, J., 2002. In: Statistical Study of Tensile Properties of A356 Aluminum Alloy, Using a New Casting Design. ASM 2nd Internat. Al Casting Technology Symposium, 7–10 October, 2002, Columbus, Ohio.
Emamy, M., Abbasi, R., Kaboli, S., Campbell, J., 2009. Int. J. Cast Met. Res. 22 (6), 430–437.
Enderle, R.J., 1979. Trans. Am. Found. Soc. 87, 59–64.
Engels, G., Schneider, G., 1986. Cast. Plant Technol. (2), 12–20.
Enright, P., Hughes, I.R., November 1996. Foundryman 390–395.
Enright, N., Lu, S.Z., Hellawell, A., Pilling, J., 2000. TAFS 108, 157–162.
Enright, P., 2001. Private communication.
Evans, J., Runyoro, J., Campbell, J., 1997a. In: Beech, J., Jones, H. (Eds.), Solidification Processing. Dept. Engineering Materials, University of Sheffield, pp. 74–78.
Evans, J., Runyoro, J., Campbell, J., 1997b. In: SP97 4th Decennial International Conf. on Solidification Processing, Sheffield, pp. 74–78.
Evans, M.-H., Richardson, A.D., Wang, L., Wood, R.J.K., 2013. Effect of hydrogen on butterfly and white etching crack (WEC) formation under rolling contact fatigue (RCF). Wear. http://dx.doi.org/10.1016/j.wear.2013.03.008i.
Evans, E.P., 1951. BCIRA J. 4 (319), 86–139.
Evans, R.W., August 2002. Mater. Sci. Technol. 18, 831–839.
Fan, Z.-T., Ji, S., 2005. Mater. Sci. Technol. 21 (6), 727–734.
Fan, Z., Ji, S., Fang, X., Liu, G., Patel, J., Das, A., 2007. In: Crepeau, P.N., Tiryakioglu, M., Campbell, J. (Eds.), Shape Casting: The 2nd Internat. Symp. TMS, pp. 299–306.
Fang, O.T., Granger, D.A., 1989. TAFS 97, 989–1000.
Fasoyinu, F.A., Sadayappan, M., Cousineau, D., Sahoo, M., 1998a. TAFS 106, 721–734.
Fasoyinu, F.A., Sadayappan, M., Cousineau, D., Zavadil, R., Sahoo, M., 1998b. TAFS 106, 327–337.
Fathallah, R., Inglebert, G., Castex, L., 1998. Mater. Sci. Technol. 14, 631–639.
Fathallah, R., Sidham, H., Braham, C., Castex, L., August 2000. Mater. Sci. Technol. 19, 1050–1056.
Faubert, G.P., Moore, D.J., Rundman, K.B., 1990. TAFS 98, 831–845.
Feest, E.A., McHugh, G., Morton, D.O., Welch, L.S., Cook, I.A., 1983. In: Solidification Technology in the Foundry and Cast House. Warwick Univ. Conf. 1980. Metals Soc. Publication, pp. 232–239.
Felberbaum, M., Dahle, A., 2011. In: Cast House Symposium. TMS Annual Congress, San Diego, CA.
Feliu, S., Flemings, M.C., Taylor, H.F., 1960. Br. Found. 53, 413–425.
Feliu, S., Luis, L., Siguin, D., Alvarez, J., 1962. TAFS 70, 838–844 (1964) 71, 145–157; (1964) 72, 129–137.
Feliu, S., 1962. TAFS 70, 838–844.
Feliu, S., 1964. TAFS 72, 129–137.
Fenn, D., Harding, R.A., 2002. University of Birmingham, UK, unpublished work.
Fernandez, E., Musalem, P., Fava, J., Flinn, R.A., 1970. TAFS 78, 308–312.
Feurer, U., 1976. Giesserei 28, 75–80 (English translation by Alusuisse).
Fischer, R.B., 1988. TAFS 96, 925–944.
Fisher, J.C., 1948. J. Appl. Phys. 19, 1062–1067.
Flemings, M.C., Mehrabian, R., 1970. TAFS 78, 388–394.
Flemings, M.C., Nereo, G.E., 1967. TMS-AIME 239, 1449–1461.
Flemings, M.C., Conrad, H.F., Taylor, H.F., 1959. TAFS 67, 496–507.
Flemings, M.C., Niiyama, E., Taylor, H.F., 1961. TAFS 69, 625–635.
Flemings, M.C., Mollard, F.R., Niyama, E.F., Taylor, H.F., 1962. TAFS 70, 1029–1039.
Flemings, M.C., Poirier, D.R., Barone, R.V., Brody, H.D., April 1970. J. Iron Steel Inst. 371–381.
Flemings, M.C., Mollard, F.R., Niyama, E.F., 1987. TAFS 95, 647–652.
Flemings, M.C., Kattamis, T.Z., Bardes, B.P., 1991. TAFS 99, 501–506.

Flemings, M.C., 1963. In: 30th Internat. Found. Cong. Prague, pp. 61–81 and Br. Found. (1964) 57, 312–325.
Flemings, M.C., 1974. Solidification Processing. McGraw-Hill, USA.
Fletcher, A.J., Griffiths, W.D., 1995. Mater. Sci. Technol. 11 (3), 322–326.
Fletcher, A.J., 1989. Thermal Stress and Strain Generation in Heat Treatment. Elsevier.
Flinn, R.A., Van Vlack, L.H., Colligan, G.A., 1986. TAFS 94, 29–46.
Flood, S.C., Hunt, J.D., 1981. Met. Sci. 15, 287–294.
Fomin, V.V., Stekol'nikov, G.A., Omel'chenko, V.S., 1965. Russ. Cast. Prod. 229–231.
Fonda, R.W., Lauridsen, E.M., Ludwig, W., Tafforeau, P., Spanos, G., 2007. Metall. Mater. Trans. A 38A (11), 2721–2726.
Ford, D.A., Wallbank, J., 1998. Int. J. Cast Met. Res. 11, 23–35.
Forest, B., Berovici, S., 1980. In: Solidification Technology in the Foundry and Casthouse. Metals Soc. Conf. Warwick, Paper 93, pp. 1–12.
Forgac, J.M., Angus, J.C., June 1981. Metall. Trans. B 12B, 413–416.
Forgac, J.M., Schur, T.P., Angus, J.C., 1979. J. Appl. Mech. 46, 83–89.
Forslund, S.H.C., 1954. In: 21st Internat. Foundry Congress, Florence, Paper 15.
Forsyth, P.J.E., 1995. Mater. Sci. Technol. 11 (3), 1025–1033.
Forsyth, P.J.E., 1999. Mater. Sci. Technol. 15 (3), 301–308.
Forward, G., Elliott, J.F., 1967. J. Met. 19, 54–59.
Fox, S., Campbell, J., 2000. Scr. Mater. 43 (10), 881–886.
Fox, S., Campbell, J., 2002. Int. J. Cast Met. Res., 14 (6), 335–340.
Franklin, A.G., Rule, G., Widdowson, R., September 1969. J. Iron Steel Inst. 1208–1218.
Fras, E., Gorny, M., Lopez, H.F., 2007a. TAFS 115, 1–17.
Fras, E., Wiencek, K., Gorny, M., Lopez, H.F., 2007b. Trans. Met. Mater. A 38A (2), 385–395.
Fras, E., Gorny, M., Lopez, H.F., 2012. Metall. Mater. Trans. 43A (11), 4204–4218.
Frawley, I.J., Moore, W.F., Kiesler, A.L., 1974. TAFS 82, 561–570.
Fredriksson, H., Lehtinen, B., 1979. In: Solidification and Casting of Metals. Metals Soc. Conf. Metals Soc, Sheffield, pp. 260–267.
Fredriksson, H., Haddad-Sabzevar, M., Hansson, K., Kron, J., 2005. Mater. Sci. Technol. 21 (5), 521–529.
Fredriksson, H., 1984. Mat. Sci. Eng. 65, 137–144.
Fredriksson, H., 1996. In: Ohnaka, I., Stefanescu, D.M. (Eds.), Solidification Science and Processing. The Minerals, Metals and Materials Society.
French, A.R., 1957. Inst. Met. Monogr. Rep. 22, 60–84.
Freti, S., Bornand, J.D., Buxmann, K., 12–16 June 1982. Light Metal Age.
Friebel, V.R., Roe, W.P., 1963. TAFS 71, 388–393.
Frommeyer, G., Derder, C., Jimenez, J.A., 2002. Mater. Sci. Technol. 18 (9), 981–986.
Fruehling, J.W., Hanawalt, J.D., 1969. TAFS 77, 159–164.
Fruehling, J.W., Hanawalt, J.D., 1969. TAFS 77, 159–590.
Fuji, N., Fuji, M., Morimoto, S., Okada, S., 1984. J. Jpn. Inst. Light Met. 34 (8), 446–453 (Met. Abstr. 51-1657).
Fuji, M., Fuji, N., Morimoto, S., Okada, S., 1986. J. Jpn. Inst. Light Met. 36 (6), 353–360 (Transl. NF 197).
Fuoco, R., Correa, E.R., Correa, A.V.O., 1995. TAFS 103, 379–387.
Fuoco, R., Correa, E.R., Andrade Bastos, M.de, 1998. TAFS 106, 401–409.
Gadd, M.A., Bennett, G.J., 1984. In: Fredriksson, H., Hillert, M. (Eds.), The Physical Metallurgy of Cast Iron. North-Holland, p. 99.
Gagne, M., Goller, R., 1983. TAFS 91, 37–46.
Gagne, M., Paquin, M.-P., Cabanne, P.-M., 2008. World Foundry Congress, 68th, pp. 101–106.
Gall, K., Horstemeyer, M.F., van Schilfgaarde, M., Baskes, M.I., 2000. J. Mech. Phys. Solids 48, 2183–2218.
Gallois, B., Behi, M., Panchal, J.M., 1987. TAFS 95, 579–590.
Gammelsaeter, R., Beck, K., Johansen, S.T., 1997. In: Light Metals (Conference), pp. 1007–1011.
Garat, M., Guy, S., Thomas, J., 1991. Foundryman 84 (1), 29–34.
Garat, M., European Patent Application 0274964 Filed 16 November 1987 (in French).
Garbellini, O., Palacio, H., Biloni, H., 1990. Cast. Met. 3 (2), 82–90.
Garcia-Garcia, O., Sanches-Araiza, M., Castro-Roman, M., Escobedo, B.J.C., 2007. In: Crepeau, P.N., Tiryakioglu, M., Campbell, J. (Eds.), Shape Casting: 2nd International Symposium. TMS, pp. 109–116.

Garnar, T.E., 1997. TAFS 85, 399–416.
Gauthier, J., Samuel, F.H., 1995a. TAFS 103, 849–857.
Gauthier, J., Louchez, P.R., Samuel, F.H., 1995b. Cast. Met. 8 (2), 91–106.
Gebelin, J.-C., Griffiths, W.D., 2007. In: Jones, H. (Ed.), Proc 5th Decennial Internat. Conf. on Solidification Processing. (SP07) Sheffield, UK, pp. 512–516.
Gebelin, J.-C., Jolly, M.R., 2002. TAFS 110, 109–119.
Gebelin, J.-C., Jolly, M.R., 2003. In: Modeling of Casting, Welding and Solidification Processing Conference X, Destin USA.
Gebelin, J-C., Jolly, M.R., Jones, S., December. INCAST 13 (11), 22–27.
Gebelin, J.-C., Jolly, M.R., Hsu, F.-Y., 2005. In: Tiryakioglu, M., Crepeau, P.N. (Eds.), Shape Casting: The John Campbell Symposium. TMS, pp. 355–364.
Gebelin, J.-C., Jolly, M.R., Hsu, F.-Y., 2006. Int. J. Cast Met. Res. 19 (1), 18–25.
Geffroy, P.-M., Lakehal, M., Goni, J., Beaugnon, E., Heintz, J.-M., Silvain, J.-F., 2006a. Metall. Mater. Trans. A 37A, 441–447.
Geffroy, P.-M., Pena, X., Lakehal, M., Goni, J., Egizabal, P., Silvain, J.-F., February 2006b. Fonderie Fondeur d'aujord'hui 252, 8–19.
Gelperin, N.B., 1946. TAFS 54, 724–726.
Genders, R., Bailey, G.L., 1934. The Casting of Brass Ingots. The British Non-Ferrous Metals Research Association.
Gentry, E.G., 1966. TAFS 74, 142–149.
Gernez, M., 1867. Philos. Mag. 33 (4), 479.
Geskin, F.S., Ling, E., Weinstein, M.I., 1986. TAFS 94, 155–158.
Ghomashchi, M.R., Chadwick, G.A., 1986. Met. Mater. 477–482.
Ghomashchi, M.R., 1995. J. Mater. Process. Technol. 52, 193–206.
Ghosh, S., Mott, W.J., 1964. Trans. Am. Found. Soc. 72, 721–732.
Giese, S.R., Stefanescu, D.M., Barlow, J., Piwonka, T.S., 1996. Part II, TAFS 104, 1249–1257.
Girshovich, N.G., Lebedev, K.P., Nakhendzi Yu, A., April 1963. Russ. Cast. Prod. 174–178.
Giuranno, D., Ricci, E., Arato, E., Costa, P., 2006. Acta Mater. 54, 2625–2630.
Glatz, J., December 1996. Materials Evaluation, 1352–1362 and INCAST June 1997, 1–7.
Glenister, S.M.D., Elliott, R., 1981. Met. Sci. 15 (4), 181–184.
Gloria, D., Hernandez, F., Valtierra, S., Cisneros, M.A., October 2000. In: 20th ASM Heat Treating Soc. Conf. Proc, St Louis, MO, pp. 674–679.
Goad, P.W., 1959. TAFS 67, 436–448.
Godding, R.G., 1962. BCIRA J. 10 (3), 292–297.
Godlewski, L.A., Zindel, J.W., 2001. TAFS 109, 315–325.
Godlewski, L.A., Su, X., Pollock Tresa, M., Allison, J.E., 2013. Metall. Mater. Trans. A 44A, 4809–4818.
Goklu, S.M., Lange, K.W., April 1986. In: Proc. Conf. Process Technology, vol. 6. Iron and Steel Society, Washington, USA, pp. 1135–1146.
Gonya, H.J., Ekey, D.C., 1951. TAFS 59, 253–260.
Goodwin, F.E., 2008. ASM Handbook. In: Casting, vol. 15, pp. 1095–1099.
Goodwin, F.E., 2009. Cast Met. Diecast. Times 11 (4), 14.
Goria, C.A., Serramoglia, G., Caironi, G., Tosi, G., 1986. TAFS 94, 589–600 (These authors quote Kobzar, A.I., Ivanyuk, E.G., 1975. Russ. Cast. Prod. 7, 302–330).
Gorshov, A.A., 1964. Russ. Cast. Prod. 338–340.
Gould, G.C., Form, G.W., Wallace, J.F., 1960. TAFS 68, 258–267.
Gouwens, P.R., 1967. TAFS 75, 401–407.
Graham, A.L., Mizzi, B.A., Pedicini, L.J., 1987. TAFS 95, 343–350.
Granath, O., Wessen, M., Cao, H., 2008. Int. J. Cast Met. Res. 21 (5), 349–356.
Grandfield, J.F., Nguyen, T.T., Rohan, P., Nguyen, V., 2007. In: Jones, H. (Ed.), Proc. 5th Decennial Internat. Conf. on Solidification Processing. (SP07). Sheffield, UK, pp. 507–511.
Grassi, J.R., Campbell, J., Shaw, C.W., 6 November 2012. Integrated Quiescent Processing of Melts. US Patent 8303890.
Graverend, J-B.Le, Cormier, J., Kruch, S., Gallerneau, F., Mendez, J., 2012. Metall. Mater. Trans. A 43A (11), 3988–3997.
Gray, R.J., Perkins, A., Walker, B., 1977. In: Sheffield Conference "Solidification & Casting of Metals", pp. 300–305 (Published by Metals Soc. 1979 Book 192).

Greaves, R.H., 1936. ISI Spec. Rep. 15, 26–42.
Grebe, W., Grimm, G.P., 1967. Aluminium 43, 673–683.
Green, N.R., Campbell, J., 1994. Trans. AFS 102, 341–347.
Green, R.A., Heine, 1990. TAFS 98, 495–503.
Green, N.R., Tomkinson, A.M., Wright, T.C., Evans, J.P., Fuchs, U., Tschegg, S., 2005. In: Tiryakioglu, M., Crepeau, P.N. (Eds.), Shape Casting: The John Campbell Symposium. TMS.
Green, R.A., 1990. TAFS 98, 947–952.
Green, N.R., 1995. University of Birmingham, UK, unpublished work.
Greer, A.L., Bunn, A.M., Tronche, A., Evans, P.V., Bristow, D.J., 2000. Acta Mater. 48 (11), 2765–3026.
Grefhorst, C., Crepaz, R., 2005. Casting Plant and Technology International 1, pp. 28–35.
Griffiths, W.D., Lai, N.-W., 2007. Metall. Mater. Trans. 38A, 190–196.
Griffiths, W.D., Cox, M., Campbell, J., Scholl, G., 2007. Mater. Sci. Technol. 23 (2), 137–144.
Grill, A., Brimacombe, J.K., 1976. Ironmak. Steelmak. 3 (2), 76–86.
Grote, R.E., 1982. TAFS 90, 93–102.
Groteke, D.E., 1985. TAFS 93, 953–960.
Groteke, D.E., 2008. Personal communication to JC.
Grube, K.R., Kura, J.G., 1955. TAFS 63, 35–48.
Grunengerg, N., Escherle, E., Sturm, J.C., 1999. TAFS 107, 153–159.
Grupke, C.C., Hunter, L.J., Leonard, C., Nath, R.H., Weaver, G.J., 2011. Quality assurance with process compensated resonant testing. TAFS paper 11-036.
Gruzleski, J., Handiak, N., Campbell, H., Closset, B., 1986. TAFS 94, 147–154.
Gruznykh, I.V., Nekhendzi, Yu A., 1961. Russ. Cast. Prod. 6, 243–245.
Gunasegaram, D., Givord, M., O'Donnell, R., Finnin, B., December 2007. Foundry Trade J. 362–365.
Guo, J., Sheng, W., Su, Y., Ding, H., Jia, J., January 2001. Int. J. Cast Met. Res.
Gupta, N., Satyanarayana, K.G., 2006. The solidification processing of metal-matrix composites: the Rohatgi symposium. JOM: J. Miner. Met. Mater. Soc. ISSN: 1047-4838 58 (11), 92–94.
Gurland, J., 1966. Trans. AIME 236, 642–646.
Gurland, J., 1979. Application of the hall-petch relation to particle strengthening in spherodized steels and aluminum-silicon alloys. In: Proceedings of the 2nd Conf. on Strength of Metals and Alloys, pp. 621–625.
Guthrie, R.I.L., 1989. Engineering in Process Metallurgy. Clarendon Press, Oxford, UK.
Guven, Y.F., Hunt, J.D., 1988. Cast. Met. 1 (2), 104–111.
Ha, M., Kim, W.-S., Moon, H.-K., Lee, B.-J., Lee, S., 2008. Metall. Mater. Trans. A 39 (5), 1087–1098.
HabibollahZadeh, A., Campbell, J., 2003. Paper 03-022 Trans. Am. Found. Soc. 10.
Habibollahzadeh, A., Campbell, J., 2004. Int. J. Cast Met. Res. 17 (3), 1–8.
Hadian, R., Emamy, M., Campbell, J., 2009. The modification of cast Al-Mg$_2$Si metal matrix composite by Li. Metall. Mater. Trans. B 40 (6), 822–832.
Haginoya, I., 1976. J. Jpn. Inst. Light Met. 26 (3), 131–138 and Ibid 1981 31, 769–774.
Hairy, P., Longa, Y., Laguerre, C., et al., June/July 2003. Fonderie Fondeur d'aujourd'hui (226), 20–30. ALFED catalogue C/27918.
Hall, F.R., Shippen, J., 1994. Eng. Failure Anal. 1 (3), 215–229.
Hallam, C.P., Griffiths, W.D., Butler, N.D., 2000. In: Sahm, P.R., Hansen, P.N., Conley, J.G. (Eds.), Modeling of Casting, Welding and Advanced Solidification Processes IX.
Halvaee, A., Campbell, J., 1997. TAFS 105, 35–46.
Hammitt, F.G., 1974. Proc. 1973 Symp. Copenhagen. In: Bjorno, L. (Ed.), Finite-Amplitude Wave Effects in Liquids. IPC Science and Technology Press, pp. 258–262. Paper 3.10.
Hammond, D.E., August 1989. Mod. Cast. 29–33.
Hangai, Y., Kato, H., Utsunomiya, T., Kitahara, S., 2010. Metall. Mater. Trans. A 41A, 1883–1886.
Hansen, P.N., Rasmussen, N.W., 1994. In: BCIRA International Conference. University of Warwick, UK.
Hansen, P.N., Sahm, P.R., 1988. In: Giamei, A.F., Abbaschian, G.J. (Eds.), Modelling of Casting and Welding Processes IV. The Mineral, Metals and Materials Society.

Hansen, P.N., Sahm, P.R., 1988. In: Giamei, A.F., Abbaschian, G.J. (Eds.), Modelling of Casting, Welding and Advanced Solidification Processes IV. The Mineral, Metals and Materials Society.
Hansen, P.N., 1975. Ph.D. thesis. Part 2, Technical University of Denmark, Copenhagen (Lists 283 publications on hot tearing.).
Harding, R.A., Campbell, J., Saunders, N.J., 1997. Inoculation of ductile iron. In: Beech, J., Jones, H. (Eds.), Solidification Processing 97 Sheffield Conference, 7–10 July, 1997.
Harding, R.A., June 2006. In: 67th World Foundry Congress, Harrogate, UK. Paper 17.
Harinath, U., Narayana, K.L., Roshan, H.M., 1979. TAFS 87, 231–236.
Harper, S., 1966. J. Inst. Met. 94, 70–72.
Harris, K.P., September 2004. Foundry Pract. (Foseco) 242, 01–07.
Harris, K.P., 2005a. In: Tiryakioglu, M., Creapeau, P.N. (Eds.), Shape Casting: The JC Symposium. TMS Annual Congress, pp. 433–442.
Harris, K.P., July 2005b. Mod. Cast. 36–38.
Harrison Steel 1999. Personal communication.
Harsem, O., Hartvig, T., Wintermark, H., 1968. In: 35th Internat. Found. Cong. Kyoto, Paper 15.
Hart, R.G., Berke, N.S., Townsend, H.E., 1984. Met. Trans. 15B, 393–395.
Hartmann, D., Stets, W., 2006. TAFS 114, 1055–1058.
Hartung, C., Ecob, C., Wilkinson, D., 2008. Cast. Plant Technol. 2, 18–21.
Hashemi, H.R., Ashoori, H., Davami, P., 2001. Mater. Sci. Technol. 17, 639–644.
Hassan, M.I., Al-Kindi, R., 2014. J. Met. 66 (9), 1603–1611.
Hassell, H.J., 1980. Br. Found. 73 (4), 95.
Haugland, E., Engh, T.A., 1997. In: Light Metals (TMS Conference), pp. 997–1005.
Hayes, K.D., Barlow, J.O., Stefanescu, D.M., Piwonka, T.S., 1998. TAFS 106, 769–776.
Hayes, J.S., Keyte, R., Prangnell, P.B., 2000. Mater. Sci. Technol. 16, 1259–1263.
Hebel, T.E., 1989. Heat Treating, Fairchild Business Publication, USA.
Hedjazi, D., Bennett, G.H.J., Kondic, V., 1975. Br. Found. 68, 305–309.
Hedjazi, Dj., Bennett, G.H.J., Kondic, V., December 1976. Met. Technol. 537–541.
Heine, R.W., Green, R.A., 1989. TAFS 97, 157–164.
Heine, R.W., Green, R.A., 1992. TAFS 100, 499–508.
Heine, H.J., Heine, R.W., 1968. TAFS 76, 470–484.
Heine, R.W., Loper, C.R., 1966a. TAFS 74, 274–280.
Heine, R.W., Loper, C.R., 1966b. TAFS 74, 421–428 and Br. Found. 1967 60, 347–353.
Heine, R.W., Rosenthal, P.C., 1955. Principles of Metal Casting. McGraw-Hill Book Co Inc + AFS, NY, USA.
Heine, R.W., Uicker, J.J., 1983. TAFS 91, 127–136.
Heine, R.W., Uicker, J.J., Gantenbein, D., 1984. TAFS 92, 135–150.
Heine, R.W., Green, R.A., Shih, T.S., 1990. TAFS 98, 245–252.
Heine, R.W., 1951. TAFS 59, 121–138.
Heine, R.W., 1968. TAFS 76, 463–469.
Heine, R.W., 1982. TAFS 90, 147–158.
Hendricks, M.J., Wang, P., Bijvoet, M., 2005. Br. Invest. Casters Assoc. Bull. 47, 4–5.
Henschel, C., Heine, R.W., Schumacher, J.S., 1966. TAFS 74, 357–364.
Hernandes-Reyes, B., 1989. TAFS 97, 529–538.
Herrara, A., Campbell, J., 1997. TAFS 105, 5–11.
Herrera, A., Kondic, V., 1979. In: Solidification and Casting of Metals. Conf. Sheffield, 1977. Metals Soc, London, pp. 460–465.
Hess, K., March 1974. AFS Cast Met. Res. J. 6–14.
Hetke, A., Gundlach, R.B., 1994. TAFS 102, 367–380.
Heusler, L., Feikus, F.J., Otte, M.O., 2001. TAFS 109, 215–223.
Heyn, E., 1914. J. Inst. Met. 12, 3.
Hill, H.N., Barker, R.S., Willey, L.A., 1960. Trans. Am. Soc. Met. 52, 657–671.
Hillert, M., Lindblom, Y., 1954. The growth of nodular graphite. J. Iron Steel Inst. 148, 388–390.

Hillert, M., Steinhauser, H., 1960. Jernkontorets Ann. 144, 520–522.
Hillert, M., 1968. In: Merchant, H.D. (Ed.), Recent Research on Cast Iron. Gordon and Breach, NY, pp. 101–127.
Hillis, J.E., November 2002. In: Internat. Conf. on SF6 and the Environment.
Hiratsuka, S., Niyama, E., Anzai, K., Hori, H., Kowata, T., 1966. In: 4th Asian Foundry Congress, pp. 525–531.
Hiratsuka, S., Niyama, E., Funakubo, T., Anzai, K., 1994. Trans. Jpn. Found. Soc. 13 (11), 18–24.
Hiratsuka, S., Niyama, E., Horie, H., Kowata, T., Anzai, K., Nakamura, M., 1998. Int. J. Cast Met. Res. 10 (4), 201–205.
Hirt, C.W., 2003. www.flow3d.com.
Ho, K., Pehlke, R.D., 1984. TAFS 92, 587–598.
Ho, P.S., Kwok, T., Nguyen, T., Nitta, C., Yip, S., 1985. Scr. Metall. 19 (8), 993–998.
Hoar, T.P., Atterton, D.V., 1950. J. Iron Steel Inst. 166, 1–7.
Hoar, T.P., Atterton, D.V., Houseman, D.H., 1953. J. Iron Steel Inst. 175, 19–29.
Hoar, T.P., Atterton, D.V., Houseman, D.H., 1956. Metallurgia 53, 21–25.
Hochgraf, F.G., 1976. Metallography 9, 167–176.
Hodaj, F., Durand, F., 1997. Acta Mater. 45, 2121.
Hodjat, Y., Mobley, C.E., 1984. TAFS 92, 319–321.
Hoff, O., Andersen, P., 1968. In: 35th Internat. Found. Cong., Kyoto, Paper 8.
Hoffman, E., Wolf, G., 2001. Giessereiforschung 53 (4), 131–151.
Hofmann, R., Wittmoser, A., 16 November 1971, US Patents 3619866 and 3620286
Hofmann, F., 1962. TAFS 70, 1–12.
Hofmann, F., 1966. AFS Cast Met. Res. J. 2 (4), 153–165.
Hoffmann, J., 2001. Foundry Trade J. 175 (3578), 32–34.
Holt, G.S., Mitchell, C.J., Simmons, R.E., 1989. BCIRA Report 1779 and BCIRA J. July 1989, 291–296.
Holtzer, M., March 1990. Foundryman 135–144.
Hong, C.P., Shen, H.F., Lee, S.M., 2000. Metall. Mater. Trans. B 31B, 297–305.
Hoover, W.R., 1991. In: Hansen, N., et al. (Eds.), Proc 12th Riso Internat. Symp. Materials Science: Metal Matrix Composites – Processing, Microstructure and Properties, pp. 387–392.
Hoult, F.H., 1979. Trans. AFS 87, 237–240 and 241–244.
Hoult, F.H., 1979a. TAFS 87, 237–240.
Hoult, F.H., 1979b. TAFS 87, 241–244.
Howe Sound Co – Superalloy Group USA, 1965. British Patent 1125124.
Hsu, W., Jolly, M.R., Campbell, J., 2006. Int. J. Cast Met. Res. 19 (1), 38–44.
Hsu, F.-Y., Jolly, M.R., Campbell, J., 2005. In: Tiryakioglu, M., Crepeau, P.N. (Eds.), Shape Casting: The John Campbell Symposium. TMS, San Francisco, CA, pp. 73–82.
Hu, D., Loretto, M.H., 2000. University of Birmingham, UK, personal communication.
Hu, Z.C., Zhang, E.L., Zeng, S.Y., 2008. Mater. Sci. Technol. 24 (11), 1304–1308.
Hua, C.H., Parlee, N.A.D., 1982. Met. Trans. 13B, 357–367.
Huang, J., Conley, J.G., 1998. TAFS 106, 265–270.
Huang, H., Lodhia, A.V., Berry, J.T., 1990. TAFS 98, 547–552.
Huang, L.W., Shu, W.J., Shih, T.S., 2000. TAFS 108, 547–560.
Huber, G., Brechet, Y., Pardoen, T., 2005. Acta Mat. 53, 2739–2749.
Hudak, D., Tiryakioğlu, M., 2009. On estimating percentiles of the Weibull distribution by the linear regression method. J. Mater. Sci. 44, 1959–1964.
Hughes, I.C.H., 1988. Metals Handbook. In: Casting, vol. 15. ASM, Ohio, USA, 647–666.
Hull, D.R., 1950. Casting of Brass and Bronze. ASM.
Hultgren, A., Phragmen, G., 1939. Trans. AIME 135, 133–244.
Hummer, R., 1988. Cast. Met. 1 (2), 62–68.
Hunsicker, H.Y., 1980. Met. Trans. 11A, 759–773.
Hunt, J.D., Thomas, R.W., 1997. In: Beech, J., Jones, H. (Eds.), Proc. 4th Decennial Internat. Conf. on Solidification Processing (SP97). University of Sheffield, UK, pp. 350–353.
Hunt, J.D., 1980. University of Oxford, personal communication.
Hurtuk, D.J., Tzavaras, A.A., 1975. TAFS 83, 423–428.

Hurum, F., 1952. TAFS 60, 834–848.
Hurum, F., 1965. TAFS 73, 53–64.
Hutchins, M., 2007. Foundry Trade J. 181 (3645), 195.
Hutchinson, H.P., Sutherland, D.S., 1965. Nature 206, 1036–1037.
IBF Technical Subcommittee TS 17, 1948. In: Symp. Internal Stresses in Metals and Alloys, London 1947. The Inst of Metals, pp. 179–188.
IBF Technical Subcommittee TS 18, 1949. Proc. IBF 42, A61–A77.
IBF Technical Subcommittee TS 32, 1952. Br. Found. 45, A48–A56. Foundry Trade J. 93, 471–477.
IBF Technical Subcommittee TS 32, 1956. Foundry Trade J. 101, 19–27.
IBF Technical Subcommittee TS 32, 1960. Br. Found. 53, 10–13 (but original work reported in Br. Found. 1952, 45, A48–A56).
IBF Technical Subcommittee TS 35, 1960. Br. Found. 53, 15–20.
IBF Technical Subcommittee TS 61, 1964. Br. Found. 57, 75–89, 504–508.
IBF Technical Subcommittee TS 71, 1969. Br. Found. 62, 179–196.
IBF Technical Subcommittee TS 71, 1971. Br. Found. 64, 364–379.
IBF Technical Subcommittee TS 71, 1976. Br. Found. 69, 53–60.
IBF Technical Subcommittee TS 71, 1979. Br. Found. 72, 46–52.
Iida, T., Guthrie, R.I.L., 1988. The Physical Properties of Liquid Metals. Clarendon Press, Oxford, p. 14.
Impey, S., Stephenson, D.J., Nicholls, J.R., 1993. In: Microscopy of Oxidation 2. Cambridge Conference, pp. 323–337.
International Magnesium Association, 2006. Alternatives to SF6 for Magnesium Melt Protection. EPA-430-R-06-007.
Ionescu, V., 2002. Mod. Cast. 92 (5), 21–23.
Isaac, J., Reddy, G.P., Sharman, G.K., 1985. TAFS 93, 29–34.
Isawa, T., Campbell, J., November 1994. Trans. Jpn. Found. Soc. 13, 38–49.
Isawa, T., 1993. The Control of the Initial Fall of Liquid Metal in Gravity Filled Casting systems (Ph.D. thesis). University of Birmingham Department of Metallurgy and Materials Science.
ISO Standard 8062, 1984. Castings – System of Dimensional Tolerances.
Isobe, T., Kubota, M., Kitaoka 5., 1978. J. Jpn. Found. Soc. 50 (11), 671–676.
Itamura, M., Yamamoto, N., Niyama, E., Anzai, K., 1995. In: Lee, Z.H., Hong, C.P., Kim, M.H. (Eds.), Proc 3rd Asian Foundry Congress, pp. 371–378.
Itamura, M., Murakami, K., Harada, T., Tanaka, M., Yamamoto, N., 2002. Int. J. Cast Met. Res. 15 (3), 167–172.
Itofugi, H., Uchikawa, H., 1990. TAFS 98, 429–448.
Itofugi, H., 1996. TAFS 104, 79–87.
Iyengar, R.K., Philbrook, W.O., 1972. Met. Trans. 3, 1823–1830.
Jackson, W.J., Wright, J.C., September 1977. Met. Technol. 425–433.
Jackson, K.A., Hunt, J.D., Uhlmann, D.R., Seward, T.P., 1966. Trans. AIME 236, 149–158.
Jackson, R.S., 1956. Foundry Trade J. 100, 487–493.
Jackson, W.J., April 1972. Iron Steel 163–172.
Jacob, S., Drouzy, M., 1974. In: Internat. Foundry Congress, 41 Liege, Belgium. Paper 6.
Jacobi, H., 1976. Arch. Eisenhuttenwes. 47, 441–446.
Jacobs, M.H., Law, T.J., Melford, D.A., Stowell, M.J., 1974. Met. Technol. 1 (11), 490–500.
Jakumeit, J., Laqua, R., Hecht, U., Goodheart, K., Peric, M., 2007. In: Jones, H. (Ed.), SP07 Proc. 5th Decennial Internat. Conf. Solidification Processing, pp. 292–296.
Jaquet, J.C., 1988. In: 8th Colloque Europeen de la Fonderie des Metaux Non Ferreux du CAEF.
Jaradeh, M.M., Carlberg, T., 2011. Metall. Mater. Trans. B 42B, 121–132.
Jay, R., Cibula, A., 1956. Proc. Inst. Br. Found. 49, A126–A140.
Jayatilaka, A. de S., Trustrum, K., 1977. J. Mater. Sci. 12, 1426.
Jeancolas, M., Devaux, H., 1969. Fonderie 285, 487–499.
Jeancolas, M., Cohen de Lara, G., Hanf, H., 1962. TAFS 70, 503–512.
Jeancolas, M., Devaux, H., Graham, G., 1971. Br. Found. 64, 141–154.
Jelm, C.R., Herres, S.A., 1946. Trans. Am. Found. Assoc. 54, 241–251.
Jennings, J.M., Griffin, J.A., Bates, C.E., 2001. TAFS 109, 177–186.
Ji, S., Roberts, K., Fan, Z., February 2002. Mater. Sci. Technol. 18, 193–197.

Jian, X., Xu, C., Meek, T., Han, Q., 2005. TAFS 113, 131–138.
Jiang, H., Bowen, P., Kntt, J.F., 1999. J. Mater. Sci. 34, 719–725.
Jianzhong, L., 1989. TAFS 97, 31–34.
Jirsa, J., October 1982. Foundry Trade J. 7, 520–527.
Jiyang, Z., Schmidt, W., Engler, S., 1990. TAFS 98, 783–786.
Jo, C.-Y., Joo, D.-W., Kim, I.-B., 2001. Mater. Sci. Technol. 17, 1191–1196.
Johnson, W.H., Baker, W.O., 1948. TAFS 56, 389–397.
Johnson, S.B., Loper, C.R., 1969. TAFS 77, 360–367.
Johnson, A.S., Nohr, C., 1970. TAFS 78, 194–207.
Johnson, R.A., Orlov, A.N., 1986. Physics of Radiation Effects in Crystals. Elsevier, North Holland.
Johnson, W.C., Smart, H.B., 1979. In: "Solidification and Casting of Metals" Sheffield Conf. 1977. Metals Soc. Publication 192, pp. 125–130.
Johnson, W.H., Bishop, H.F., Pellini, W.S., 1954. Foundry 102–107 and 271–272.
Johnson, T.V., Kind, H.C., Wallace, J.F., Nieh, C.V., Kim, H.J., 1989. TAFS 97, 879–886.
Jolly, M.J., Lo, H.S.H., Turan, M., Campbell, J., Yang, X., 2000. In: Sahm, P.R., Hansen, P.N., Conley, J.G. (Eds.), Modeling of Casting, Welding and Advanced Solidification Processing IX, Aachen.
Jolly, M.J., 2008. Engineering School, University of Birmingham, UK, personal communication.
Jones, D.R., Grim, R.E., 1959. TAFS 67, 397–400.
Jones, C.A., Fisher, J.C., Bates, C.E., 1974. TAFS 82, 547–559.
Jones, S.G., April 1948. Am. Found. 139.
Jones, S., June 2005. Foundry Trade J. 179, 156–157.
Jones, S., October 2006. Foundry Trade J. 180 (3638), 267–270.
Jones, S., April 2006. INCAST 18–21.
Joo, S.-H., Jung, J., Chun, M.S., Moon, C.H., Lee, S., Kim, H.S., August 2014. Metall. Mater. Trans. A 45A, 4002–4011.
Jorstad, J.L., 1971. TAFS 79, 85–90.
Jorstad, J.L., 1996. TAFS 104, 669–671.
Kahl, W., Fromm, E., 1984. Aluminium 60 (9), E581–E586.
Kahl, W., Fromm, E., 1985. Met. Trans. B 16B (3), 47–51.
Kahn, P.R., Su, W.M., Kim, H.S., Kang, J.W., Wallace, J.F., 1987. TAFS 95, 105–116.
Kaiser, W.D., Groenveld, T.P., 1975. In: 8th Internat. Die Casting Congress, Detroit, Michigan, USA, pp. 1–9. Paper Number G-T75-084.
Kaiser, G., 1966. Ber. Bunsenges. Phys. Chem. 70 (6), 635–639.
Kallbom, R., Hamberg, K., Wessen, M., Bjorkegren, L.-E., 2005. Mater. Sci. Eng. A A413–A414, 346–351.
Kallbom, R., Hamberg, K., Bjorkegren, L.-E., 2006. In: World Foundry Congress, Harrogate, UK, paper 184/1-10.
Kang, X., Li, D.-Z., Xia, L., Campbell, J., Li, Y.Y., 2005. In: Tiryakioglu, M., Crepeau, P.N. (Eds.), Shape Casting: The John Campbell Symposium. TMS, pp. 377–384.
Karsay, S.I., 1971. Ductile Iron II; Engineering Design Properties Applications. Quebec Iron & Titanium (QIT) Corporation, Canada.
Karsay, S.I., 1980. Ductile Iron: The State of the Art 1980. QIT-Fer et Titane Inc, Canada.
Karsay, S.I., 1985. Ductile Iron Production Practices. AFS.
Karsay, S.I., 1992. Ductile Iron Production Practices; the State of the Art 1992. QIT Fer et Titane Inc., Canada and "Ductile Iron; the essentials of gating and risering system design" revised 2000 published by Rio Tinto Iron & Titanium Inc.
Kasch, F.E., Mikelonis, P.J., 1969. TAFS 77, 77–89.
Kaspersma, J.H., Shay, R.H., 1982. Met. Trans. 13B, 267–273.
Katashima, S., Tashima, S., Yang, R.-S., 1989. TAFS 97, 545–552.
Katgerman, L., 1982. J. Met. 34 (2), 46–49.
Kato, E., Metall. Mater. Trans. A 30A, 2449–2453.
Kay, J.M., Nedderman, R.M., 1974. An Introduction to Fluid Mechanics and Heat Transfer, third ed. CUP. pp. 115–119.
Kearney, A.L., Raffin, J., 1987. Hot Tear Control Handbook for Aluminium Foundrymen and Casting Designers. American Foundrymen's Soc., Des Plaines Illinois, USA.
Khalili, A., Kromp, 1991. J. Mater. Sci. 26, 6741–6752.

Khan, P.R., Su, W.M., Kim, H.S., Kang, J.W., Wallace, J.F., 1987. TAFS 95, 105–116.
Khorasani, A.N., 1995. TAFS 103, 515–519 and Mod. Cast. 1996 86, 36–38.
Kiessling, R., 1987. Non-metallic Inclusions in Steel. The Metals Society.
Kihlstadius, D., 1988. ASM Metals Handbook. In: Casting, vol. 15, 273–274.
Kilshaw, J.A., 1963. BCIRA J. 11, 767.
Kilshaw, J.A., 1964. BCIRA J. 12, 14.
Kim, S.B., Hong, C.P., 1995. In: Modeling of Casting, Welding and Advanced Solidification Processes VII. TMS, pp. 155–162.
Kim, M.H., Loper, C.R., Kang, C.S., 1985. TAFS 93, 463–474.
Kim, M.H., Moon, J.T., Kang, C.S., Loper, C.R., 1993. TAFS 101, 991–998.
Kim, D., Han, K., Lee, B., Han, I., Park, J.H., Lee, C., 2014. Metall. Mater. Trans. A 45A, 2046–2054.
Kirner, J.F., Anewalt, M.R., Karwacki, E.J., Cabrera, A.L., 1988. Met. Trans. 19A, 3045–3055.
Kita, K., 1979. AFS Int. Cast Met. J. 4 (4), 35–40.
Kitaoka, S., 2001. In: Conference: Light Metals (Metaux Legers): International Symposium on Light Metals as Held at the 40th Annual Conference of Metallurgists of CIM (COM 2001), Toronto, Ontario, Canada, 26–29 August 2001. Canadian Institute of Mining, Metallurgy and Petroleum, pp. 13–24. Xerox Tower Suite 1210 3400 de Maisonneuve Blvd. W, Montreal, PQ, Quebec H3Z 3B8, Canada, 2001.
Klemp, T., 1989. TAFS 97, 1009–1024.
Knott, J.F., Elliott, D., 1979. Worked Examples in Fracture Mechanics. Inst. Metals Monograph 4, Inst. Metals, London.
Kokai, K., 28 March 1985. Japan patent application by Toyota 60 54244.
Kolorz, A., Lohborg, K., 1963. In: 30th Internat. Found. Congress, pp. 225–246.
Kolsgaard, A., Brusethaug, S., 1994. Mater. Sci. Technol. 10 (6), 545–551.
Kondic, V., 1959. Foundry 87, 79–83.
Kono, R., Miura, T., 1975. Br. Found. 69, 70–78.
Koster, W., Goebring, K., 1941. Giesserei 28 (26), 521.
Kotschi, T.P., Kleist, O.F., 1979. AFS Int. Cast. Met. J. 4 (3), 29–38.
Kotschi, R.M., Loper, C.R., 1974. TAFS 82, 535–542.
Kotschi, R.M., Loper, C.R., 1975. TAFS 83, 173–184.
Kotsyubinskii, O. Yu, Gerchikov, A.M., Uteshev, R.A., Novikov, M.I., 1961. Russ. Cast. Prod. 8, 365–368.
Kotsyubinskii, O Yu, Oberman, Ya I., Gerchikov, A.M., 1962. Russ. Cast. Prod. 4, 190–191.
Kotsyubinskii, O. Yu, Oberman, Ya I., Gini, E.Gh, 1968. Russ. Cast. Prod. 4, 171–172.
Kotsyubinskii, O. Yu, 1961–1962. Russ. Cast. Prod. 269–272.
Kotsyubinskii, O. Yu, 1963. In: 30th Internat. Found. Cong, pp. 475–487.
Kotzin, E.L., 1981. Metalcasting and Molding Processes. American Foundrymen's Soc, Des Plaines, Illinois, USA.
Kraft, G., 1978. Metall. ISSN: 0075-2819 32 (6), 560–562.
Krinitsky, A.I., 1953. TAFS 61, 399–410.
Kron, J., Bellet, M., Ludwig, A., Pustal, B., Wendt, J., Fredriksson, H., 2004. Int. J. Cast Met. Res. 17 (5), 295–310.
Kruse, B.L., Richards, V.L., Jackson, P.D., 2006. TAFS 114, 783–795.
Kubo, K., Pehlke, R.D., 1985. Met. Trans. 16B, 359–366.
Kubo, K., Pehlke, R.D., 1986. Met. Trans. 17B, 903–911.
Kujanpaa, V.P., Moisio, T.J.I., 1980. In: Solidification Technology in the Foundry and Cast House, Warwick Conf, pp. 372–375 (Metals Soc. 1983).
Kulas, M.-A., Green, W.P., Taleff, E.M., Krajewski, P.E., McNelly, T.R., 2006. Metall. Mater. Trans. A 37A, 645–655.
Kunes, J., Chaloupka, L., Trkovsky, V., Schneller, J., Zuzanak, A., 1990. TAFS 98, 559–563.
Kuyucak, S., 2002. TAFS, 110.
Kuyucak, S., February 2008. In: 68th World Foundry Congress, pp. 483–487.
Lagowski, B., Meier, J.W., 1964. TAFS 72, 561–574.
Lagowski, B., 1967. TAFS 75, 229–256.
Lagowski, B., 1979. TAFS 87, 387–390.
Lai, N.-W., Griffiths, W.D., Campbell, J., 2003. In: Stefanescu, D.M., Warren, J.A., Jolly, M.R., Krane, M.J.M. (Eds.), Modeling of Casting, Welding and Solidification Processes-X, pp. 415–422.
Laid, E., 16 February 1978. US patent application number 878309.

Lalpoor, M., Eskin, D.G., Katgerman, L., 2009. Metall. Mater. Trans. A 40A (13), 3304–3313.
Lalpoor, M., Eskin, D.G., Katgerman, L., 2010. Metall. Mater. Trans. A 41A (9), 2425–2434.
Lane, A.M., Stefanescu, D.M., Piwonka, T.S., Pattabhi, R., October 1969. Mod. Cast. 54–55.
Lang, G., 1972. Aluminium 48 (10), 664–672.
Lansdow, P., 1997. Personal communication regarding casting plant from KWC, Unterkulm, Switzerland, seen at Maynal Castings limited, Wolverhampton, UK.
Larranaga, P., Asenjo, I., Sertucha, J., Suarez, R., Ferrer, I., Lacaze, J., 2009. Trans. Met. Mater. A 40A, 654–661.
Lashkari, O., Ghomashchi, R., April 2007. Mater. Sci. Eng. A 454–455, 30–36.
Laslaz, G., Laty, P., 1991. TAFS 99, 83–90.
Latimer, K.G., Read, P.J., 1976. Br. Found. 69, 44–52.
LaVelle, D.L., 1962. TAFS 70, 641–647.
Lawrence, M., February 1990. Mod. Cast. 51–53.
Ledebur, A., 1882. Stahl Eisen 2, 591.
Leduc, L., Nadarajah, T., Sellars, C.M., July 1980. Met. Technol. 269–273.
Lee, J.-K., Kim, S.K., March 2007. Mater. Sci. Eng. A 449–451, 680–683.
Lee, R.S., 1987. TAFS 95, 875–882.
Lees, D.C.G., 1946. J. Inst. Met. 72, 343–364.
Lerner, Y., Aubrey, L.S., 2000. TAFS 108, 219–226.
Lerner, S.Y., Kouznetsov, V.E., May 2004. Mod. Cast. 37–41.
Leth-Olsen, H., Nisancioglu, K., 1998. Corros. Sci. 40, 1179–1194 and 1194–1214.
Lett, R.L., Felicelli, S.D., Berry, J.T., Cuesta, R., Losua, D., 2009. TAFS 8, paper 09-043.
Levelink, H.G., van den Berg, H., 1962. TAFS 70, 152–163.
Levelink, H.G., van den Berg, H., 1968. TAFS 76, 241–251.
Levelink, H.O., van den Berg, H., 1971. TAFS 79, 421–432.
Levelink, H.O., Julien, F.P.M.A., June 1973. AFS Cast Met. Res. J. 56–63.
Levelink, H.O., 1972. TAFS 80, 359–368.
Leverant, G.R., Gell, M., 1969. TAIME 245, 1157–1172.
Levinson, D.W., Murphy, A.H., Rostoker, W., 1955. TAFS 63, 683–686.
Lewis, G.M., 1961. Proc. Phys. Soc. (London) 71, 133.
Lewis, R., Ransing, R., 1998. Metall. Mater. Trans. B 29B (2), 437–448.
Li, Y.X., Liu, B.C., Loper, C.R., 1990. TAFS 98, 483–488.
Li, Y., Jolly, M.R., Campbell, J., 1998. In: Thomas, B.G., Beckermann, C. (Eds.), Modeling of Casting, Welding and Advanced Solidification Processes VIII. The Minerals, Metals and Materials Soc., pp. 1241–1253.
Li, D.Z., Campbell, J., Li, Y.Y., 2004. J. Process. Technol. 148 (3), 310–316.
Li, J., Wang, B., Li, J., July 2014. ICASP 4: The 4th Internat. Conf. on Advances in Solidification Processes, Windsor, UK, pp. 8–11.
Li, Q.L., Xia, T.D., Lan, Y.F., Li, P.F., 2014. Mater. Sci. Technol. 30 (7), 835–841.
Liass, A.M., Borsuk, 18 December 1962. French Patent 1342529.
Liass, A.M., 1968. Foundry Trade J. 124, 3–10.
Liddiard, E.A.G., Baker, W.A., 1945. TAFS 53, 54–65.
Lin, H.J., Hwang, W.-S., 1988. TAFS 96, 447–458.
Lin, D.-Y., Shih, M.C., 2003. Int. J. Cast Met. Res. 16 (6), 537–540.
Lin, J., Sharif, M.A.R., Hill, J.L., 1999. Alum. Trans. 1 (1), 72–78.
Lind, M., Holappa, L., 2010. Metall. Mater. Trans. B 41B, 359–366.
Ling, Y., Mampaey, F., Degrieck, J., Wettinck, E., 2000. In: Sahm, P.R., Hansen, P.N., Conley, J.G. (Eds.), Modeling of Casting, Welding and Advanced Solidification Processes IX, pp. 357–364.
Liu, S., Loper, C.R., 1990. TAFS 98, 385–394.
Liu, P.C., Loper, C.R., Kimura, T., Park, H.K., 1980. TAFS 88, 97–118.
Liu, P.C., Li, C.L., Wu, D.H., Loper, C.R., 1983. TAFS 91, 119–126.
Liu, S.L., Loper, C.R., Witter, T.H., 1992. TAFS 100, 899–906.
Liu, F., Zhao, D.W., Yang, G.C., 2001. Metall. Mater. Trans. B 32B, 449–460.

Liu, L., Samuel, A.M., Samuel, F.H., Dowty, H.W., Valtierra, S., 2002. The role of Sr oxide on porosity. Trans. AFS 110, 449–462.
Liu, L., Samuel, A.M., Samuel, F.H., Doty, H.W., Valtierra, S., 2003a. The role of Fe in Sr-modified 319 and 356. J. Mater. Sci. 38, 1255–1267.
Liu, L., Samuel, A.M., Samuel, F.H., Doty, H.W., Valtierra, S., 2003b. Int. J. Cast Met. Res. 16 (4), 397–408.
Liu, Z., Wang, F., Qiu, D., Taylor, J.A., Zhang, M., September 2013. Metall. Mater. Trans. A 44A, 4025–4030.
Livingston, H.K., Swingley, C.S., 1971. Surf. Sci. 24, 625–634.
Llewelyn, G., Ball, J.I., 1962. GB Patent Application, Complete Published 1965, 987060.
Lo, H., Campbell, J., 2000. In: Sahm, P.R., Hansen, P.N., Conley, J.G. (Eds.), Modeling of Casting, Welding and Advanced Solidification Processes IX, pp. 373–380.
Locke, C., Ashbrook, R.L., 1950. TAFS 58, 584–594.
Locke, C., Ashbrook, R.L., 1972. TAFS 80, 91–104.
Locke, C., Berger, M.J., 1951. Steel Founders Soc. America Research Report 25 "The Flow of Steel in Sand Molds Part II".
Longden, E., 1931–32. Proc. IBF 25, 95–145.
Longden, E., 1939–40. Proc. IBF 33, 77–107.
Longden, E., 1947–48. Proc. IBF 41, A152–A165.
Longden, E., 1948. TAFS 56, 36–56.
Loper, C.R., Fang, K., 2008. TAFS 8, paper 08-065 (05).
Loper, C.R., Heine, R.W., 1961. TAFS 69, 583–600.
Loper, C.R., Heine, R.W., 1968. TAFS 76, 547–554.
Loper, C.R., Kotschi, R.M., 1974. TAFS 82, 279–284.
Loper, C.R., LeMahieu, D.L., 1971. TAFS 79, 483–492.
Loper, C.R., Miskinis, 1985. TAFS 93, 545–560.
Loper, C.R., Newby, M.R., 1994. TAFS 102, 897–901.
Loper, C.R., Saig, A.G., 1976. TAFS 84, 765–768.
Loper, C.R., Javaid, A., Hitchings, J.R., 1996. TAFS 104, 57–65.
Loper, C.R., 1992. TAFS 100, 533–538.
Loper, C.R., 1999. TAFS 107, 523–528.
Low, J.R., 1969. Trans. AIME 245, 2481–2494.
Lu, Y.H., Chen, Z.R., Zhu, X.F., Shoji, T., 2014. Mater. Sci. Technol. 30 (15), 1944–1950.
Lubulin, I., Christensen, R.J., 1960. TAFS 68, 539–550.
Ludwig, T.H., Schaffer, P.L., Arnberg, L., December 2013. Metall. Mater. Trans. 44A, 5796–5805.
Lukens, M.C., Hou, T.X., Pehlke, R.D., 1990. TAFS 98, 63–70.
Lukens, M.C., Hou, T.X., Purvis, A.L., Pehlke, R.D., 1991. TAFS 99, 445–449.
Lumley, R.N., Sercombe, T.B., Schaffer, G.B., 1999. Metall. Mater. Trans. 30A, 457–468.
Lumley, R.N., O'Donnell, R.G., Gunasegaram, D.R., Givord, M., 2007. Metall. Mater. Trans. A 38A, 2565–2574.
Lumley, R., May 2013. Foundry Trade J. 106–107.
Lynch, R.F., Olley, R.P., Gallagher, P.C.J., March 1977. AFS Int. Cast Met. J. 61–86.
Ma, Z., Samuel, A.M., Samuel, F.H., Doty, H.W., Valtierra, S., 2008. Mater. Sci. Eng. A 490, 36–51.
Mae, Y., Sakonooka, A., 1985. Met. Soc. AIME, TMS, Paper 0197 7 (US Patent 4808243, 1987).
Maeda, Y., Nomura, H., Otsuka, Y., Tomishige, H., Mori, Y., 2002. Int. J. Cast Met. Res. 15, 441–444.
Maidment, L.J., Walter, S., Raul, G., 1984. In: 8th Heat Treating Conf., Detroit.
Majumdar, I., Raychaudhuri, B.C., 1981. Int. J. Heat Mass Transf. 24 (7), 1089–1095.
Malzer, G., Hayes, R.W., Mack, T., Eggler, G., 2007. Metall. Mater. Trans. A 38A (2), 314–327.
Mampaey, F., Beghyn, K., 2006. TAFS 114, 637–656.
Mampaey, F., Xu, Z.A., 1997. TAFS 105, 95–103.
Mampaey, F., Xu, Z.A., 1999. TAFS 107, 529–536.
Mampaey, F., Habets, D., Plessers, J., Seutens, F., 2008. Int. Found. Res./Geissereiforschung 60 (1), 2–19.
Mampaey, F., 1999. TAFS 107, 425–432.
Mampaey, F., 2000. TAFS 108, 11–17.
Mandal, B.P., 2000. Indian Foundry J. 46 (4), 25–28.

Mansfield, T.L., 1984. In: Proc. Conf. "Light Metals". Met. Soc. AIME, pp. 1305–1327.
Manzari, M.T., Lewis, R.W., Gethin, D.T., 2000. Optimum design of chills in sand casting process. In: Proc. IMECE 2000 International Mech. Eng. Congr., Florida, USA. DETC98/DAC-1234 ASME, p. 8.
Marck, C.T., Keskar, A.R., 1968. TAFS 76, 29–43.
Marin, T., Utigard, T., 2010. Metall. Mater. Trans. B 41B (3), 535–542.
Mariotto, C.L., 1994. TAFS 102, 567–573.
Martin, L.C.B., Keller, C.T., Shivkumar, S., 1992. In: AFS 3rd Internat. Conf. on Molten Aluminum Processing. Orlando, Florida, pp. 79–91.
Mashl, S.J., 2008. ASM Handbook. In: Casting, vol. 15, pp. 408–416.
Maske, F., Piwowarsky, E., March 1929. Foundry Trade J. 28, 233–243.
Masuda, S., Toda, H., Aoyama, S., Orii, S., Ueda, S., Kobayashi, M., 2009. J. Jpn. Found. Eng. Soc. 81, 312–322.
Mathier, V., Grasso, P.-D., Rappaz, M., 2008. Metall. Mater. Trans. 39A, 1399–1409.
Mathier, V., Vernede, S., Jarry, P., Rappaz, M., 2009. Metall. Mater. Trans. A 40A (4), 943–957.
Matsubara, Y., Suga, S., Trojan, P.F., Flinn, R.A., 1972. TAFS 80, 37–44.
Matsuda, M., Ohmi, M., 1981. AFS Int. Cast. Met. J. 6 (4), 18–27.
Mazed, S., Campbell, J., 1992. Univ. Birmingham, UK, unpublished work.
Mbuya, T.O., Oduari, M.F., Rading, G.O., Wekesa, M.S., 2006. Int. J. Cast Met. Res. 19 (6), 357–360.
McCartney, D.G., 1989. Int. Mater. Rev. 34 (5), 247–260.
McClain, S.T., McClain, A.S., Berry, J.T., 2001. TAFS 109, 75–86.
McDavid, R.M., Dantzig, J., 1998. Metall. Mater. Trans. 29B, 679–690.
McDonald, R.J., Hunt, J.D., 1969. Trans. AIME 245, 1993–1997.
McDonald, R.I., Hunt, J.D., 1970. Trans. AIME 1, 1787–1788.
McGrath, C., Fischer, R.B., 1973. TAFS 81, 603–620.
McKim, P.E., Livingstone, K.E., 1977. TAFS 85, 491–498.
McParland, A.J., 1987. In: Sheffield Conf. "Solidification Processing" Institute of Metals, London, 1988, pp. 323–326.
Medvedev Ya, I., Kuzukov, V.K., 1966. Russ. Cast. Prod. 263–266.
Mehta, R., 1 April 2008. Mater. World.
Mejia, I., Altamirano, G., Bedolla-Jacuinde, A., Cabrera, J.M., November 2013. Metall. Mater. Trans. A 44A, 5165–5176.
Melendez, A.J., Carlson, K.D., Beckermann, C., 2010. Int. J. Cast Met. Res. 23 (5), 278–288.
Mendelson, S.J., 1962. Appl. Phys. 33 (7), 2182–2186.
Mertz, J.M., Heine, R.W., 1973. TAFS 81, 493–495.
Merz, R., Marincek, B., 1954. In: 21st International Foundry Congress. Paper 44, pp. 1–7.
Metcalf, G.J., 1945. J. Inst. Met. 1029, 487–500.
Meyers, C.W., 1986. TAFS 94, 511–518.
Mi, J., Harding, R.A., Campbell, J., 2002. Int. J. Cast Met. Res. 14 (6), 325–334.
Micks, F.W., Zabek, V.J., 1973. TAFS 81, 38–42.
Middleton, 1. M., Canwood, B., 1967. Br. Found. 60, 494–503.
Middleton, J.M., 1953. TAFS 61, 167–183.
Middleton, J.M., 1964. Br. Found. 57, 1–19.
Middleton, J.M., 1965. Br. Found. 58, 13–24.
Middleton, J.M., 1970. Br. Found. 64, 207–223.
Midea, A.C., 2001. TAFS 109, 41–50 and Foundryman March 2003 60–63.
Midson, S.P., September 2008. Die Cast. Eng. Semi solid casting of aluminum alloys - an update.
Miguelucci, E.W., 1985. TAFS 93, 913–916.
Mihaichuk, W., February 1986. Mod. Cast. 36–38 and (March) 33–35.
Mikkola, P.H., Heine, R.W., 1970. TAFS 78, 265–268.
Miles, G.W., 1956. Proc. Inst. Br. Found. 49, A201–A210.
Miller, G.F., 1967. Foundry 95, 104–107.
Minakawa, S., Samarasekera, I.V., Weinberg, F., 1985. Met. Trans. 16B, 823–829.
Minetola, P., Iuliano, L., Argentieri, G., 2012. Int. J. Cast Met. Res. 25 (1), 38–46.
Mintz, B., Yue, S., Jonas, J.J., 1991. Int. Mater. Rev. 36 (5), 187–217.

Mirak, A.R., Divandari, M., Boutorabi, S.M.A., Campbell, J., 2007. Int. J. Cast Met. Res. 20 (4), 215–220.
Mirak, A.R., Divandari, M., Boutorabi, S.M.A., August 2010. Mater. Sci. Technol.
Miresmaeili, S.M., Campbell, J., Shabestari, S.G., Boutorabi, S.M.A., 2005. Metall. Mater. Trans. A 36A, 2341–2349.
Miresmaeili, S.M., 2006. University of Birmingham, UK, unpublished work.
Miyagi, Y., Hino, M., Tsuda, O., 1985. "R&D" Kobe Steel Engineering Reports 1983 33 July (3). Also published in Kobelco Technical Bulletin, 1076 (1985).
Mizoguchi, T., Perepezko, J.H., Loper, C.R., 1997. TAFS 105, 89–94 and Mater. Sci. Eng. A 1997 A226–A228, 813–817.
Mizumoto, M., Sasaki, S., Ohgai, T., Kagawa, A., 2008. Int. J. Cast Met. Res. 21 (1–4), 49–55.
Mizuno, K., Nylund, A., Olefjord, I., 1996. Mater. Sci. Technol. 12, 306–314.
Mohanty, P.S., Gruzleski, J.E., 1995. Acta Metall. Mater. 43, 2001–2012.
Mohla, P.P., Beech, J., 1968. Br. Found. 61, 453–460.
Molaroni, A., Pozzesi, V., 1963. In: 30th Internat. Found. Congress, pp. 145–161.
Molina, J.M., Voytovych, R., Louis, E., Eustathopoulos, N., 2007. Int. J. Adhes. Adhes. 27, 394–401.
Mollard, F.R., Davidson, N., 1978. TAFS 78, 479–486.
Momchilov, E., 1993. Institute for Metal Science, Bulgarian Academy of Sciences, Sofia J. Mater. Sci. Technol. 1 (1), 5–12.
Mondloch, P.A., Baker, D.W., Euvrard, L., 1987. TAFS 95, 385–392.
Monroe, R.W., Blair, M., 1995. Am. Found. Soc. Trans. 103, 633–640.
Morgan, A.D., 1966. Br. Found. 59, 186–204.
Morgan, P.C., 1989. Met. Mater. 5 (9), 518–520.
Morthland, T.E., Byrne, P.E., Tortorelli, D.A., Dantzig, J.A., 1995. Metall. Mater. Trans. B 26B, 871–885.
Moumeni, E., Stefanescu, D.M., Tiedje, N.S., Larranaga, P., Hattel, J.H., November 2013. Metall. Mater. Trans. 44A, 5134–5146.
Mountford, N.D.G., Calvert, R., 1959–1960. J. Inst. Met. 88, 121–127.
Mountford, N.D.G., Sommerville, I.D., 1993. Steel Technology International. London, UK, pp. 155–169.
Mountford, N.D.G., Sommerville, I.D., From, L.E., Lee, S., Sun, C., 1992/1993. In: A Measuring Device for Quality Control in Liquid metals. Scaninject Conf., Lulea, Sweden.
Mountford, N.D.G., Sommerville, I.D., Simionescu, A., Bai, C., 1997. TAFS 105, 939–946.
Mroz, S.S., Goodrich, G.M., 2006. TAFS 114, 493–505.
Mufti, N.A., Webster, P.D., Dean, T.A., 1995. Mater. Sci. Technol. 11 (8), 803–809.
Muhmond, H.M., Frederiksson, H., 2013. Metall. Mater. Trans. 44B, 283–298.
Mukherjee, M., Garcia-Moreno, F., Banhart, J., 2010. Metall. Mater. Trans. B 41B (3), 500–504.
Mulazimoglu, M.H., Handiak, N., Gruzleski, J.E., 1989. TAFS 97, 225–232.
Mulazimoglu, M.H., Tenekedjiev, N., Closset, B.M., Gruzleski, J.E., 1993. Cast. Met. 6 (1), 16–28.
Muller, W., Feikus, F.J., 1996. TAFS 104, 1111–1117.
Muller, F.C.G., 1887. Zeit. ver dent. Ingenieure 23, 493.
Munoz-Moreno, R., Perez-Prado, M.T., Llorca, J., Ruiz-Navas, E.M., Boehlert, C.J., April 2013. Metall. Mater. Trans. A 44A, 1887–1896.
Mutharasan, R., Apelian, D., Romanowski, C., 1981. J. Met. 83 (12), 12–18.
Mutharasan, R., Apelian, D., Ali, S., 1985. Met. Trans. 16B, 725–742.
Muthukumarasamy, S., Seshan, S., 1992. TAFS 100, 873–879.
Myllymaki, R., 1987. In: Young, W.B. (Ed.), Residual Stress in Design, Process and Materials Selection. ASM Conf., USA, pp. 137–141.
Nadell, R., Eskin, D., Katgerman, L., 2007. Mater. Sci. Technol. 23 (11), 1327–1335.
Nai, S.M.L., Gupta, M., 2002. Mater. Sci. Technol. 18, 633–641.
Nakae, H., Shin, H., 1999. Int. J. Cast Met. Res. 11 (5), 345–349.
Nakae, H., Takai, K., Okauchi, K., Koizumi, H., 1991. IMONO 63, 692–703.
Nakae, H., Koizumi, H., Takai, K., Okauchi, K., 1992. IMONO Either 64, 34–39.
Nakagawa, Y., Momose, A., 1967. Tetsu-to-Hagane 53, 1477–1508.
Naro, R.L., Pelfrey, R.L., 1983. TAFS 91, 365–376.
Naro, R.L., Tanaglio, R.O., 1977. TAFS 85, 65–74.

Naro, R.L., Wallace, J.F., 1967. TAFS 75, 741–758.
Naro, R.L., Wallace, J.F., 1992. TAFS 100, 797–820.
1992. TAFS 100, 797–820.
Naro, R.L., http://www.asi-alloys.com/DocsPDF/Naro.pdf, the first 2 paragraphs of p. 6.
Naro, R.L., 1974. TAFS 82, 257–266.
Naro, R.L., 2004. TAFS 112, 527–545.
Nath, R.H., Grupke, C.C., Leonard, C., Johnson, M.K., 2011. In: Crepeau, P., et al. (Eds.), Shape Casting the John Berry Honorary Symposium. TMS, San Diego.
Navisi, S., 2010. Mod. Cast. 100 (9), 35–38.
Nayal, G.El, Beech, J., 1987. In: 3rd Internat. Conf. Solidification Processing, Sheffield, pp. 384–387 (Inst Metals Publication 421).
Nayal, G.El, 1986. Mater. Sci. Technol. 2, 803.
Nazar, A.M.M., Cupini, N.L., Prates, M., Daview, G.J., 1979. Metall. Trans. B 10B, 203–210.
Neiswaag, H., Deen, H.J.J., 1990. In: 57 World Foundry Congress, Osaka, Japan.
Nereo, G.E., Polch, R.F., Flemings, M.C., 1965. TAFS 73, 1–13.
Neumeier, L.A., Betts, B.A., Crosby, R.L., 1976. TAFS 84, 437–448.
Newell, M., D'Souza, N., Green, N.R., 2009. Int. J. Cast Met. Res. 22 (1–4), 66–69.
Nguyen, T., Carrig, J.F., 1986. TAFS 94, 519–528.
Nguyen, T., Loose, G.de, Carrig, J., Nguyen, V., Cowley, B., 2006. TAFS 114, 695–706.
Ni, P., Jonsson, L.T.I., Ersson, M., Johsson, P.G., December 2014. Metall. Mater. Trans. B 45B, 2414–2424.
Nicholas, K.E.L., Roberts, R.W., 1961. BCIRA J. 9 (4), 519.
Nicholas, K.E.L., 1972. Br. Found. 65, 441–451.
Nikolai, M.F., 1996. TAFS 104, 1017–1029.
Nishida, Y., Droste, W., Engler, S., 1986. Met. Trans. 17B, 833–844.
Niu, J.P., Yang, K.N., Sun, X.F., Jin, T., Guan, H.R., Hu, Z.Q., 2002. Mater. Sci. Technol. 18 (9), 1041–1044.
Niu, J.P., Sun, X.F., Jin, T., Yang, K.N., Guan, H.R., Hu, Z.Q., 2003. Mater. Sci. Technol. 19 (4), 435–439.
Niyama, E., Ishikawa, M., 1966. In: 4th Asian Foundry Congress, pp. 513–523.
Niyama, E., Uchida, T., Morikawa, M., Saito, S., 1982. In: Internat. Found. Congress 49, Chicago, paper 10.
Noesen, S.J., Williams, H.A., 1966. In: 4th National Die Casting Congress, Cleveland, Ohio, paper number 801.
Noguchi, T., Kamota, S., Sato, T., Sakai, M., 1993. TAFS 101, 231–239.
Noguchi, T., Kano, J., Noguchi, K., Horikawa, N., Nakamura, T., 2001. Int. J. Cast Met. Res. 13, 363–371.
Noguchi, T., Horikawa, N., Nagate, H., Nakamura, T., Sato, K., 2005. Int. J. Cast Met. Res. 18 (4), 214–220.
Nolli, P., Cramb, A.W., 2008. Metall. Mater. Trans. 39B, 56–65.
Nordland, W.A., 1967. Trans. AIME 239, 2002–2004.
Nordlien, J.H., Davenport, A.J., Scamans, G.M., 2000. In: ASST 2000 (2nd Internat. Symp. Al Surface Science Technology) UMIST Manchester, UK. Alcan + Dera, pp. 107–112.
Nordlien, J.H., 1999. Alum. Extrus. 4 (4), 39–41.
Noreo, G.E., Polich, R.F., Flemings, M.C., 1965. TAFS 73, 1–13 and 28–33.
Northcott, L., 1941. J. Iron Steel Inst. 143, 49–91.
Novikov, I.I., Grushko, O.E., 1995. Mater. Sci. Technol. 11, 926–932.
Novikov, I.I., Portnoi, V.K., 1996. Russ. Cast. Prod. 4, 163–166.
Novikov, I.I., Novik, F.S., Indenbaum, G.V., 1966. Izv. Akad. Nauk. Met. 5, 107–110 (English translation in Russ. Metall. Min., 1966 5, 55–59).
Novikov, I.I., April 1962. Russ. Cast. Prod. 167–172.
Nyahumwa, C., Green, N.R., Campbell, J., 1998. TAFS 106, 215–223.
Nyahumwa, C., Green, N.R., Campbell, J., 2000. Metall. Mat. Trans. A 31A, 1–10.
Nyamekye, K., An, Y.-K., Bain, R., Cunningham, M., Askeland, D., Ramsay, C., 1994a. TAFS 102, 127–131.
Nyamekye, K., Wei, S., Askeland, D., Voigt, R.C., Pischel, R.P., Rasmussen, W., Ramsay, C., 1994b. TAFS 102, 869–876.
Nyichomba, B.B., Campbell, J., 1998. Int. J. Cast Met. Res. 11 (3), 163–186.
O'Hara, P., February 1990. Engineering 41–42.
Ocampo, C.M., Talavera, M., Lopez, H., 1999. Metall. Mater. Trans. A 30A, 611–620.

Oelsen, W., 1936. Stahl Eisen 56, 182.
Ohnaka, 2003. Mould Coat Doubles the Back-Pressure in Moulds.
Ohno, A., 1987. Solidification: The Separation Theory and its Practical Applications. Springer-Verlag.
Ohsasa, K.-I., Takahash, T., Kobori, K.J., 1988. Jpn. Inst. Met. 52 (10), 1006–1011 and Ohsasa, K.-I., Ohmi, T., Takahashi, T., 1988. Bull. Fac. Eng. Hokkaido Univ. 143 (in Japanese).
Okada, Y., Fujii, N., Goto, A., Morimoto, S., Yusuda, Y., 1982. TAFS 90, 135–146.
Okamoto, S., August 2013. Personal Communication to JC 01. Ehime University, Matsuyama, Japan.
Oki, S., 1970. IMONO J. Jpn. Found. Soc. 42 (11), 9. English Translation in Bulletin of the Faculty of Engineering, Yokohama National University, March 1969, vol. 18, pp. 127–136.
Omura, N., Tada, S., TMS Annual Meeting, 2012.
Oper, C.R., 1981. TAFS 89, 405–408.
Osborn, D.A., 1979. Br. Found. 72, 157–161.
Ostrom, T.R., Trojan, P.K., Flinn, R.A., 1974. TAFS 82, 519–524.
Ostrom, T.R., Trojan, P.K., Flinn, R.A., 1975. TAFS 83, 485–492.
Ostrom, T.R., Frasier, D.J., Trojan, P.K., Flinn, R.A., 1976. TAFS 84, 665–674.
Ostrom, T.R., Trojan, P.K., Flinn, R.A., 1981. TAFS 89, 731–736.
Ostrom, T.R., Biel, J., Wager, W., Flinn, R.A., Trojan, P.K., 1982. TAFS 90, 701–709.
Outlaw, R.A., Peterson, D.T., Schmidt, F.A., 1981. Met. Trans. 12A, 1809–1816.
Owen, M., 1966. Br. Found. 59, 415–421.
Owusu, Y.A., Draper, A.B., 1978. TAFS 86, 589–598.
Pakes, M., Wall, A., 1982. Zinc Development Assoc., UK.
Pan, E.N., Hu, J.F., 1996. In: 4th Asian Foundry Congress, pp. 396–405.
Pan, E.N., Hu, J.F., 1997. TAFS 105, 413–418.
Pan, E.N., Hsieh, M.W., Jang, S.S., Loper, C.R., 1989. TAFS 97, 379–414.
Panchanathan, V., Seshadri, M.R., Ramachandran, A., 1965. Br. Found. 58, 380–384.
Pandee, P., Gourley, G.M., Beyakov, S.A., Ozaki, R., Yasuda, H., Limmanaevichit, C., September 2014. Metall. Mater. Trans. A 45A, 4549–4560.
Papworth, A., Fox, P., 1998. Mater. Lett. 35, 202–206.
Paray, F., Kulunk, B., Gruzleski, J.E., 2000. Int. J. Cast Met. Res. 13, 147–159.
Parkes, T.W., Loper, C.R., 1969. TAFS 77, 90–96.
Parkes, W.B., 1952. TAFS 60, 23–37.
Pattabhi, R., Lane, A.M., Piwonka, T.S., 1996. Part III, TAFS 104, 1259–1264.
Patterson, W., Koppe, W., 1962. Giesserel 4, 225–249.
Patterson, W., Engler, S., Kupfer, R., 1967a. Giesserei-Foschung 19 (3), 151–160.
Pattyn, R., 1967. In: 34th Internat. Foundry Cong., Paris, paper 26.
Paxton D.M., Dudder G.J., Reynolds J., Charron W., Cleaver T., TMS Annual Meeting Light Metals Division Session III, 2006.
Pechiney Aluminium, 26 April 1977. British Patent, 1574321.
Pekguleryuz, M.O., Lin, S., Ozbakir, E., Temur, D., Aliravci, C., 2010. Int. J. Cast Met. Res. 23 (5), 310–320.
Pelleg, J., Heine, R.W., 1966. TAFS 74, 541–544.
Pellerier, M., Carpentier, M., April 1988. Hommes et Fonderie 184, 7–14. In French. Abstract English translation BCIRA 1989 Abstr. 27.
Pellini, W.S., 1952. Foundry 80, 124–133, 194, 196, 199.
Pellini, W.S., 1953. TAFS 61, 61–80 and 302–308.
Pell-Walpole, W.T., 1946. J. Inst. Met. 72, 19–30.
Pennors, A., Samuel, A.M., Samuel, F.H., Doty, H.W., 1998. TAFS 106, 251–264.
Perbet, D., 1988. Hommes Fonderie (Mars) 15. French plus English Translation.
Perets, S., Arbel, A., Ariely, A., Venkert, A., Schneck, R.Z., 2004. Mater. Sci. Technol. 20 (12), 1519–1524.
Peters, T.M., Twarog, D.L., 1992. TAFS 100, 1005–1023.
Peterson, W.M., Blanke, J.E., 1980. TAFS 88, 503–506.
Petro, A., Flinn, R.A., 1978. TAFS 86, 357–364.
Petrzela, L., 31 October 1968. Foundry Trade J. 693–696.

Pettersson, H., 1951. TAFS 59, 35–55.
Phillion, A.B., Vernede, S., Rappaz, M., Cockcroft, S.L., Lee, P.D., 2009. Int. J. Cast Met. Res. 22 (1–4), 1–5.
Pillai, R.M., Mallya, V.D., Panchanathan, V., 1976. TAFS 84, 615–620.
Pitcher, P.D., Forsyth, P.J.E., November 1982. The Influence of Microstructure on the Fatigue Properties of an Al Casting Alloy. Royal Aircraft Establishment Technical Report 82107.
Piwonka, T.S., Flemings, M.C., 1966. Trans. Met. Soc. AIME 236, 1157–1165.
Plyatskii, V.M., 1965. "Extrusion Casting" Primary Sources. New York, USA.
Poirier, D.R., Yeum, K., 1987. In: Solidification Processing Conference, Sheffield. Institute of Metals, London.
Poirier, D.R., Yeum, K., 1988. In: Light Metals Conf., USA, pp. 469–476.
Poirier, D.R., Yeum, K., Maples, A.L., 1987. Met. Trans. 18, 1979–1987.
Poirier, D.R., Sung, P.K., Felicelli, S.D., 2001. TAFS 105, 139–155.
Poirier, D.R., 1987. Met. Trans. 18B, 245–255.
Pokorny, M.G., Monroe, C.A., Beckermann, C., 2009. In: Campbell, J., Crepeau, P.N., Tiryakioglu, M. (Eds.), Shape Casting: The 3rd International Symposium. TMS.
Polich, R.F., Flemings, M.C., 1965. TAFS 73, 28–33.
Pollard, W.A., 1964. TAFS 72, 587–599.
Pollard, W.A., 1965. TAFS 73, 371–379.
Polmear, I.J., 2006. Light Alloys, fourth ed. Butterworth Heinemann.
Polodurov, N.N., 1965. Russ. Cast. Prod. 5, 209–210.
Pope, J.A., 1965. Br. Found. 58, 207–224.
Popel, G.I., Esin, O.A., 1956. Zh. Fiz. Khim 30, 1193.
Porter, L.F., Rosenthal, P.C., 1952. Trans. Am. Found. Soc. 60, 125–136.
Portevin, A., Bastien, P., 1934. J. Inst. Met. 54, 45–58.
Portevin, A., Bastien, P., 1936. Inst. Br. Found. In: 33rd Annual Conf. Glasgow, pp. 88–116.
Pouly, P., Wuilloud, E., 1997. In: Light Metals (TMS Conference), pp. 829–835.
Pozdniakov, A.V., Zolotorevskiy, V.S., 2014. Int. J. Cast Met. Res. 27 (4), 193–198.
Prakash, M., Cleary, P., Grandfield, J., 2009. J. Mater. Process. Technol. 209 (7), 3396–3407.
Prates, M., Biloni, H., 1972. Met. Trans. 3A, 1501–1510.
Prible, J., Havlicek, F., 1963. In: 30th Internat. Foundry Congress, pp. 394–410.
Prodham, A., Carpenter, M., Campbell, J., 1999. CIATF Technical Forum.
Puhakka, R., Campbell, J., June 2009. Mod. Cast. 27–29.
Puhakka, R., 2009. Personal communication.
Puhakka, R., June 2009. Mod. Cast. 27–29.
Puhakka, R., November 2010. Foundry Trade J. 184 (3679), 277–279.
Puhakka, R., 2011. Shape Casting; the John Berry Honorary Symposium. TMS Annual Meeting, San Diego, CA.
Puhakka, R., 2011. www.castdifferently.com.
Pulkonik, K.J., Lee, W.E., Rosenberg, R.A., 1967. TAFS 75, 38–41.
Pumphrey, W.I., Lyons, J.V., 1948. J. Inst. Met. 74, 439–455.
Pumphrey, W.I., Moore, D.C., 1949. J. Inst. Met. 75, 727–736.
Pumphrey, W.I., 1955. Researches into the Welding of Aluminium and its Alloys. Research Report 27. Aluminium Development Association, UK.
Purdom, P., March 1992. Met. Mater. 169.
Purvis, A.L., Hanslits, C.R., Diehm, R.S., 1994. TAFS 102, 637–644.
Qian, M., Graham, D., Zheng, L., St John, D.H., Frost, M.T., 2003. Mater. Sci. Technol. 19 (2), 156–162.
Qingchung, L., Kuiying, C., Chi, L., Songyan, Z., 1991. TAFS 99, 245–253.
Rabinovich, A., March 1969. AFS Cast Met. Res. J. 19–24.
Ragab, Kh A., Bournane, M., Samuel, A.M., Al-Ahmari, A.M.A., Samuel, F.H., Doty, H.W., 2013. Mater. Sci. Technol. 29 (4), 412–425.
Ragone, D.V., Adams, C.M., Taylor, H.F., 1956. TAFS 64, 640–652 and 653–657.
Rahmani, Kh, Nategh, S., 2010. Metall. Mater. Trans. A 41A (1), 125–137.
Raiszadeh, R., Griffiths, W.D., 2008. Metall. Mater. Trans. 39B, 298–303.

Raiszadeh, R., Griffiths, W.D., 2011. Metall. Mater. Trans. 43B, 133–143.
Raiszadeh, R., Nateghian, M., Doostmohammadi, H., December. Metall. Mater. Trans. B 43B, 1540–1549.
Raiszadeh, R., Amirinejhad, S., Doostmohammadi, H., 2013. Int. J. Cast Met. Res. 26 (6), 330–338.
Ramesh, G., Prabhu, K.N., August 2013. Metall. Mater. Trans. B 44B, 797–799.
Ramrattan, S.N., Guichelaar, P.J., Palukunnu, A., Tieder, R., 1996. TAFS 104, 877–886.
Ramseyer, J.C., Gabathuler, J.P., Feurer, U., 1982. Aluminium 58 (10), E192–E194, 581–585.
Ransley, C.E., Neufeld, H., 1948. J. Inst. Met. 74, 599–620.
Ransing, R., Pao, W.K.S., Lin, C., Snood, M.P., Lewis, R.W., 2004. Int. J. Cast Met. Res. 18, 1–12.
Rao, Y.K., Lee, H.G., 1984. Met. Trans. 15B, 396–400.
Rao, G.V.K., Panchanathan, V., 1973. Cast. Met. Res. J. 19 (3), 135–138.
Rao, T.S.V., Roshan, H.Md, 1988. TAFS 96, 37–46.
Rao, G.V.K., Srinivasan, M.N., Seshadri, M.R., 1975. TAFS 83, 525–530.
Rapoport, D.B., 1964. Foundry Trade J. 116, 169.
Rappaz, M., Drezet, J.-M., Gremaud, M., 1999. Metall. Mater. Trans. A 30A (2), 449–455.
Rappaz, M., Drezet, J.-M., Mathier, V., Vernede, S., July 2006. Materials Science Forum, vols. 519–521. Trans Tech Publications, Switzerland p.1665–1674.
Rappaz, M., 1989. Int. Mater. Rev. 34 (3), 93–123.
Rashid, A.K.M.B., Campbell, J., 2004. Metall. Mater. Trans. 35A (7), 2063–2071.
Rashid, M.S., Hanna, M.D., 1993. North American Die casting Association (NADCA) Conf., Cleveland, paper T93-041. pp. 105–111 (US Patent 4990310, 1989).
Rassenfoss, J.A., 1977. TAFS 85, 583–596.
Rauch, A.H., Peck, J.P., Thomas, G.F., 1959. TAFS 67, 263–266.
Rault, L., Allibert, M., Prin, M., Dubus, A., 1996. Light Met. 345–355.
Reddy, D.C., Murty, S.S.N., Chakravorty, P.N., 1988. TAFS 96, 839–844.
Reding, J.N., 1968. TAFS 76, 92–98.
Rege, R.A., Szekeres, E.S., Foreng, W.D., 1970. Met. Trans. 1, 2652–2653.
Ren, X.C., Zhou, Q.J., Shan, G.B., Chu, W.Y., Li, J.X., Su, Y.J., Qiao, L.J., 2008. Metall. Mat. Trans. 39A, 87–97.
Ren, Y., Zhang, L., Yang, W., Duan, H., December 2014. Metall. Mater. Trans. B 45B, 2057–2071.
Revankar, V., Baker, P., Schultz, A.H., Brandt, H., 2000. Light Met. 51–55.
Rezvani, M., Yang, X., Campbell, J., 1999. TAFS 107, 181–188.
Richins, D.S., Wetmore, W.O., 1951. In: AFS Symposium on Principles of Gating, pp. 1–24.
Richmond, O., Tien, R.H., 1971. J. Mech. Phys. Solids 19, 273–284.
Richmond, O., Hector, L.G., Fridy, J.M., 1990. Trans. ASME J. Appl. Mech. 57, 529–536.
Rickards, P.J., 1975. Br. Found. 68, 53–60.
Rickards, P.J., 1982. Br. Found. 75, 213–223.
Ricken, H., Ostermeier, M., Hoffmann, H., Fent, A., 2009. Cast. Plant Technol. (1), 30–32.
Riposan, I., Chisamera, M., Stan, S., Toboc, P., Ecob, E., White, D., 2008. Mater. Sci. Technol. 24 (5), 579–584.
Rishel, L.L., Pollock, T.M., Cramb, A.W., February 1999. In: Proc. 1999 Internat. Symp. Liquid Metal Processing and Casting; Santa Fe, NM, pp. 287–299.
Rivas, R.A.A., Biloni, H., 1980. Zeit. Met. 71 (4), 264–268.
Rivera, G., Boeri, R., Sikora, J., 2002. Mater. Sci. Technol. 18, 691–697.
Roberts, T.E., Kovarik, D.P., Maier, R.D., 1979. TAFS 87, 279–298.
Roberts, R.J., 1996. TAFS 104, 523–526.
Roedter, H., September 1986. Foundry Trade J. Int. 6.
Roehrig, K., 1978. TAFS 86, 75–92.
Rogberg, B., 1980. Solidification technology in the foundry and cast house. Warwick Conf, pp. 365–371 (Metals Soc, 198).
Rogers, K.P., Heathcock, C.J., 1990. US Patent 5316070. Date: 31 May 1994.
Rohatgi, P., 1990. Adv. Mater. Process. 2, 39–44.
Romankiewicz, F., 1976. AFS Int. Cast Met. J. 1 (4), 13–17.
Romero, J.M., Smith, R.W., Sahoo, M., 1991. TAFS 99, 465–468.
Rometsch, P.A., Schaffer, G.B., Taylor, J.A., 2001. Int. J. Cast Met. Res. 14 (1), 59–69.

Rong De, L., Xiang, Y.J., 1991. TAFS 99, 707–712.
Rooy, E.L., Fischer, E.F., 1968. Am. Found. Soc. Trans. 76, 237–240.
Rosenberg, R.A., Flemings, M.C., Taylor, H.F., 1960. TAFS 68, 518–528.
Rossmann, M., February 1982. Giesserei 69 (4), 102–103.
Rostoker, W., Berger, M.J., July 1953. Foundry 81, 100–105 and 260–265.
Roth, M.C., Weatherly, G.C., Miller, W.A., 1980. Acta Met. 28, 841–853.
Rouse, J., 1987. 54th Internat. Foundry Congress, New Dehli, p. 16. Paper 30.
Roviglione, A.N., Hermida, J.D., 2004. Trans. Metall. Mater. B 35B, 313–330.
Ruddle, R.W., Cibula, A., 1957. Inst. Met. Monogr. Report. 22, 5–32.
Ruddle, R.W., Mincher, A.L., 1949–1950. J. Inst. Met. 76, 43–90.
Ruddle, R.W., 1956. The Running and Gating of Sand Castings. Monograph & Report Series No 19, Institute of Metals, London.
Ruddle, R.W., 1960. TAFS 68, 685–690.
Runyoro, J., Boutorabi, S.M.A., Campbell, J., 1992. TAFS 100, 225–234.
Runyoro, J., 1992. Design of the Gating System (Ph.D. thesis). University of Birmingham.
Ruxanda, R., Sanchez, L.B., Massone, J., Stefanescu, D.M., 2001. Trans. Am. Found. Soc. 109, 37–48 (Cast iron division).
Sabatino Di, M., Syvertsen, F., Arnberg, L., Nordmark, A., 2005. Int. J. Cast Met. Res. 18 (1), 59–62.
Sabatino Di, M., Arnberg, L., Brusethaug, S., Apelian, D., 2006. Int. J. Cast Met. Res. 19 (2), 94–97.
Sadayappan, M., Fasoyinu, F.A., Thomson, J., Sahoo, M., 1999. TAFS 107, 337–342.
Sadayappan, M., Sahoo, M., Liao, G., Yang, B.J., Li, D., Smith, R.W., 2001. TAFS 109, 341–352.
Saeger, C.M., Ash, E.J., 1930. TAFS 38, 107–145.
Safaraz, A.R., Creese, R.C., 1989. TAFS 97, 863–870.
Sahoo, M., Whiting, L.V., 1984. TAFS 92, 861–870.
Sahoo, M., Worth, M., 1990. TAFS 98, 25–33.
Sahoo, M., Whiting, L.V., White, D.W.G., 1985. TAFS 93, 475–480.
Saia, A., Edelman, R.E., 1968. TAFS 76, 189–195.
Saigal, A., Berry, J.T., 1984. TAFS 92, 703–708.
Sakakibara, Y., Suzuki, T., Hayashi, H., et al. 1988. Japan Patent 17578/88, Europe patent EP0306841 A2.
Sakamoto, M., Akiyama, S., Ogi, K., 1996. In: 4th Asian Found. Cong. Proc., Australia, pp. 467–476.
Samarasekera, I.V., Anderson, D.L., Brimacombe, J.K., 1982. Met. Trans. 13B, 91–104.
Sambasivan, S.V., Roshan, H.Md, 1977. TAFS 85, 265–270.
Samuel, A.M., Samuel, F.H., 1993. Met. Trans. A 24A, 1857–1868.
Samuel, F.H., Samuel, A.M., Doty, H.W., Valtierra, S., 2001. Metall. Mater. Trans. A 32A, 2061–2075.
Sandford, P., March 1988. Foundryman 110–118.
Sandford, P., 1993. TAFS 101, 817–824.
Santos, R.G., Garcia, A., 1998. Int. J. Cast Met. Res. 11, 187–195.
Sarazin, J.R., Hellawell, A., 1987. In: Beech, J., Jones, H. (Eds.), Solidification Processing, Sheffield University, UK, pp. 94–97.
Sare, I.R., 1989. Cast. Met. 1 (4), 182–190.
Saucedo, I.G., Beech, J., Davies, G.J., 1980. In: Conf. "Solidification technology in foundry and cast house". Warwick Univ., Metals Society Publication 1983, pp. 461–468.
Scarber, P., Bates, C.E., 2006. TAFS 114, 37–44.
Scarber, P., Bates, C.E., Griffin, 2006. TAFS 114, 435–445.
Schaffer, P.L., Dahle, A.K., 2009. Metall. Mater. Trans. A 40A (2), 481–485.
Schaffer, P.L., Dahle, A.K., Zindel, J.W., 2004. Light Met. 821–826.
Schaffer, G.B., 2004. Mater. Forum 28, 65–74.
Scheffer, K.D., 1975. TAFS 83, 585–592.
Schilling, H., Reithmann, M., 1992. Cast. Plant Technol. (3), 12–14.
Schilling, H., 1987. Patent PCT WO 87/07543.
Schmidt, D.G., Jacobson, A.E., 1970. TAFS 78, 332–337.
Schmied, H.-J., 1988. Giessereitechnik 34 (4), 133–135.
Schneider, W., 2006. Cast. Plant Technol. (1), 30–37.
Schrey, A., June 2007. Foundry Pract. (Foseco) 246, 01–07.

Schumacher, P., Greer, A.L., 1993. Key Eng. Mater. 81-83, 631.
Schumacher, P., Greer, A.L., 1994. Mater. Sci. Eng. A181/A182, 1335–1339.
Schurmann, E., 1965. Arch. Eisenh. 36, 619–631 (BISI Translation 4579).
Schwandt, C., Fray, D.J., December 2014. Metall. Mater. Trans. B 45B, 2145–2152.
Sciama, G., 1974. TAFS 82, 39–44.
Sciama, G., 1975. TAFS 83, 127–140.
Sciama, G., 1993. TAFS 101, 643–651.
Scott, W.D., Bates, C.E., 1975. TAFS 83, 519–524.
Scott, D., Smith, T.J., 1985. Personal communication.
Scott, A.F., et al., 1948. J. Chem. Phys. 16, 495–502.
Scott, W.D., Goodman, P.A., Monroe, R.W., 1978. TAFS 86, 599–610.
SCRATA, 1981. Hot Tearing – Causes and Cures, Tech. Bull. No. 1. Steel Castings Research and Trade Assoc, Sheffield, UK.
Seetharamu, S., Srinivasan, M.W., 1985. TAFS 93, 347–350.
SERF, June 2002. Foundry Trade J. 34.
Sexton, A.H., Primrose, J.S.G., 1911. The Principles of Ironfounding. The Technical Publishing Co, London.
SFSA (Steel Founders Society of America), May 2000. Foundry Trade J. 40–41.
Shabestari, S.G., Gruzleski, J.E., 1995. TAFS 103, 285–293.
Shafaei, A., Raiszadeh, R., December 2014. Metall. Mater. Trans. B 45B, 2486–2494.
Shamsuzzoha, M., Nastac, L., Berry, J.T., 2012. TAFS 116, paper 12-101.
Shen, P., Zhen, X.-H., Lin, Q.-L., Zhang, D., Jiang, Q.-P., 2009. Metall. Mater. Trans. 40A (2), 444–449.
Shendye, S.B., Gilles, D.J., 2009. TAFS 117, 305–311.
Sherby, O.D., May 1962. Met. Eng. Q. (ASM) 3–13.
Shi, Q., Ding, X., Chen, J., Zhang, X., Zheng, Y., Feng, Q., 2014. Metall. Mater. Trans. A 45A, 1665–1669.
Ship Department Publication 18, 1975. Design & Manufacture of Nickel-Aluminium-Bronze Sand Castings. Ministry of Defence (Procurement Executive), Foxhill, Bath, UK.
Shirey, D.R., Williams, D.C., 1968. TAFS 76, 661–674.
Shivkumar, S., Wang, L., Steenhoff, B., 1989. TAFS 97, 825–836.
Showman, R.E., Aufderheide, R.C., 2003. 111, paper 145, p. 12.
Showman, R.E., Aufderheide, R.C., Yeomans, N.P., 2006. TAFS 114, 391–399.
Showman, R.E., Aufderheide, R.C., Yeomans, N.P., August 2007. Foundry Trade J. 224–227.
Sicha, W.E., Boehm, R.C., 1948. TAFS 56, 398–409.
Sidhu, R.K., Richards, N.L., Chaturvedi, M.C., 2005. Mater. Sci. Technol. 21 (10), 1119–1131.
Sieurin, S.I., June 1974. Foundry, page 94 and following.
Sieurin, S.I., March 1975. Paper G-T75-T045. In: 8th Soc. Die Casting Engineers Internat. Congress, Detroit, MI (4 pages).
Sigworth, G.K., Engh, T.A., 1982. Met. Trans. 13B, 447–460.
Sigworth, G.K., Kuhn, T.A., 2007. Int. J. Metalcast. 1 (1), 31–40.
Sigworth, G.K., Wang, C., Huang, H., Berry, J.T., 1994. TAFS 102, 245–261.
Sigworth, G.K., Howell, J., Rios, O., Kaufman, M.J., 2006. Int. J. Cast Met. Res. 19 (2), 123–129.
Sigworth, G., Jorstad, J., Campbell, J., 2009. AFS Int. J. Metalcast. Corresp. 3 (1), 65–77.
Sigworth, G., 1984. Met. Trans. 15A, 227–282.
Sikorski, S., Groteke, D.E., October 2005. In: AFS Internat. Conf. High Integrity Light Metal Castings.
Simard, A., Proulx, J., Paquin, D., Samuel, F.H., Habibi, N., November 2001. In: Am. Found. Soc. Molten Al Processing Conf., Orlando, Florida.
Simensen, C.J., 1993. Zeit Met. 84 (10), 730–733.
Simensesn, C.J., 1981. Metall. Trans. 12B, 733–743.
Simmons, W., Trinkl, G., 1987. In: BCIRA Conf. British Cast Iron Research Assoc., UK.
Sin, S.L., Dube, D., Tremblay, R., 2006. Mater. Sci. Technol. 22 (12), 1456–1463.
Singh, S.N., Bardes, B.P., Flemings, M.C., 1970. Met. Trans. 1, 1383.
Sinha, N.P., Kondic, V., 1974. Br. Found. 67, 155–165.
Sinha, N.P., 1973. Ph.D. thesis. University of Birmingham, UK.
Sirrell, B., Campbell, J., 1997. TAFS 105, 645–654.

Skaland, T., 2001. TAFS 105, 77–88.
Skarbinski, M., 1971. Br. Found. 44, 126–140.
Skelly, H.M., Sunnucks, D.C., 1954. TAFS 62, 481–491.
Smith, D.D., Aubrey, L.S., Miller, W.C., 1988. In: Welch, B. (Ed.), "Light Metals" Conf., The Minerals, Metals and Materials Soc., pp. 893–915.
Smith, T.J., Lewis, R.W., Scott, D.M., 1990. Foundryman 83, 499–507.
Smith, C.S., 1948. Trans. AIME 175, 15–51.
Smith, C.S., 1949. Trans. AIME 185, 762–768.
Smith, C.S., 1952. Metal Interfaces. ASM, Cleveland, Ohio, pp. 65–113.
Smith, D.M., 1981. UK Patent application GB 2085920 A.
Smith, R.A., 1986. UK Patent GB2187984 A; priority date 21 February 1986.
Sokolowski, J.H., Kierkus, C.A., Brosnan, B., Evans, W.J., 2000. TAFS 108, 497–503.
Sokolowski, J.H., Kierkus, C.A., Brosnan, B., Evans, W.J., 2003. Mod. Cast. 93 (1), 39–42.
Sokolowski, J.H., 2006. University of Windsor, Canada, unpublished work.
Solberg, J.K., Onsoien, M.I., 2001. Mater. Sci. Technol. 17, 1238–1242.
Song, T., Cooman, B.C.de, April 2013. Metall. Mater. A 44A, 1686–1705.
Song, H., Hellawell, A., 1989. Light Met. 819–823.
Sontz, A., 1972. TAFS 80, 1–12.
Sosman, R.B., 1927. The properties of silica. In: Am. Chem. Soc. Monogr. Ser. The Chemical Catalogue Co., USA, pp. iv-45.
Southam, D.L., 1987. Foundry Manage. Technol. 7, 34–38.
Southin, R.T., Romeyn, A., 1980. In: Warwick Conf. "Solidification Technology in the Foundry and Cast House" Metals Soc. 1983, pp. 355–358.
Southin, R.T., 1967. The solidification of metals. In: Brighton Conf. UK. ISI Publication 110, pp. 305–308.
Spada, A.T., 20–24 February 2004. Mod. Cast. 94. June 2004, 48; July 2005, 18–22.
Speidel, M.O., 1982. In: Sixth European Non-ferrous Metals Industry Colloquium CAEF, pp. 65–78.
Spenser, D., Mehrabian, R., Flemings, M.C., 1972. Met. Trans. 3, 1925.
Spitaler, P., 1957. Giesserei 44, 757–766.
Spittle, J.A., Brown, S.G.R., 1989a. J. Mater. Sci. 23, 1777–1781.
Spittle, J.A., Brown, S.G.R., 1989b. J. Mater. Sci. 5, 362–368.
Spittle, J.A., Cushway, A.A., 1983. Met. Technol. 10, 6–13.
Spraragen, W., Claussen, G.E., 1937. J. Am. Weld. Soc. 16 (11), 2–62 (Supplement: Welding Research Committee).
Srimanosaowapak, S., O'Reilly, Keyna, 2005. In: Tiryakioglu, M., Crepeau, P.N. (Eds.), Shape Casting: The John Campbell Symposium. TMS, pp. 41–50.
Srinagesh, K., 1979. AFS Int. Cast Met. J. 4 (1), 50–63.
Srinivasan, A., Pillai, U.T.S., Pai, B.C., 2006. TAFS 114, 737–746.
Stahl, G.W., 1961. TAFS 69, 476–478.
Stahl, G.W., 1963. TAFS 71, 216–220.
Stahl, G.W., 1986. TAFS 94, 793–796.
Stahl, G.W., 1989. The gravity tilt pour process. In: Proc. AFS Internat Conf: Permanent Mold Castings, Miami. Paper 2.
Staley, J.T., 1981. Metals Handbook. In: Heat Treating, ninth ed., vol. 4. American Society for Metals, USA. pp. 675–718.
Staley, J.T., 1986. Aluminium technology. In: Inst. Metals UK Conference, vol. 86, pp. 396–407.
Stanford, N., Sabirov, I., Sha, G., La Fontaine, A., Ringer, S.P., Barnett, M.R., 2010. Metall. Mater. Trans A 41A (3), 734–743.
Starobin, A., Goettsch, D., Walker, M., 2011. AFS Int. J. Met. 5 (3), 57–64.
Steel Founders Society of America (Anon), May 2000. Foundry Trade J. 40–41.
Steel Founders' Society of America (Anon), 1970. Steel Castings Handbook, fourth ed.
Steen, H.A.H., Hellawell, A., 1975. Acta Met. 23, 529–535.
Stefanescu, D.M., Hummer, R., Nechtelberger, E., 1988. Metals Handbook. In: Casting, ninth ed., vol. 15. ASM, Ohio, USA. pp. 667–677.
Stefanescu, D.M., Giese, S.R., Piwonka, T.S., Lane, A.M., Barlow, J., Pattabhi, R., 1996. TAFS 104, 1233–1264.
Stefanescu, D.M., Wills, S., Massone, J., 2009. TAFS, pp. 20, paper 09-120.
Stefanescu, D.M., 1988. Metals Handbook. In: Casting, ninth ed., vol. 15. ASM, Ohio, USA. pp. 168–181.

Stefanescu, D.M., 2007. Trans. Metall. Mater. A 38A (7), 1433–1447.
Stefanescu, D.M., 2015. Thermal analysis – theory and applications in metalcasting. Int. J. Metalcast. 9 (1), 7–22.
von Steiger, R., 1913. Stahl Eisen 33, 1442.
Stein, H., Iske, F., Karcher, D., 1958. Giesserei Technisch Wissenschaftliche Beihefte 21, 115–1124.
StJohn, D.H., Qian, M., Easton, M.A., Cao, P., Hildebrand, Z., 2005. Metall. Mater. Trans. A 36A, 1669–1679.
StJohn, D.H., Easton, M.A., Cao, P., Qian, M., July 2007. Int. J. Cast Met. Res. 20 (3), 131–135 and In: Jones, H. (Ed.), SP07 Proc. 5th Decennial Internat. Conf. Solidification Processing. July 2007, Sheffield, UK, pp. 99–103.
StJohn, D.H., Easton, M.A., Qian, M., Taylor, J.A., July 2013. Metall. Mater. Trans. A 44A, 2935–2949.
Stolarczyk, J.E., 1960. Br. Found. 53, 531–548.
Stoltze, P., Norskov, I.K., Landman, U., 1988. Phys. Rev. Lett. 61 (4), 440–443.
Storaska, G.A., Howe, J.M., 2004. Mater. Sci. Eng. A 368, 183–190.
Stratton, P., June 2010. Mater. World 28–29.
Street, A.C., 1986. The Diecasting Book, second ed. Portcullis Press, Redhill, UK. (Chapter 8) on zinc alloys pp. 170–194.
Su, J.Y., Chow, C.T., Wallace, J.F., 1982. TAFS 90, 565–574.
Sugden, A.A.B., Bhadeshia, H.K.D.H., June 1988. Metall. Trans. A 19A. University of Cambridge, UK. pp.1597–1610.
Sullivan, E.J., Adams, C.M., Taylor, H.F., 1957. TAFS 65, 394–401.
Sumiyoshi, Y., Ito, N., Noda, T., 1968. J. Cryst. Growth 3 and 4, 327–339.
Sun, L., Campbell, J., 2003. TAFS 111 (paper 03-018).
Sun, G.X., Loper, C.R., 1983. TAFS 91, 841–854.
Surappa, M.K., Blank, E., Jaquet, J.C., 1986a. In: Sheppard, T. (Ed.), Conference "Aluminium technology". Inst. Metals, pp. 498–504.
Surappa, M.K., Blank, E., Jaquet, J.C., 1986b. Scr. Met. 20, 1281–1286.
Sutton, T., June 2002. Foundryman 95 (6), 223–231.
Suutala, N., 1983. Met. Trans. 14A, 191–197.
Suzuki, K., Nishikawa, K., Watakabe, S., 1996. Mater. Trans. Jpn. Inst. Met. 37 (12), 1793–1801.
Suzuki, S., October 1989. Mod. Cast. 38–40.
Svensson, I.L., Dioszegi, A., 2000. In: Sahm, P., Hansen, P., Conley (Eds.), Modelling of Casting, Welding and Advanced Solidification Processes IX, pp. 102–109.
Svensson, I., Villner, L., 1974. Br. Found. 67, 277–287.
Svoboda, J.M., Geiger, G.H., 1969. TAFS 77, 281–288.
Svoboda, J.M., Monroe, R.W., Bates, C.E., Griffin, J., 1987. TAFS 95, 187–202.
Svoboda, J.M., 1994. TAFS 102, 461–471.
Sweeney, V.D., 1964. TAFS 72, 911–913.
Swift, R.E., Jackson, J.H., Eastwood, L.W., 1949. TAFS 57, 76–88.
Swing, E., 1962. TAFS 70, 364–373.
de Sy, A., 1967. TAFS 75, 161–172.
Syvertsen, M., 2006. Metall. Mater. Trans. B 37B (6), 495–504.
Syvertsen, M., Engh, T.A., 2001. Light Met. 957–963.
Szklarska-Smialowska, Z., 1999. Corros. Sci. 41, 1743–1767.
Tadayon, M.R., Lewis, R.W., 1988. Cast. Met. 1, 24–28.
Tadayon, M.R., 28 September 1982. Finite Element Modelling of Heat Transfer and Solidification in the Squeeze Forming Process (Ph.D. thesis). University College, Swansea, UK.
Tafazzoli-Yadzi, M., Kondic, V., 1977. AFS Internat. Cast. Met. J. 2 (4), 41–47.
Taft, D.J., 1968. Br. Found. 61, 69–75.
Takeuchi, E., Brimacombe, J.K., 1984. Metall. Mater. Trans. B 15, 493–509.
Talbot, D.E.G., Granger, D.A.J., 1962. Inst. Met. 91, 319–320.
Talbot, D.E., 1975. J. Int. Metall. Rev. 20, 166.
Taleyarkhan, R.P., Kim, S.H., Gulec, K., 1998. TAFS 106, 619–624.
Tasaki R., Noda Y., Terashima K., Hashimoto K., Proc. 28th World Foundry Cong., February 2008, Chennai, India, pp. 121–126.
Taylor, K.C., Baier, A., 2003. Cast. Plant Technol. Int. 123 (2), 36–46.
Taylor Jr., C.M., Taylor, H.F., 1953. Fundamentals of riser behavior. AFS Trans. 686–693.

Taylor, L.S., 1960. Foundry Trade J. 109 (2287), 419–427.
Teague, J., Richards, V., 2010. AFS Int. J. Metalcast. 4 (2), 45–57.
Terashima K., Noda Y., et al. Proc. 28th World Foundry Cong., February 2008, Chennai, India, pp. 113–119.
Theuwissen, K., Lafont, M.-C., Laffont, L., Viguier, B., Lacaze, J., 2012. Microstructural characterization of graphite spheroids in ductile iron. Trans. Indian Inst. Met. 65 (6), 627–631.
Theile von, W., 1962. Alum. Ger. 38, 707–715, 780–786. (English Translation by Nercessian, H., 29 April 1963, 5229, Alcan Banbury, UK and Electicity Council Research Centre, UK.).
Thomas, B.G., 1995. J. Iron Steel Inst. Jpn. Int. 35 (6), 737–743.
Thomas, B.G., Parkman, J.T., 1997. In: Conf. "Solidification" Indianapolis Indiana. The Minerals, Metals and Materials Society, 1998, pp. 509–520.
Thomason, P.F., 1968. J. Inst. Met. 96, 360–365.
Thornton, D.R., 1956. J. Iron Steel Inst. 183 (3), 300–315.
Tian, C., Irons, G.A., Wilkinson, D.S., 1999. Metall. Mater Trans. B 30B (2), 241–252.
Tiberg, L., 1960. Jerkont. Ann. 144 (10), 771–793.
Tiekink, W., Boom, R., Overbosch, A., Kooter, R., Sridar, S., 2010. Ironmak. Steelmak. 37 (7), 488–495.
Tien, R.H., Richmond, O., 1982. Trans. ASME J. Appl. Mech. 49, 481–486.
Tigges, U., December 2010. Foundry Trade J. 310–311.
Timelli, G., Bonollo, F., 2007. Int. J. Cast Met. Res. 20 (6), 304–311.
Timmons, W.W., Spicgelberg, W.D., Wallace, J.F., 1969. TAFS 77, 57–61.
Tiryakioglu, M., Campbell, J., 2007. Int. J. Cast Met. Res. 20 (1), 25–29.
Tiryakioglu, M., Campbell, J., 2009. Mater. Sci. Technol. 25 (6), 784–789.
Tiryakioglu, M., Hudak, D., 2008. J. Mater. Sci. 43, 1914–1919.
Tiryakioglu, M., Askeland, D.R., Ramsay, C.W., 1993. TAFS 101, 685–691.
Tiryakioğlu, M., Tiryakioğlu, E., Askeland, D.A., 1997a. Int. J. Cast Met. Res. 9, 259–267.
Tiryakioğlu, M., Tiryakioğlu, E., Askeland, D.A., 1997b. TAFS 105, 907–915.
Tiryakioğlu, M., Tiryakioğlu, E., Campbell, J., 2002. Int. Cast Met. Res. J. 14 (6), 371–375.
Tiryakioglu, M., Hudak, D., 2008. Mater. Sci. Eng. A 498, 501–503.
Tiryakioglu, M., Hudak, D., Okten, G., 2009a. Mater. Sci. Eng. A 527, 397–399.
Tiryakioglu, M., Campbell, J., Alexopoulos, N.D., 2009b. Metall. Mater. Trans. A 40A, 1000–1007.
Tiryakioglu, M., Campbell, J., Alexopoulos, N.D., 2009c. Mater. Sci. Eng. A 506, 23–26.
Tiryakioglu, M., Campbell, J., Nyahumwa, C., 2011. Fracture surface facets and fatigue life potential of castings. In: Crepeau, P., et al. (Eds.), Shape Casting: The John Berry Honorary Symposium. TMS, San Diego, CA.
Tiryakioğlu E., 1964. A study of the dimensioning of feeders for sand castings (Ph.D. thesis). University of Birmingham, UK.
Tiryakioglu, M., 2001. Personal communication.
Tiryakioglu, M., 2008. Statistical distributions for the size of fatigue-initiating defects. A comparative study. Mater. Sci. Eng. A 497, 119–125.
Tiryakioglu, M., 2009a. The size of fatigue-initiating defects. Mater. Sci. Eng. 520, 114–120.
Tiryakioglu, M., Unpublished Work on the Derivation of Linear Equation for M.
Tiryakioglu, M., 2009b. Metall. Mater. Trans. 40A, 1000–1007.
Tiryakioglu, M., 2011. In: Shape Casting; "John Berry Honorary Symposium" IV. TMS Annual Congress, San Diego, CA, USA.
Tiwara, S.N., Gupta, A.K., Maihotra, S.L., 1985. Br. Found. 78 (1), 24–27.
Toda, H., Minami, K., Koyama, K., Ichitani, K., Kobayashi, M., Uesugi, K., et al., 2009. Acta Mater. 57, 4391.
Toda, H., Oogo, H., Horikawa, K., Uesugi, K., Takeuchi, A., Suzuki, Y., et al., February 2014. Metall. Mater. Trans. A 45A, 765–776.
Tokarev, A.I., 1966. Iz.v VUZ Chern. Met. 3, 193–200. BCIRA Translation T1190, January 1966.
van Tol, R., Katerman, L., van der Akker, H.E.A., 1997. Solidification processing. In: Beech, J., Jones, H. (Eds.), Conference Sheffield, UK (SP97). pp. 79–82.
Tomono, H., Ackermann, P., Kurz, W., Heinemann, W., 1980. Solidification technology in foundry and cast house. In: Warwick Univ. Conf. Metals Society Publication 1983, pp. 524–531.
Torabi Rad, M., Kotas, P., Bekermann, C., September 2013. Metall. Mater. Trans. 44A, 4266–4281.
Tordoff, E.G., Wolfgram, T., Talwar, V., Hysell, M., 1996. TAFS 104, 461–466.

Townsend, D.W., May 1984. Foundry Trade J. 24, 409–414.
Travena, D.H., 1987. Cavitation and Tension in Liquids. Adam Huger, Inst. Phys, Bristol.
Trbizan Katerina, 2001. Casting Simulation. World Foundry Organisation (paper 4) pp. 83–97.
Trikha, S.K., Bates, C.E., 1994. TAFS 102, 173–180.
Trojan, P.K., Guichelaar, P.J., Flinn, R.A., 1966. TAFS 74, 462–469.
Tsai, H.L., Chiang, K.C., Chen, T.S., 1988. In: Giamei, A.F., Abbaschian, O.J. (Eds.), Modeling of Casting and Welding Processes IV. The Minerals, Metals and Materials Society, USA, 1989.
Tschopp, M.A., Ramsay, C.W., Askeland, D.R., 2000. TAFS 108, 609–614.
Tsuruya, S., Ishikawa, Y., Sono, K., 1974. Trans. AFS 82, 27–34.
Tucker, S.P., Hochgraph, F.G., 1973. Metallography 6, 457–464.
Turchin, A.N., Eskin, D.G., Katgerman, L., 2007. Int. J. Cast Met. Res. 20 (6), 312–318.
Turchin, A.N., Eskin, D.G., Katgerman, L., 2007. Metall. Mater. Trans. A 38A (6), 1317–1329.
Turkdogan, E.T., 1986. Foundry processes, their chemistry and physics. In: Katz, S., Landefeld, C.F. (Eds.), International Symposium, Warren, Mich. USA. Plenum Press 1988, pp. 53–100.
Turner, A., Owen, F., 1964. Br. Found. 57, 55–61, 355–356.
Turner, G.L., 1965. Br. Found. 58, 504–505.
Twitty, M.D., 1960. BCIRA I., Report 575. In: Walton, C.F. (Ed.) (1971), Gray and Ductile Iron Casting Handbook. Gray and Ductile Iron Founders' Soc. Inc, Cleveland, Ohio, pp. 844–856.
Tyberg, B., Granehult, O., 1970. In: 37th Internat. Foundry Congress, p. 4.
Tyndall, J., 1872. The Forms of Water. D. Appleton and Co, New York.
Tynelius, K., Major, J., Apelian, D., 1993. Am. Found. Soc. Trans. 101, 401–413.
Unsworth, W., February 1988. Met. Mater. 83–86.
Uto, Y., Yamasaki, D., 1967. UK Patent Specification 1198700.
Valdes, J., King, P., Liu, X., 2010. Metall. Mater. Trans. A 41A (9), 2408–2416.
Valdez, M.E., Uranga, P., Cramb, A.W., 2007. Metall. Mater. Trans. B 38B, 257–266.
Van Ende, M.A., Guo, M., Proost, J., Blanpain, B., Wollants, P., 2010. Ironmak. Steelmak. 37 (7), 496–501.
Vandenbos, S.A., 1985. TAFS 93, 871–878.
Venturelli, G., Sant'Unione, G., February 1981. Alluminio 100–106.
Vernede, S., Jarry, P., Rappaz, M., 2006. Acta Mater. 54, 4023–4034.
Vernede, S., Dantzig, J.A., Rappaz, M., 2009. Acta Mater. 57, 1554–1569.
Versnyder, F.I., Shank, M.E., 1970. Mater. Sci. Technol. 6 (4), 213–247.
Vigh, L., Bennett, G.H.J., 1989. Cast Met. 2 (3), 144–150.
Villner, L., 1969. Br. Found. 62, 458–468. Also published in Giesserei 1970 57 (27), 837–844 and Cast Met. Res. J. 1970 6 (3) 137–142.
Vincent, R.S., Simmons, G.H., 1943. Proc. Phys. Soc. (London) 376–382.
Vogel, A., Doherty, R.D., Cantor, B., 1977. In: Univ. Sheffield Conf. "The Solidification and Casting of Metals". Metals Soc. 1979, pp. 518–525.
Voigt, R.C., Holmgren, S.D., 1990. TAFS 98, 213–225.
Vorren, O., Evensen, J.E., Pedersen, T.B., 1984. TAFS 92, 459–466.
Wakefield, G.R., Sharp, R.M., 1992. J Mater. Sci. Technol. 8, 1125–1129.
Wakefield, G.R., Sharp, R.M., 1996. J Mater. Sci. Technol. 12, 518–522.
Walker, J.L., 1961. Physical Chemistry of Process Metallurgy. II. Interscience, NY, p. 845.
Wall, A.J., Cocks, D.L., 1980. Br. Found. 73, 292–300.
Wallace, J.F., Hrusovsky, J.P., TAFS 1979. 87, 269–278.
Wallace, J.F., Kissling, R.J., December 1962. Foundry 36–39. January 1963, 64–68.
Wallace, J.F., 1988. TAFS 96, 261–270.
Walther, W.D., Adams, C.M., Taylor, H.F., 1954. TAFS 62, 219–230.
Wan, L., Nakashima, T., Kato, E., Nomura, H., 2002. Int. J. Cast Met. Res. 15, 187–192.
Wang, Q.G., Apelian, D., Lados, D.A., 2001a. Part I. J. Light Met. 1 (1), 73–84.
Wang, Q.G., Apelian, D., Lados, D.A., 2001b. Part II. J. Light Met. 1 (1), 85–97.
Wang, R.-Y., Lu, W.-H., Ma, Z.-Y., 2007. TAFS 111, 8, paper 124.

Wang, J., Lee, P.D., Hamilton, R.W., Li, M., Allison, J., 2009. Scr. Met. 60 (7), 516–519.
Wang, C., Gao, H., Dai, Y., Ruan, X., Wang, J., Sun, B., 2010. Metall. Mater. Trans. A 41A (7), 1616–1620.
Wang, L., Lett, R., Felicelli, S.D., Berry, J.T., 2011. TAFS, paper 11-039.
Wang, Q., Geng, H., Zhang, S., Jiang, H., Zuo, M., March 2014. Metall. Mater. Trans. A 45A, 1621–1630.
Wannasin, J., Schwam, D., Wallace, J.F., 2007. J. Mater. Process. Technol. 191 (1–3), 242–246.
Ward, C.W., Jacobs, I.C., 1962. TAFS 70, 332–337.
Wardle, G., Billington, J.C., October 1983. Met. Technol. 10, 393–400.
Warrick, R.J., 1966. TAFS 74, 722–733.
Warrington, D., McCartney, D.G., 1989. Cast Met. 2 (3), 134–143.
Watmough, T., 1980. TAFS 88, 481–488.
Waudby, P.E., George, G.H., June 1986. In: Twentieth EICF Conf. on Investment Casting Brussels, p. 19, paper 12.
Way, L.D., 2001. Mater. Sci. Technol. 17 (10), 1175–1190.
Weber, J.A., Rearwin, E.W., 1961. Foundry 2, 69–71.
Webster, P.D., 1964. Br. Found. 57, 520–523.
Webster, P.D., 1966. Br. Found. 59, 387–393.
Webster, P.D., 1967. Br. Found. 60, 314–319.
Webster, P.D., 1980. In: Fundamentals of Foundry Technology. Portcullis Press.
Weibull, W., 1951. J. Appl. Mech. 18, 293–297.
Weiner, J.H., Boley, B.A., 1963. J. Mech. Phys. Solids 11, 145–154.
Weins, M.J., Bottom, J.L.S.de, Flinn, R.A., 1964. TAFS 72, 832–839.
Weiss, D.J., Rose, D., 1993. TAFS 101, 1065–1066.
Wells, M.B., Oleka, J.T., 1988. TAFS 96, 913–918.
Welter, V.G., 1931. Z. Met. 23.
Wen, S.W., Jolly, M.R., Campbell, J., 1997a. In: Beech, J., Jones, H. (Eds.), Proc. 4th Decennial Internat. Conf. on Solidification Processing, Sheffield, pp. 66–69.
Wen, S.W., Jolly, M.R., Campbell, J., 7–10 July 1997b. In: Solidification Processing Conference SP97, Sheffield (Paper on "Promotion of directional solidification…").
West, T.D., 1882. American Foundry Practice, 11th ed. Wiley NY, Chapman & Hall London. 1910, p. 107.
Whittenberger, E.J., Rhines, F.N., 1952. J. Met. 4 (4), 409–420 and Trans AIME, 194, 409–420.
Wieser, P.F., Dutta, I., 1986. TAFS 94, 85–92.
Wieser, P.F., Wallace, J.F., 1969. TAFS 77, 22–26.
Wieser, P.F., 1983. TAFS 91, 647–656.
Wightman, G., Fray, D.J., 1983. Met. Trans. 14B, 625–631.
Wildermuth, J.W., Lutz, R.H., Loper, C.R., 1968. TAFS 76, 258–263.
Wile, L.E., Strausbaugh, K., Archibald, J.J., Smith, R.L., Piwonka, T.S., 1988. Metals Handbook. In: Casting, ninth ed., vol. 15. ASM International. pp. 240–241.
Williams, J.A., Singer, A.R.E., 1968. J. Inst. Met. 96, 5–12.
Williams, S.J., Bache, M.R., Wilshire, B., 2010. Mater. Sci. Technol. 26 (11), 1332–1337.
Williams, D.C., 1970. TAFS 78, 374–381, 466–467.
Wilshire, B., Scharning, P.J., 2008. Mater. Sci. Technol. 24, 1–9.
Winardi, L., Littleton, H.E., Bates, C.E., 2007. TAFS 04, 10, paper 062.
Winter, B.P., Ostrom, T.R., Sleder, T.A., Trojan, P.K., Pehlke, R.D., 1987. TAFS 95, 259–266.
Winzer, N., Cross, C.E., 2009. Metall. Mater. Trans. A 40A (2), 273–274.
Wittmoser, A., Hofmann, R., 1968. In: 35th Foundry Congress, Kyoto, Japan.
Wittmoser, A., Steinack, K., Hofmann, R., February 1972. Br. Found. 65, 73–84.
Wittmoser, A., 1975. Trans. AFS 83, 63–72.
Wlodawer, R., 1966. Directional Solidification of Steel Castings. English translation by Hewit, L.D., Riley, R.V. Pergamon Press.
Wojcik, W.M., Raybeck, R.M., Paliwoda, E.J., December 1967. J. Met. 19, 36–41.
Wolf, A., Steinhauser, T., 2004. Cast. Plant Technol. Int. 3, 6–11.
Woodbury, K.A., Chen, Y., Parker, J.K., Piwonka, T.S., 1998. TAFS 106, 705–711.

Woodbury, K.A., Piwonka, T.S., Ke, Q., 2000. In: Sahm, P.R., Hansen, P.N., Conley, J.G. (Eds.), Modeling of Casting, Welding and Advanced Solidification Processes IX, pp. 270–277.
Woodbury, K.A., Ke, Q., Piwonka, T.S., 2000. TAFS 108, 259–265.
Woolley, J.W., Woodbury, K.A., 2007. TAFS 115, 18 (Paper 075(01)).
Woolley, J.W., Woodbury, K.A., 2009. TAFS 117, 31–40.
Worman, R.A., Nieman, J.R., 1973. TAFS 81, 170–179.
Wray, P.J., 1976a. Acta Met. 76, 125–135.
Wray, P.J., 1976b. Met. Trans. 7B, 639–646.
Wright, T.C., Campbell, J., 1997. TAFS 105, 639–644 and Mod. Cast. June 1997 (see also Dimmick, T., March 2001. Mod. Cast. 91 (3), 31–33).
Wu, R., Sandstrom, R., 1995. Mater. Sci. Technol. 11 (6), 579–588.
Wu, C., Sahajwalla, Pehlke, R.D., 1997. TAFS 105, 739–744.
Wu, M., 1997. TAFS 105, 693–702.
Wu, M., Sahm, P.R., 1997. Trans. Am. Found. Soc. 105, 693–702.
Wuilloud, E., 1994. Light Met. 1079–1082.
Wurker, L., Zeuner, Th, 2004. Aluminium 80 (11), 1207–1213.
Wurker, L., Zeuner, Th, 2006. Cast. Plant Technol. (1), 38–42.
Xiao, L., Anzai, K., Niyama, E., Kimura, T., Kubo, H., 1998. Int. J. Cast Met. Res. 11 (2), 71–81.
Xiao, B., Wang, Q.G., Jadhav, P., Li, K., 2010. J. Mater. Process. Technol. 210, 2023–2028.
Xiong, S.-M., Liu, X.-L., 2007. Metall. Mater. Trans. A 38A, 248–434.
Xu, Z.A., Mampaey, F., 1994. TAFS 102, 181–190.
Xu, Z.A., Mampaey, F., 1997. TAFS 105, 853–860.
Xu, J., Liu, F., Xu, X., Chen, Y., 2012. Metall. Mater. Trans. A 43A, 1268–1276.
Xu, Z.A., August 2007. TMS U.S.A.
Xue, X., Thorpe, R., 1995. TAFS 103, 743–747.
Yamaguchi, K., Healy, G., 1974. Met. Trans. 5, 2591–2596.
Yamamoto, N., Kawagoishi, N., 2000. Trans. AFS 108, 113–118.
Yamamoto, S., Kawano, Y., Murakami, Y., Chang, B., Ozaki, R., 1975. TAFS 83, 217–226.
Yamamoto, Y., Iwahori, H., Yonekura, K., Nakmura, M., 1980. AFS Int. Cast Met. J. 5 (2), 60–65.
Yan, Y., Yang, G., Mao, Z., 1989. J. Aeronaut. Mater. (China) 9 (3), 29–36.
Yang, X., Campbell, J., 1998. Pouring basin. Int. J. Cast Met. Res. 10, 239–253.
Yang, X., Din, T., Campbell, J., 1998. Offset sprue. Int. J. Cast Met. Res. 11, 1–12.
Yang, X., Jolly, M.R., Campbell, J., 2000. Vortex flow runner. Alum. Trans. 2 (1), 67–80 and Sahm, P., Hansen, P., Conley, (Eds.), Modelling of Casting, Welding and Advanced Solidification Processing IX. 2000, pp. 420–427.
Yaokawa, J., Miura, D., Anzai, K., Yamada, Y., Yoshi, H., 2007. Jpn. Foundry Eng. Soc. Mater. Trans. 48 (5), 1034–1041.
Yarborough, W.A., Messier, R., 1990. Science 247, 688–696.
Yazzie, K.E., Williams, J.J., Kingsbury, D., Peralta, P., Jiang, H., Chawla, N., 2010. J. Met. 62 (7), 16–21.
Yonekura, K., et al., 1986. TAFS 94, 277–284.
Yoshimi, K., Nakamura, J., Kanekon, D., Yamamoto, S., Maruyama, K., Katsui, H., Goto, T., 2014. J. Met. 66 (9), 1930–1938.
Youdelis, W.V., Yang, C.S., 1982. Metal. Sci. 16, 275–281.
Young, P., April 2002. Quoted Anon. Foundry Trade J. 27–28.
Young, K.P., Kirkwood, D.H., 1975. Met. Trans. 6A, 197–205.
Yu, X.Q., Sun, Y.S., 2004. Mater. Sci. Technol. 20 (3), 339–342.
Yu, L.X., Sun, Y.R., Sun, W.R., Zhang, W.H., Liu, F., Xin, X., Qi, F., Jia, D., Sun, X.F., Guo, S.R., Hu, Z.Q., 2013. Mater. Sci. Technol. 29 (12), 1470–1477.
Yue, T.M., Chadwick, G.A., 1991. Personal communication relating to the effect of grain size on the 0.2PS of 7010 Al alloy, AZ91 Mg alloy and Al-4.5Cu alloy.
Yurko, J.A., Martinez, R.A., Flemings, M.C., June 2003. Metall. Sci. Technol. (Teksid Alum.) 21 (1).
Zadeh, A.H., Campbell, J., 2002. TAFS Paper 02-020, pp. 1–17.
Zang, Z., Bian, X., Liu, X., 2001. Int. J. Cast Met. Res. 14 (1), 31–35.
Zeitler, H., Scharfenberger, W., 1984. Aluminium (Germany) 60 (12), E803–E808.

Zemcik, L., 2015. Formation of oxide films in castings from nickel-base superalloys. AFS Int. J. Metalcast., in press.
Zhang, D.L., Zheng, L.H., St John, D.H., 1998. Mater. Sci. Technol. 14 (7), 619–625.
Zhang, C., Mucciardi, F., Gruzleski, J., Burke, P., Hart, M., 2003. Trans AFS paper 03-010, pp. 1–11.
Zhang, E., Wnag, G.J., Xu, J.W., Hu, Z.C., 2010. Mat. Sci. Technol. 26 (8), 956–961.
Zhao, L., Baoyin, Wang, N., Sahajwalla, V., Pehlke, R.D., 2000. Int. J. Cast Met. Res. 13, 167–174.
Zhong, H., Rometsch, P.A., Estrin, Y., 2013. Metall. Mater. Trans. A 44A, 3970–3983.
Zhou, Y.Z., Volek, A., 2007. Mater. Sci. Technol. 23 (3), 297–302.
Zhu, P., Sha, R., Li, Y., Effect of twin/tilt on the growth of graphite. In: Fredriksson, H., Hillert, M. (Eds.), The Physical Metallurgy of Cast Iron. Proc. Materials Research Soc., vol. 34. p. 3.
Zildjian Avedis Company, May 2002. Mod. Cast. 68.
Ziman, J., 2001. Non-Instrumental Roles of Science. Physics Department, University of Bristol.
Zuehlke, H.B., 1943. Trans. Am. Found. Assoc. 51, 773–797.
Zuithoff, A.1, 1964. Paper 29; and in Geisserei, (1965) 52 (9). In: 31st Internat. Foundry Congress, Amsterdam, pp. 820–827.
Zurecki, Z., Best, R.C., 1996. TAFS 104, 859–864.
Zurecki, Z., 1996. In: Saha, D. (Ed.), Gas Interactions in Nonferrous Metals Processing. TMS, pp. 79–93.

Index

Note: 'Page numbers followed by "f" indicate figures and "t" indicate tables'.

0-9

1.5 factor (initial/average fill rates), 947–949
10 rules for good castings. *See* Rules for good castings
10 test bar mould, 448

A

Ablation casting (RS), 888–891
Accuracy. *See* Dimensional accuracy
Acetylene black and fluidity, 114
Additions to melts, 787–788
ADI. *See* Austenitic ductile iron
Aero chocolate structure (zinc alloys), 865
Aerofoil fluidity tests for investment casting, 120f
Aerospace applications:
    aluminium alloys, 541
    location points, 634
AFS fineness number for aggregates, 802
Ag_3Sn (intermetallic), 223
Agglomeration, 31, 222, 321, 323
Aggregate moulding materials:
    AFS fineness number, 802
    carbon, 806
    chromite sand, 804
    description, 821
    hollow ceramic spheres, 805–806
    minerals, 805–806
    olivine, 803–805
    silica sand, 803–804
    synthetic aggregates, 806
    zircon sand, 803, 805
Aggregate moulds:
    coatings, 159–160
    Cosworth Process, 924
    mould accuracy, 904–907
    residual stress, 616–617
Aided flotation and detrainment, 784
Aided sedimentation and detrainment, 784–785
Air:
    entrainment, 677, 822–823
    gap and metal–mould interface, 167–170
    melting and casting of nickel alloys, 331–332
    quenching, 623–624
Air gap and rapid solidification, 887–888
Air inclusion, term, 724
Air lock, 741
Alcan, 379, 792–793
Alkaline phenolic (A-P) binders, 809, 814–815
Allen, Alec, 772
Alotech, 775, 784
Alpha phase Silica, 803
Alpha Fe, 55–56, 260f, 471f, 473
Alumina:
    bifilms, 78, 323, 781
    bismuth, 77
    furling and unfurling, 74, 74f
    inclusions, 239
    liquid aluminium, 19
    liquid steel, 31–32
    slag and flux inclusions, 31–33
    stringers, 31–32, 34f
    vertical filling of casts, 51, 52f
    zinc alloys, 226
    γ-alumina, 237
Aluminides (titanium alloys), 240–242, 242f, 336–337, 339–340, 403
Aluminium:
    bifilms, 469–470, 476–477
    bubble trails analysis, 63
    contact pouring, 660–661
    critical velocity, 41, 43f, 545
    crucible melting, 771
    diffusion coefficients, 10f
    ductility, 476–477
    expansion and contraction, 899
    fcc structure, 482
    feeders, 574, 586–587, 599f
    filtration, 785–786
    fins, 187–188, 187f–188f, 190f
    fluidity, 117
    freezing range, 577–579
    hexachloroethane degassing, 780
    holding, 774
    HPDC, 862–863
    hydrogen, 6, 6f, 267–269, 782, 788
    inserts, 185
    L-junctions, 188, 188f
    magnesium alloys, 230, 233
    melting, 769
    melts, 3
    mould gases, 144–145
    Ni-base alloys, 330–335
    oxides, 25, 230, 231f–232f, 789
    reclamation and re-cycling of aggregates, 814–816
    residual stress (quenching), 617–618
    rotary degassing, 780–783

Aluminium: (*Continued*)
  rotary furnaces, 771
  shot, 805
  shot blasting, 927–928
  siphons, 776
  solidification, 347–348
  strength of castings, 545
  strontium, 914
  structure, 3
  surface tension, 655, 659
  T-junctions, 187, 187f
  thermal conductivity, 770–771
  vapour zones in greensand mould, 140, 140f
  wall plaques, 821
  water vapour, 8
  zinc alloys, 223–227. *See also* Liquid aluminium
Aluminium alloys:
  7000 series, 328, 444, 618
  7050 high strength, 497
  ablation casting, 888–891
  accuracy, 932
  aerospace, 541
  alumina films, 525
  aluminium bronze
    chills and fins, 191–192, 192f
    critical velocity, 44
    Durville casting, 829
    feeder neck constriction, 542, 586
    feeding distance, 587–590
    hydrogen solubility, 7f
    penetration barriers, 153
    pouring, 47
    surface films, 13
  Al-0.2Cu porosity/interdendritic segregation, 211, 212f
  Al-1Cu equiaxed grains, 433, 435f
  Al-1Sn hot tear, 421f
  Al-3Cu-5Si modulus, 113
  Al-3Mg grain size, 483f
  Al-4Mg oxidation and corrosion resistance, 512
  Al-4.5Cu
    ablation casting, 890, 891f
    $CuAl_2$, 263
    DAS, grain size and t_f, 207–209, 208f
    elongation to fracture, 477, 478f
    eutectic and time, 930
    furling and unfurling, 73–74, 73f
    grain refinement, 482
    inclusions, 326
    macrosegregation, 611
    quality indices, 526–527
    tilt casting, 832
  Al-4.5Cu-0.7Ag tear initiation, 426
  Al-4.5Cu-1.5Mg ductility, 475–476, 476f, 491–492, 493f

Al-4.5 Mg bifilms, 80
Al-7Si-0.4Mg elongation v. yield strength, 479f
Al-5Mg
  deactivation of entrained films, 77
  freezing range, 577, 589f
  oxidation, 225f, 229
Al-5Si, penetration barriers, 154
Al-5Si-1Cu slip band direction, 486f
Al-5Si-3Cu
  chills, 182f
  residual stress, 509
Al-6Zn-2.7Mg-1.7Cu aging time, 477, 480f
Al-6.6Cu hot tear, 419f
Al-7Si
  cellular growth, 248, 248f
  chills, 181–182, 183f
  magnesium, 398
  modulus, 113
  oxide bifilms, 198–199, 200f
  phenolic-urethane binders, 397
  superheat and fluidity, 114, 115f
  thermal analysis, 264–267
Al-7Si-0.3Mg
  channel defects, 222
  external chills, 219
  gas porosity, 404
  viscosity, 111
Al-7Si-0.4Mg
  ablation casting, 889–890, 890f–891f
  bifilms, 86, 538, 539f
  bubble trails, 63, 64f
  deactivation of entrained films, 77–80
  engineering, 236
  flow channel behaviour, 194–197
  fluidity, 112
  furling and unfurling, 69f
  grain refinement, 482
  heat treatment, 931–932
  HIPping, 936–937, 936f
  hot tear, 418f
  hydrogen porosity, 406–409
  incipient melting, 933
  β-iron, 487, 488f
  liquid aluminium holding time, 775
  mechanical properties, 75
  pore initiation on bifilms, 379
  pouring, 788
  reduced pressure tests, 84, 85f
  sand inclusions, 35, 36f
  squeeze casting, 869
  strontium, 238–239
  surface films, 23f–24f, 26–28, 29f–30f
  uniform contraction, 914

INDEX

Al-7Si-0.4Mg-0.4Fe eutectic phase, 490, 492f
Al-7Si-0.5Mg
  fatigue/fracture, 498
  shot blasting, 928
  uniform contraction, 914
Al-7Si-0.6Mg fracture toughness, 495
Al-7Si-Mg
  10 test bar mould, 448
  elongation to fracture, 477, 478f
  fatigue, 501
  quality indices, 526
Al-8Si-3.3Cu-0.2Mg uniform contraction, 914
Al-9Si-4Mg oxide skin, 279, 279f
Al-10Mg
  batch melting, 769–770
  beryllium, 238
  HIPping, 937
Al-10Si lost foam castings, 877
Al-11Si
  dendritic segregation, 219
  surface films, 26–28, 29f
Al-11.5Mg layer porosity, 386, 387f
Al-11.5Si bifilms evidence, 83f
Al-12Si
  feeder size, 349–350, 349f
  freezing range, 577, 589f
  oxide bifilms, 260f–261f
  shrinkage porosity, 362, 363f
Al-17Si
  engine blocks, 119
  uniform contraction, 914, 916f
Al-21Cu channel segregation, 222
Al-Ag bonding, 185
Al-Cd hot tear, 420–421
Al-Cu
  casting, 237
  channel segregation, 222
  continuous fluidity, 132–134
  gravity segregation, 222
  hot tearing, 418f, 429f, 430, 432f
  hydrostatic tension, 357, 358f
  silicon, 263
  stress concentration, 424
  susceptibility prediction of hot tearing, 429f, 430, 432f
Al-Cu-Fe/Mg alloying and hot tears, 440
Al-Cu-Si fluidity and latent heat, 106, 118
Al-Li elastic modulus, 507
Al-Mg
  barriers to diffusion, 413
  beryllium, 81, 238
  blister formation, 932
  casting, 236–237
  eutectics and interdendritic feeding, 356–357
  hot rolling, 941
  intergranular corrosion, 518
  properties, 230
  surface films, 11–13
  susceptibility prediction of hot tearing, 430
Al-Mg$_2$Si system, 263f, 793
Al-Pb hot tear, 421
Al-Si
  bifilms, 67, 199, 200f, 248, 258–259, 289, 538–540
  casting, 236
  chills, 179
  continuous fluidity, 131–134
  CSC parameters, 432
  eutectic Si, 244–259
  extended fluidity, 128–130, 129f
  feeding theory, 598
  filter contamination, 744
  flowability, 123–124
  fluidity, 96
  furling and unfurling, 67–77
  gates, 711
  Hall–Petch relation, 480
  hot tear, 418f
  in situ MMC, 793
  interdendritic phases, 487
  iron-rich intermetallics, 260f
  latent heat and fluidity, 115–119, 118f
  machining, 939
  magnesium, 237
  micrographs, 302
  microjetting, 58
  modulus, 113
  mould explosions, 147
  nucleation, 199, 200f, 248–249
  phase diagram, 250–253
  phosphorus, 246f
  sand blasting, 929
  shrinkage porosity initiation, 365–381
  silicon particles, 199, 244, 245f, 248–249
  sodium, 238–239
  strontium, 227, 238–239, 914
  susceptibility prediction of hot tearing, 430, 432
  tensile properties, 448, 471
  titanium, 243, 263
  uniform contraction, 911–912
Al-7Si-0.4Mg reverberatory furnaces, 770
Al-6Si-3.5Cu heat treatment, 931–932
Al-Si-Mg ductility, 477
Al-Sn fluidity, 99, 100f
Al-Ti phase diagram, 240, 241f
Al-TiB$_2$ metal/matrix composite, 27f

Aluminium alloys: (*Continued*)
  Al-Zn
    AZ91, 229–230, 233, 235
    fluidity, 100–101, 102f
    superheat, 101, 102f
  Al-Zn-Mg-Cu
    ductility, 477
    surface, 933f
  argon, 775
  automatic greensand moulding, 822
  automotive systems, 570
  beryllium, 81
  bifilms
    Al-Si, 473, 477, 538–540
    bismuth, 77
    evidence, 83–87
    furling and unfurling, 67–77
    hydrogen precipitation, 68–70, 256–257
    intermetallics in Al-Si alloys, 71
    mechanism, 295f
    melting, 775
    oxides, 72–73
    pouring and turbulence, 239
    shrinkage, 61, 70
  binders, 807–813
  bismuth, 77
  borides, 239
  boron, 802
  bubble trails, 62–63, 65
  bubbles, 61
  carbides, 239
  carbon black (thin-walled parts), 125
  castings, 167, 169, 236–269
  'cell count/size', 212
  ceramic foam filters, 733
  chills, 179
  'Chinese script', 262
  clamping points, 637–638
  confluence welds, 50–51, 55–56
  contact pouring, 823
  contraction, 776
  corrosion, 481, 511–519
  counter-gravity casting, 932
  cracks, 41, 42f, 444–445
  critical fall heights, 642
  crucible melting, 771
  cylinder heads, 835
  datums, 631–633
  deactivation of entrained films, 77
  degassing, 777–783
  delivery, 769
  dendrite arm spacing, 75, 76f, 244
  dendrites, 201–202, 204–205, 206f

detrainment, 784, 787
dies, 798–799
direct chill, 328
dry hearth furnaces, 47, 541, 773–774
entrained inclusions, 239
expansion and contraction, 899
external chills, 179–182
feeding, 579
filtration, 785–786
fins, 883
flow, 91–134
fluidised beds, 934–935
furling and unfurling, 67–77
grain refinement, 158, 201, 239–244, 482
gravity die castings, 771
grit blasting, 927–928, 939
growth restriction parameter, 242, 243f
Hall–Petch relation, 479
heat pipes, 192–193
heat treatment, 623–624
high volume, 660
hot isostatic pressing, 935–938
HPDC, 798, 862–863, 900, 907–908
hydrogen, 5–9, 6f–7f, 11–13, 267–269, 407, 409, 782
hydrostatic tensions, 346–348, 352, 354f, 360
inclusions, 3–5, 4t, 8, 239, 536
internal chills, 182–185
iron-rich intermetallics, 259–262
liquid, 238–239, 769, 776
liquid fluxes, 32–33, 82
long freezing range (feeding distance), 588
lost foam casting, 875–878
lost-wax casting, 914
magnetic moulding, 802
markets, 236
melting, 769–796
melts, 769, 771, 777–790, 793
metal/matrix composites, 26, 35
Mitsubishi, 224t, 226
mushy state testing, 436
non-feeding role of feeders, 602–603
oxide bifilms, 780, 782–784, 786–787, 793
oxides, 47, 56, 228, 237–239
painting, 940
pouring, 47–50
quenching, 617–618
rapid solidification, 887, 889–891, 889f–891f
residual stress, 613–615, 617–619, 622–623, 625–626, 628–630
reverberatory furnaces, 771
rolling waves, 92
rotary degassing, 780–783
rubber moulds, 814
salt cores, 799–800

sand castings, 53
sedimentation, 82–83
shrinkage porosity, 60
SiC MMC, 793
Sn addition, 939
solid feeding, 360–365
steel
   gauze filters, 744
   titanium inserts, 185
strength, 624–626, 734
strontium, 238, 525, 914
structure, 3
superplastic forming, 510
surface tension controlled filling, 651–656
surfaces, 17, 22–25, 29f–30f, 46, 49
temperature distribution in greensand mould, 139–140, 139f
thermal analysis, 264–267, 266f
'tilt pouring', 829–831, 833
Ti-rich grain refiners, 541
titanium, 239, 241–242, 242f, 482, 857
ultimate tensile strength, 490–493
ultra-clean, 541
ultrasonics, 4, 86, 481, 537–538
unzipping waves, 94
vacuum moulding, 878–880
yield strength, 477–490
zircon sand moulds, 928
Aluminium carbide in magnesium alloys, 233
Aluminium nitride, 322, 782
Aluminium phosphide (AlP), 244, 245f–246f, 246–247, 249–251, 249f, 253, 255, 259, 262
Aluminium silicate (mullite), 800
Ammonia and nitrogen porosity, 412–413
Anorthosite (aggregate moulding materials), 805
AOD (argon/oxygen decarburisation), 777, 779, 783
Area ratios in pressurised/unpressurised filling, 754t
Areal pore density (parameter), 379, 380f
Argon:
   aluminium alloys, 775
   degassing, 775
   inert gas shroud, 788, 789f
   magnesium alloys, 229
Asarco type furnace, 774
ASTM radiographic standards, 235–236
Asymmetric films, 50
Austenite:
   cast iron, 275–314
   matrix structure, 197
   pearlite transformation, 330
   residual stress, 616
   steels, 179, 328
Austenitic ductile iron (ADI), 312
Automatic bottom pouring (ABP), 772, 823
Automotive cylinder heads:
   accuracy, 924
   air quenching, 624
   Cosworth Process, 851
   die casting, 842
   gravity dies, 825
   residual stress, 618–619
   strength reduction by heat treatment, 624–626
'Avoid bubble damage' rule:
   bubble damage
      counter-gravity systems, 560–561
      gravity filling systems, 559–560
   bubble trail, 557
'Avoid convection damage' rule:
   academic background, 603
   convection
      countering, 610
      damage and casting section thickness, 608–609
   engineering imperatives, 603–607
'Avoid core blows' rule:
   background, 561–565
   core blow model study, 569
   micro-blows, 397–398
   outgassing pressure in cores, 564–568
   prevention of blows, 569–573
'Avoid laminar entrainment of the surface film' rule:
   hesitation and reversal, 555–556
   horizontal stream flow, 553–555
   meniscus, 550–551
   oxide lap defects, 556–557
   waterfall: oxide flow tube, 552–553
'Avoid shrinkage damage' rule:
   definitions/background, 573
   feeding to avoid shrinkage problems, 574–575
   seven feeding rules, 575–596
'Avoid turbulent entrainment' (critical velocity) rule:
   maximum velocity requirement, 543–546
   'no-fall', 547–550

B

Back-diffusion and microsegregation, 216
Backing sands, 818–819
Base for sprue, 678–679
Batch melting (liquid metals), 769–772
Bcc. *See* Body-centred-cube
Bells and loam, 812
Bernoulli, Daniel, 951
Bernoulli equation, 951–953
Berthelot's experiment, 346–347
Beryllium:
   aluminium alloys, 81
   Al-Mg alloys, 238
   iron-rich intermetallics, 259–262

Beta-Fe particle/inclusion/platelet, 250, 258, 260, 440, 473
Beta silica, 803
'Bifilm crack', 23
Bifilm-free properties, 527–528
Bifilms:
    alumina, 321, 781
    aluminium, 469–470, 476–477
    aluminium alloys, 775, 786, 790, 792
    blister formation, 932–933
    bubbles, 25–28, 63f, 87, 780
    buoyancy, 25
    castings, 223
    chills, 179
    copper alloys, 274
    corrosion, 514, 514f
    Cosworth Process, 89
    cracks, 335f, 505, 796
    creep, 510
    DAS and tensile properties, 489
    deactivation of entrained films, 77–80
    defects in magnesium alloys, 228
    density, 25
    description, 21–25
    detrainment, 82
    directional solidification, 528, 883–885
    double
        entrainment defects, 21–37
        evidence for bifilms, 83–87
        furling and unfurling, 67–77
        grey iron, 280
        oxide flow tubes defects, 50, 57
        surface flooding, 50
        term, 21
    ductile iron, 280–282
    ductility, 473–477, 489, 490f
    elastic modulus, 507–508
    entrainment, 21–25, 45, 74–77, 80, 86
    flow regimes, 46, 46f
    flux and slag inclusions, 30
    furling and unfurling, 67–77
    gold, 87
    grain refinement, 239, 482
    graphite, 294–296
    grey iron, 280
    Griffith's cracks, 89, 444
    heat treatment of Al alloys, 631
    HIPping, 935–938
    hot tearing, 442, 443f
    hydrogen, 618
    importance, 88–89
    initiation of porosity, 365–381
    iron, 790
    leaks, 943–944, 944f
    magnesium alloys, 234–235
    melts, 537–541, 539f–540f
    micro-structure, 481–482
    Ni-base alloys, 333
    nitride, 283, 306
    non-destructive testing, 89
    pitting corrosion, 515–516
    pore initiation, 378–381
    pronunciation, 21
    reduced pressure test, 83–84, 85f
    residual stress, 617–618
    scanning electron microscopy, 83, 86–87
    shrinkage, 70
    silicon, 244
    single crystal solidification, 528, 886–887
    sources of, 89–90
    steels, 329, 499, 519, 790
    tear growth, 426–428
    titanium alloys, 339–340
    ultrasonics, 4, 86, 537–538
    viscosity, 110 112
    waterfall effect, 552–553
    welding cracks, 444
    white iron, 312–313
    X-rays, 942–943
    zinc alloys, 225–227. *See also* Aluminium alloys; Double bifilms; Oxide bifilms
Bimodal distribution, 456f, 459
Binary alloys, 203f, 240, 241f
Binders:
    alkaline phenolics, 809
    cement, 813
    chemical, 808–812
    Croning shell process, 811–812
    dry sand, 808
    Effset process, 812
    fluid (castable) sand, 813
    furans, 809
    gravity dies, 825
    greensand, 807–808
    hot box and warm box processes, 811
    light-metal casting, 811
    loam, 812
    phenol-urethane, 809–810
    silicates, 810–811, 817–818, 906
    sodium polyphosphate glass, 810
    sodium silicate, 810–811, 817–818, 906
Bismuth, 77, 204–205
Blind casters, 590–591
Blind feeders, 580–581, 590–591
Blister formation, 932–933
Blow, term, 389
Blow defect, 389, 561

INDEX

Blow holes and gas porosity, 389–398
Blow hole, term, 389
Blows. *See* Core blows
Body-centred-cube (bcc) structure, 197, 261, 445, 483
Boron, 159, 321–322
Boron nitride and Mg-Zr alloys, 234
Boron oxide, 321–322
Bottom-gating:
    vs. counter-gravity mode, 642
    filling system, 650, 651f, 657, 658f
    gravity pouring of open moulds, 645, 822
    moulds, 548
Bottom-pour ladle, 666–667, 826, 957–958
Box shaped castings, 649f
Branched columnar zone, 211
Brasses, 270–272
Brief case handles (zinc alloys), 226
Bronze:
    bubble trails, 63–64, 65f
    cannon, 397
    foundries, 154–155
    'Freedom Bell', 554, 554f
    grain refinement, 483–484
    'skeins of geese', 63–64
    wall plaques, 821. *See also* Aluminium alloys, aluminium bronze
Bubble damage:
    'avoid bubble damage' rule, 557–561
    copper, 64, 66f
    description, 62
    entrapment of small bubbles, 60–61
    inclusion control, 724
    leaks, 943
    radiography, 61
    runner, 686–687
    shrinkage porosity, 60
    surface turbulence in filling system, 389
    term, 60, 62, 724. *See also* Rule 4 for good castings
Bubble trails, 941–942
    aluminium alloys, 62–63
    aluminium and X-ray video, 60, 63
    'avoid bubble damage' rule, 557
    bifilms, 87
    cast iron, 566f–567f
    castings in vacuum, 67
    collapse, 59f
    copper, 64, 66f
    core blows, 566f–567f
    description, 58
    iron, 60, 65
    length, 67
    low-pressure casting, 65
    structure, 59

Bubble traps, 732, 737f
Bubbles:
    bifilms, 25–28, 780
    collapse, 28f
    core blows, 561, 562f–563f, 563–564
    degassing, 779–780
    detachment, 391, 391f
    diameters, 390f, 396
    entrainment, 544–545, 674
    grey iron, 566f–567f
    large, 25
    leaks, 522
    outgassing of cores, 391–392, 396
    shape, 28
    small, 25, 60–61
    trail, 25
Buoyancy forces and liquid metals, 899
Burst feeding (shrinkage porosity), 357–360
Bypass designs (surge control), 716

C

Cake core (mould design), 900, 901f
Calcium cyanamide and steels, 327
Calcium in magnesium alloys, 229
Calcium silicide, 291, 321
Capillary attraction, 653
Capillary repulsion, 653
Carbon:
    aggregate moulding, 806
    black/soot (casting thin-walled parts), 125
    cast iron, 275
    ferrous alloys, 806
    metal surface reactions (pick up and loss), 155–156
    steels, 315, 319–320, 433f, 445
Carbon boil, 315
Carbon dioxide in magnesium alloys, 229
Carbon equivalent (C_E), 275
Carbon equivalent value (CEV), 275, 292
Carbon films (lustrous carbon), 148–149, 284–288
Carbon monoxide:
    cast iron, 276
    gas porosity, 401
    pressure in steel castings, 404–405
Carbon tetrachloride and tensile strength, 347
Carbon-based filters for steels, 744
Carburisation in low carbon steels, 156
Cast iron:
    bubble trails, 25, 60, 566f–567f
    carbon, 275
    carbon monoxide, 276
    carbonyl, 150
    CaSi, 541–542
    chills, 182, 185

Cast iron: (*Continued*)
   chunky graphite, 310–312
   core blows, 393–396
   decarburisation, 156
   dies, 824
   dimensional accuracy, 900
   Ellingham diagram, 277
   expansion and contraction, 899
   gases, 275–277
   graphite films, 285–288
   graphite flakes, 199
   gravity dies, 824
   holders, 775
   inoculation, 290–298
   machining, 938–940
   microstructures
      eutectic growth graphite/austenite, 300–302
      flake graphite iron, 290–298
      graphite, 290
      nucleation and austenite matrix, 299–300
   nitrogen, 276–277
   nodularity, 806
   pressure requirement (feeding), 596
   pressurised systems, 752
   sand mould penetration, 153
   solidification, 165, 166f
   spheroidal graphite iron, 302–307
   structure hypothesis summary, 314
   surface films
      bifilms in ductile iron, 280–282
      carbon films, 284–288
      nitride bifilms, 283
      oxide bifilms in grey iron, 280
      oxides, 280
      plate fracture defect in ductile iron, 282–283
      temperature, 52
   white iron, 312–313
Cast material:
   just-in-time slurry production, 792
   liquid metal, 790–791
   metal matrix composites, 792–793
   Mg alloy slurry production from granules, 792
   partially solid mixtures, 791–793
   rheocasting, 791
   strain-induced melt activation, 792
   thixocasting, 791
Cast preforms for forging, 871
Casting:
   centrifugal, 856–861
   constraint, 917–922
   counter-gravity, 838–856
   dimensions statistics, 893, 894f
   gravity, 821–827

   horizontal transfer, 827–838
   lost foam, 875–878
   lost wax/ceramic mould processes, 871–875
   pressure assisted, 861–871
   vacuum melting and casting, 881–882
   vacuum moulding, 878–880
   vacuum-assisted, 880–881
Casting accuracy:
   nonuniform contraction, 915–922
   process comparison, 923–924
   shrinkage, 912f
   uniform contraction, 910–915
Casting alloys:
   aluminium, 236–269
   cast iron, 275–314
   copper, 269–274
   magnesium, 227–236
   nickel, 330–335
   steels, 314–330
   titanium, 335–340
   zinc, 230–234
Casting rules. *See* Rules for good castings
Cast-on fins, 191
Cast-resin patterns, 909
Cavitation (superelastic forming), 510
Cavitation, description, 595
Cell, term, 212
Cell count, 212
Cell size, 212
Cellular front growth, 202, 248
Cement binders, 813
Cementite (carbide), 313
Central versus external systems (gates and filling), 710–711
Centrifugal casting, 856–861
Centrifugal pumps for liquid metals, 847
Centrifuge casting, 856–861
Ceramic block filters, 733–735, 744
Ceramic foam filters, 729f, 733–735, 736f, 744
Ceramic moulds accuracy, 907
Ceramic moulds and cores:
   investment shell (lost-wax) moulds, 800
   investment 2-part block moulds, 801
   magnetic moulding, 802
   plaster investment block moulds, 801
CFD. *See* Computational fluid dynamics
CGI. *See* Compacted graphite iron
Channel segregation in aluminium alloys, 222
Chaplets, 185, 899
"Chattanooga Choo Choo", 291
Chemical binders, 808–812
Chilling power, 176t
Chills:
   benefits, 179

bifilms, 179
chaplets, 185
copper, 181, 182f
external, 179–182
feeding, 600
fins comparison, 191–192, 192f
heat transfer, 178–179
internal, 182–185
iron, 182, 185
relative diffusivities, 181f
thermal conductivity, 192
thickness, 192
Chinese script, 235, 262
Choke in pressurised/unpressurised systems, 750–755
Chromite sand (moulding aggregate), 803–804, 817–818
Chunky graphite (CHG), 310–312
Chvorinov's rule, 113, 173–174, 177, 576, 583, 597
Clamping points for machining, 637–638
Close-packed cubic symmetries (fcc, bcc), 261
Cloth filters siting, 730, 731f
CO_2 process (cores), 906
Coal pyrolysis, 147–148
Cobalt:
　aluminate and grain refinement, 158
　Co-base alloys, 331–332, 542
　　ceramic moulds and cores, 800
　　vacuum casting, 881
　moulds, 158
COD. *See* Crack opening displacement
Cold cracking, 442–445
　crack growth, 444–445
　crack initiation, 444
　nucleation, 443–444
'Cold crucible' technique, 857
Cold lap defects, 555
Columnar dendrites, 204
Columnar-to-equiaxed transition and solidification, 210
Compacted graphite iron (CGI), 308–310, 312
Computational fluid dynamics (CFD), 128–129
Computer simulation:
　convection, 601
　feeding, 601
　filling systems, 747, 749f
　moulds and cores, 571–573
　solidification of castings, 576
Cone test for hot tearing, 436, 441f
Confluence welds:
　aluminium alloys, 55–56
　Cosworth Process, 53
　ductile iron casts, 56
　filling instability, 54f
　mechanism, 54f
　surface films, 53, 55

Conical basin:
　pouring, 657–660
　sprue, 681, 682f
Connective loops and convection, 610
Constitutional undercooling, 202, 203f, 204
Contact pouring, 660–661, 759, 823
Continuous fluidity, 131–134
Continuous fluidity length, 96, 132
Continuous melting, 772–774
Contraction:
　allowance, 342, 913f, 914
　dimensional accuracy, 899. *See also* Uniform contraction
Controlled solidification:
　conventional shaped castings, 883
　cooling fins, 883
　directional solidification, 528, 883–885
　rapid solidification casting, 887–891
　single crystal solidification, 885–887
Controlled tilt casting (Durville method):
　description, 823
Convection:
　connective loops, 610
　Cosworth Process, 699
　countering, 610
　damage and casting section thickness, 608–609
　heat transfer, 193. *See also* Rule 7 for good castings
Conventional shaped castings (controlled solidification), 883
Cooling of castings, 883, 921
Cope, term, 821
Cope surface evaporation, 142, 143f
Cope-drag arrangement, 902, 904
Copper:
　Asarco type furnace, 774
　bubble damage, 64, 66f
　chills, 181, 182f
　diffusion coefficients, 10f
　ductility, 464f, 465, 473–474
　hot rolling, 941
Copper alloys:
　aluminium bronze modulus, 577
　bifilms, 274
　brass, 270–272
　chills, 179
　Cu-10Al
　　critical velocity, 41, 546
　　hydrogen porosity, 406f
　Cu-10Sn porosity, 354f
　Cu-Ni carbon, 410
　Cu-Zn vaporisation, 14
　ductility and grain refinement, 484
　Ellingham diagram, 273
　gas porosity, 409–410
　gases, 270–274

Copper alloys: (*Continued*)
   grain refinement, 274
   granulated charcoal, 273
   gunmetal, 270
   lead, 15, 270
   melting, 272–273
   ounce metal, 270
   penetration barriers, 153–154
   phosphorus, 272–273
   reduced pressure test, 274
   steam reaction, 271
   subsurface pinholes, 274
   surface films, 270
   zinc, 271–272. *See also* Bronze
Core blows:
   bubble trails, 392–393, 566f–567f
   bubbles, 561, 562f–563f, 563–564
   cast iron, 393–396
   condensation, 561
   gas porosity, 389, 397, 564
   grey iron, 566f–567f
   metal chills, 561
   moulds, 395–396
   prevention, 569–573
   steel boxes, 564. *See also* Rule 5 for good casting
Core-package system (CPS), 838
Cores:
   assembly, 903
   dimensional problems, 902–903
   making, 905
   running system, 917
   venting system, 392. *See also* Moulds and cores
Corrosion
   filiform corrosion, 517–518
   hydrogen embrittlement, 398, 541
   inter-granular corrosion, 316, 518
   pitting corrosion, 515–517
   stress corrosion cracking (SCC), 519
Cosworth Process:
   accuracy and aggregate moulds, 924
   bifilms, 89
   bubble damage in counter gravity systems, 560
   confluence welds, 53
   convection, 605
   critical velocity, 545–546
   electromagnetic pumps, 847–849
   entrainment of bubbles, 560
   grain size in aluminium alloys, 481
   incipient melting, 933–934
   ingates, 699
   leak elimination, 941–942
   leaks, 523
   liquid aluminium alloys, 775
   location points, 633–637
   pouring (Al alloys), 790
   roll-over after casting, 449, 838
   tensile strength, 490
Cothias Process (squeeze casting), 869
Counter-gravity casting:
   10 test bar mould, 449
   aluminium alloys, 932
   vs. bottom-gating technique, 642
   counter-pressure, 845
   description, 838
   direct vertical injection, 853–854, 854f
   failure modes for low-pressure casting, 856
   feedback control for counter gravity casting, 856
   filling system design, 642
   gravity problems, 646
   liquid metal pumps, 846–853
   low-pressure casting, 842–846
   medium-pressure die-casting, 845
   programmable control, 855
   surface turbulence, 648
   T-Mag process, 846
   vacuum riserless casting/pressure riserless casting, 846
Counter-gravity filling:
   bubble damage, 560–561
   filling defects, 555
   Griffin Process, 790, 797–798, 806
   lost foam castings, 877
   magnesium alloys, 862
Counter-pressure casting (CPC), 845
CPS. *See* Core-package system
Crack opening displacement (COD), 495–496
Cracking, term, 443
Cracks and tears, 417–446
Creep, 509–511
Cristobalite and investment shell moulds, 800
Critical fall heights:
   aluminium alloys, 645
   filling systems, 646
   liquids, 39, 48–50
Critical velocities:
   aluminium, 41, 43f
   aluminium bronze, 44
   copper alloys, 41
   equations, 38–39
   filling systems, 646
   good casting, 542–550
   liquid aluminium, 39, 40f, 44–46
   liquids, 41t
   metals, 544
   Weber number, 41–44
Croning, Johannes, 811
Croning shell process, 906

Crucible melting (electric resistance/gas heated), 771
Cryolite (flux), 31
CSC (cracking susceptibility coefficient) for hot tearing, 430, 432
Cupola shaft furnace for iron, 772–773
Cylindrical systems and location points, 636
C-Cr steels and microsegregation, 216, 217f

D

Darby, Abraham, 821
DAS. *See* Dendrite arm spacing
Datums and location points, 631–633
Deactivation of entrained films, 77–80
Decarburisation of iron castings, 156
Defects:
    alloys, 88
    blow, 391–392, 561
    castings
        gas porosity, 465
        inclusion types, 464–465
        quantification, 3
        shrinkage porosity, 466–467
        tears, cracks and bifilms, 467–470
    cold lap, 555
    'dents', 397
    dross, 565f
    entrainment, 269–270, 811–812, 857
    exfoliation, 391, 565f
    melts, 536–537
    oxide lap, 555–557
    pouring, 642–646
    wormholes, 405
Degassing (melts):
    aluminium alloys, 780–783
    bubbles, 763–765
    gaseous argon shield, 778–779
    liquid argon shield, 778
    passive, 777
    vacuum degassing, 783
Dendrite arm spacing (DAS):
    ablation casting, 890
    aluminium alloys, 75, 76f, 244
    cast materials, 3, 77
    chills, 179
    defects, 463
    directional solidification, 528, 881–882
    grain size, 207–209, 208f, 211
    growth, 207–209
    homogenisation and solution treatments, 929–930
    nucleation, 198–199
    pore initiation on bifilms, 378–381
    solidification structure, 163
    yield strength
        bifilms, 489
        description, 477–478, 484–485
        heat treatment, 487
        interdendritic spaces, restricted growth, 487
        residual Hall–Petch hardening, 485
    zinc alloys, 227
Dendrites:
    aluminium nitride, 322
    columnar, 204
    grains, 204, 207f
    graphite, 300
    growth, 201–202, 204–205, 205f–206f, 207–209, 211
    hot tears, 419
    instability condition, 204
    δ-iron, 329–330
    microsegregation, 216–218
    segregation, 218–219, 611
Dendrites, term, 201–202
Dendritic segregation, 218–219, 869
'Dent' defects, 397
Detrainment (cleaning):
    aided flotation, 784
    aided sedimentation, 784–785
    description, 81–83
    filters, 787
    filtration, 82, 785–786
    natural flotation and sedimentation, 783–784
    packed beds, 786–787
    practical aspects, 787
    rotary degassing, 82
    sedimentation, 82
Diamond films, 288
Die casting:
    contraction allowance, 914
    expansion and contraction, 899
    exterior shapes, 898
    interior shapes, 898–899
    low-pressure, 842
    maximum velocity requirement, 544–546
    medium pressure, 845
    zinc alloys, 909
Differentiation of solid (grain multiplication), 209–212
'Diffraction mottle', 481
'Diffuser' term (runner), 688
Diffusion coefficients of elements, 9, 10f
Diffusion distance, 216, 230, 292, 300, 374, 400–401, 405, 407, 409–410, 619, 779
Diffusivity (heat), 113, 173, 176t, 180, 181f
Diffusivity (thermal), 165, 176t, 180, 619, 625, 629, 801, 805
Digital laser-reading systems (metrology), 925
'Dilute air' (vacuum), 67
Dimensional accuracy of castings:
    accuracy, 910–924
    alloys, 900

Dimensional accuracy of castings: (*Continued*)
  description, 893
  expansion and contraction, 899
  ISO, 893, 897
  metrology, 924–926
  moulds, 900–909
  net shape concept, 897–900
  non-uniform contraction, 915–922
  tolerances, 896f
  tooling, 909–910
Direct chill (DC):
  aluminium alloys, 328, 525
  casting, 81
  grain refinement, 785
  packed beds, 786
  residual stress, 618
Direct gates, 690
Direct pour casting, 824
Direct pour filters, 742–744, 743f
Direct squeeze casting, 868
Direct vertical injection, 853–855
Direct-acting piston displacement pump, 853
Directional solidification (DS), 528, 883–885
Distortion. *See* Non-uniform contraction
Distortion and residual stress, 626
Division of sprue, 677–678
Double bifilms, term, 21
Down-runner. *See* Sprue
DPI. *See* Dye penetrant inspection
Drag, term, 821
Dross defects, 563
'Dross stringers' in ductile iron, 280–282
Dross trap (slag trap), 717f, 724–725
Dry coatings for moulds and cores, 161
Dry hearth furnaces, 47, 541, 773–774
DS. *See* Directional solidification
Ductile iron:
  brittleness, 306
  casting, 279
  confluence welds, 56
  critical velocity, 41, 43f
  distortion, 920
  'dross stringers', 280–282
  ductility, 314
  feeding, 582
  fluidity, 116f
  gas porosity, 411
  gravity casting, 821–827
  gravity dies, 824
  growth, 299f
  machining, 938–939
  melts, 541–542
  oxidation, 512, 513f
  plate fracture defect, 282–283
  pouring, 47
  pyrolysis, 147
  reclamation and recycling of aggregates, 816–817
  strength and filters, 734
  stringers, 280–282
  sulphur, 157
  surface films, 13
  vaporisation, 14. *See also* Spheroidal graphite iron
Ductility:
  bifilms, 469–470, 476–477, 489, 490f
  copper, 464f, 473–474
  failure, 474–475, 475f
  freezing of castings, 477
  layer porosity, 492
  pores and cracks, 477
Duplex stainless steels, 316
Duralcan, 793
Durville casting (controlled tilt), 829
Dye penetrant inspection (DPI), 332, 335, 943

E

Ease of removal (castings), 648
ECAP. *See* Equal channel angular pressing
Effset process (ice binder), 812
Elastic modulus, 507–508
Elastic stiffness (Young's modulus):
  damping capacity, 507–508
  Mitsubishi alloys, 226
  residual stress from casting, 613–617
  resonant ultrasound spectrometry, 945
Elastic–plastic model, 921–922, 922f
Electrical conductivity inclusions, 537
Electro magnetic (EM) pumps for liquid metals, 847–849
Electro-slag re-melting (ESR), 179, 794–795
Ellingham diagram:
  cast iron, 277
  copper, 273
  oxides, 11–13, 12f, 238
EM. *See* Electro magnetic
Entrained inclusions:
  aluminium alloys, 239
  aluminium nitride, 322
  description, 319
  liquid surface oxide films, 321–322
  partly liquid surface oxide films, 321
  pourings, 326
  solid oxide surface films, 320–321
Entrainment:
  air, 822, 857
  air bubbles and gas porosity, 349–350
  defects

bifilms, 21–25, 45, 545
bubbles, 25–28, 26f–30f, 544–545, 724
Cosworth Process, 838
critical heights, 642
extrinsic inclusions, 30–37
gravity filling, 827
liquid surface, 544
description, 21–25
filling, 18, 18f
films
  elimination, 13
  oxide, 19–20, 30, 33–35
  viscosity, 110–112
oxide films, 549
processes
  bifilms evidence, 83–87
  bifilms importance, 88–89
  bubble trails, 58–67
  deactivation of entrained films, 77–80
  detrainment, 81–83
  furling and unfurling, 67–77
  microjetting, 57–58
  oxide flow tube, 56–57
  oxide lap defects
    confluence weld, 52–56
    surface flooding, 50–52
  oxide skins from melt charge materials, 46–47
  pouring, 47–50
  soluble, transient films, 80–81
  surface turbulence, 37–46
surface films, 4, 13, 19–20, 37–38. *See also* Rules 2, 3 for good castings
Equal channel angular pressing (ECAP), 792
Equilibrium gas pressure concept, 7–8
ESR. *See* Electro-slag remelting
Eutectic Si in aluminium alloys:
  Al-Si
    hypereutectic, 244–247
    hypoeutectic, 247–248
    phase diagram, 250–253
  mechanical properties, 254
  non-chemical modification, 258–259
  nucleation of Si, 248–250
  Sr modification
    application, 253
    porosity, 254–257
Eutectics:
  fluidity, 102, 103f–105f, 106–107
  growth of graphite/austenite, 300–302
  interdendritic feeding, 355–357
  latent heat, 115–116
  superheat, 102, 102f, 104f
  temperature, 102, 102f, 104f

Evaporation, 139–144
Exfoliation defect, 391, 565f
Expanded runner, 688–690
Expansion and dimensional accuracy, 899
Exploded nodular graphite, 307
Extended fluidity, 128–131
External chills, 179–182, 219
External porosity (surface links), 367–368
External running system, 685f
Extrinsic inclusions:
  flux and slag, 30–37
  old oxides, 33–35
  sand, 35–37
Eye-dropper ladle. *See* Snorkel ladle

F

Face-centred cube (fcc) structure, 197, 483
Facets, 72–74, 135, 235, 257–258, 282–283, 322, 501–504, 503f, 537–538
Facing sands, 818–819
Fatigue:
  high cycle
    defects, 499–500
    initiation, 498–504
    performance, 504
    steel solidification, 499
  low cycle and thermal, 506–507
Fayalite (ferrous orthosilicate), 804
Fcc. *See* Face-centred cube
Fe_3Al (intermetallic) and tensile properties, 510–511
Fe-3.25%Si steel, 329
FeCr powder and steels, 327
Feed path requirement:
  criteria functions, 590–593
  directional solidification towards feeder, 584–587
  feeding distance, 587–590
  minimum temperature gradient requirement, 587
Feedback control for counter-gravity casting, 855–856
Feeders:
  absence, 574
  blind, 580–581, 590–591
  neck constriction, 542, 586
  non-feeding role, 602–603
  reverse taper, 579
  theory, 577–579
  thin-walled castings, 575–576
  volume, 579–580, 594f
Feeding:
  blind feeders, 580–581
  computer modelling, 601
  criteria and shrinkage porosity, 348–352, 590
  definition, 573
  ductile iron, 281f, 579, 582

Feeding: (Continued)
    five mechanisms, 600–601
    freeze volume, 579–580
    freezing systems design, 600
    graphite, 582
    Heuvers circles, 584, 595f
    new logic, 596–600
    non-feeding role of feeders, 602–603
    padding, 585, 595f
    random perturbations, 602
    reverse tapered feeder, 579
    rule 1: do not feed (unless necessary), 575–576
    rule 2: heat-transfer requirement, 576–577, 641
    rule 3: mass transfer (volume) requirement, 577–582
    rule 4: junction requirement, 582–583
    rule 5: feed path requirement, 583–590
    rule 6: pressure gradient requirement, 590–593
    rule 7: pressure requirement, 593–596
    rules, introduction, 575
    solid feeding, 600–601
    system, 901, 917
    thin-walled castings, 575–576
Feldspars and silica sand, 804
Feliu, S, 96, 131–134, 131f, 133f
Ferrite:
    matrix structure (bcc), 197, 197f
    residual stress, 616
    Schoefer diagram, 316, 317f
    titanium, 328
Ferrite potential (FP), 328
Ferromanganese (liquid), 31, 33f
Ferrosilicon (FeSi), 291–293, 293f–294f, 541–542
Ferrous alloys:
    carbon, 806
    oxide films, 546
    residual stress, 616
    silica sand, 803–804
Ferrous castings, heat treatment, 929
Fe/Si competition for sites (intermetallics), 262
Filiform corrosion, 517–518
Fillability and surface tension, 123–124
Filling:
    bubble damage and, 563
    definition, 573
    entrainment elimination, 13
    flow channel structure, 194f, 195
    moulds, 542–543
Filling system components:
    filters, 728–745
    gates, 690–716
    inclusion control: filters and traps, 724–728
    pouring basin, 657–669
    runner, 683–690
    sprue, 669–683
    surge control system, 716–719
    vortex systems, 719–723
Filling system design:
    bottom-gating, 648, 650, 651f
    casting constraint, 917–922
    description, 821
    down-runner, 669
    ease of removal, 648
    economy of size, 647
    filling speed, 647
    gravity pouring
        closed moulds, 822–825
        'no fall' conflict, 642–646
    liquid metal and mould cavity, 648
    maximum fluidity, 95–128
    maximum velocity requirement, 642
    mould design, 900–904
    pouring speed, 645
    practice
        fill rate, 758–759
        gates, 765–766
        layout, 748
        methoding approach, 748
        pouring basin, 759
        pouring time, 755–757, 764
        pressurised/unpressurised, 750–755
        runner, 763–765
        sprue, 759–763
        thin sections and slow filling, 758
        weight and volume estimate, 748–750
    reduction/elimination of gravity problems, 646–651
    surface turbulence elimination, 646
    surface tension controlled filling, 651–656
    targets, 686
    top-gating, 643
    turbulence, 645
    vacuum-assisted, 653–654, 654f
Filters:
    bubble traps, 737f
    ceramic block, 733–735, 744
    detrainment, 783–787
    direct pour, 742–744, 743f
    flow control, 733–734
    gravity dies, 825
    leakage control, 735–739
    runner systems, 686
    running systems, 739–741
    siting, 737–740
    sprue, 738f
    strainers, 729
    tangential filter gate, 714, 714f–715f
    tangential placement, 741–742
    woven cloth or mesh, 729–733
Filtration and detrainment, 82, 783–784

Fins:
  cast-on, 191
  chills comparison, 191–192, 192f
  cooling, 883
  description, 178–179
  feeding, 600
  junctions, 185–187, 186f–187f
  L-junctions, 188, 188f
  length, 188, 190f
  T-junctions, 185–187, 187f
  thermal conductivity, 191–192
  thickness, 188, 190f
Fischer, R.B., 274, 407
Flake graphite iron and inoculation:
  eutectic grain, 301f
  graphite growth, 294–296
  history, 291
  mechanism, 291–294
  practical experience, 296–298
Flemings, Merton C.:
  fluidity, 96, 99
  growth of solid, 203–204
  heat transfer, 163–197
  homogenisation and solution treatments, 929–930
  resistance and heat transfer, 164–165, 166f
  rheocasting, 791
  solidification
    metals, 198
    mode, 97–98
    time, 112–113
Flow:
  channel structure
    gates, 706–707
    heat transfer, 194–197
  control and filters, 733–734
  maximum fluidity (unrestricted flow), 95–128
  microjetting, 91
  surface forms and filling, 91–95
    rolling waves, 92
    surface tension, 91–92
Flow-off device, 716
Flowability and heat transfer, 123–124
Fluid beds, 934–935
Fluid (castable) sand binders, 813
Fluidity:
  acetylene black, 114
  advantages, 119
  alloys, 106
  critical length, 96
  definition, 96–97
  eutectics, 102, 103f–105f, 106–107
  extended, 128–131
  freezing, 106–107
  intermetallics, 103, 105f, 106

  phosphorus, 106–107, 109f
  solidification, 97–107
  term, 95
  tests
    comparison, 125–128
    description, 125
    spiral, 125–128
    superheat data, 123–124, 126, 126f
    VK strip, 122f, 125–127. See also Maximum fluidity
Fluidity to modulus ratio, 114
Fluorides (flux), 31
Fluxes:
  inclusions, 30–33
  liquids, 31–32, 34f, 82
  melt cleaning, 782
FNB. See Furan no-bake
Forrest, Reginald, 291
Foseco Porotec test (France), 83–84
FP. See Ferrite potential
Fracture pressure of liquids, 348, 348t
Fracture toughness:
  applied stress/critical defect size, 497, 498f
  description, 493
  yield strength, 496–497, 497f
Frankel Alloys Limited, 401–402
'Freckle defects', 222
'Freckles' in Ni-base superalloys, 197
Free energy of formation (surface films), 11, 12f, 13
'Freedom Bell', Washington DC, US, 554, 554f
Freezing:
  description, 343–345
  ductility of castings, 477
  fluidity, 106–107
  ice cubes, 343
  mild steel liquid, 185
  plate shaped castings (alloys), 173, 174f
  skins, 97–98
  systems design for feeding, 600–601
  times for pouring, 755, 757f
Freight wheel market in US, 790
Froude number (Fr), 44–45
Furan no-bake (FNB) binders, 805, 809
Furane resin in moulds, 157
Furane/sulfonic acid (FS), 397
Furans binders, 809
Furfuryl alcohol, 809
Furling and unfurling of bifilms, 67–77
Furnaces:
  dry hearth, 47, 541, 773–774
  holding, 784
  iron cupola shaft, 772–773
  liquid aluminium, 238–239
  reverberatory, 770–771
  rotary, 771

G

'Gas inclusion', term, 724
'Gas' porosity:
    blow holes, 389–398
    castings
        entrained air bubbles, 350
    core blows, 387, 564
    diagnosis, 413, 414t
    entrained pores (air bubbles), 388–389
    initiated in situ
        barriers to diffusion, 413
        carbon, oxygen and nitrogen, 409–412
        description, 398–413
        hydrogen, 406–409
    layer porosity, 465, 466f, 467
    moulds, 564
Gases:
    cast iron, 275–277
    Cu alloys, 270–274
Gates (filling system):
    central versus external systems, 710–711
    design, 765–766
    direct, 690, 706
    flow channel structure, 706–707
    gating ratio, 691–692
    horizontal velocity in the mould, 695–696
    horn, 703
    indirect gating, 690, 707–710
    junction effect, 696–700
    knife, 701–702, 701f
    L-junction, 699
    multiple, 692, 693f–694f
    pencil, 702–703
    premature filling problem via early gates, 692–695
    priming, 712–714
    sequential, 711
    siting, 690
    slot, 699, 707–711, 708f, 723f
    theory, 699
    T-junction, 696, 697f–698f, 698–699
    total area, 691
    touch, 700
    vertical, 703–706
    vortex, 723
Gating:
    joint line, 548, 645
    ratios, 753–755
Gauges (metrology), 925–926, 926f
Geffroy (metal mould), 802
Glass cloth filters, 729–730, 732f
Glissile (gliding) drops, 38, 544
Gold, 3, 87
'Go/no go' gauges, 925

Good fluidity, 96
Grain boundary wetting by liquid (hot tearing), 420–421
Grain refinement:
    aluminium alloys, 239–244
    amplitude/frequency of vibration, 205, 208f
    bifilms, 482
    copper alloys, 274
    direct chill, 786
    grain multiplication/fragmentation, 158, 209–212, 244, 481–482
    Hall–Petch relation, 482
    hot tearing control, 439
    magnesium alloys, 233
    mass feeding, 355
    melt cleaning, 243–244
    metal surface reactions, 158–159
    mould coats, 211
    nucleation of solid, 198–201
    porosity, 355
    rotary degassing, 785
    steels, 328–329
    titanium in Al alloys, 239, 241–242, 242f
    zirconum, 233
Grain size:
    dendrite arm spacing, 207–209
    hot tearing, 440, 443f
    vibration, 211
    yield strength, 479, 481–484
Granulated charcoal and copper alloys, 273
Graphite:
    bifilms, 297–298
    cast iron, 288–290
    dies, 843, 909
    feeding, 582
    films in Fe-C alloys, 285–288
    growth mechanisms, 313
    machining of, 843
    nodule structure, 302–304, 303f
    nucleation, 309–310
    oxide bifilms, 297–298
    titanium, 825. *See also* Flake graphite iron
Gravity casting:
    cost of, 642
    dies, 150, 570, 686–687, 798, 908
    dies (permanent moulds), 824
    gravity pouring
        closed moulds, 822–825
        open moulds, 821–822
    horizontal stack moulding, 826–827
    postscript to gravity filling, 827
    sand, 646
    two-stage filling, 825–826
    vertical stack moulding, 826

INDEX

Gravity filling:
 air bubbles, 559
 disadvantages, 827
 gate design, 765
 horizontal, 50–51, 57
 segregation, 219–222, 609
Gravity pouring of closed moulds:
 automatic bottom pouring, 823
 contact pouring, 823
 description, 822
 direct pour, 824
 gravity dies, 824
 shrouded pouring, 822
Gravity pouring of open moulds, 821–822
Grey iron:
 bubble trails, 65
 carbides, 313
 carbon films, 285, 286f
 casting, 278–279
 cooling fins, 883
 core blows and bubbles, 566f–567f
 cylinder blocks, 65
 decarburisation, 157f
 distortion, 920, 921f
 eutectic grain, 301f
 flake graphite iron, 297–298
 fluidity, 106, 109f, 116f
 gas porosity, 410–411
 graphite flakes, 295–296
 gravity dies, 824
 industrial revolution, 275
 machining, 938
 magnesium oxide surface film, 525
 mould accuracy, 909
 nitrogen, 156
 oxide bifilms, 280
 permanent moulds, 160
 phosphorus, 106, 109f, 275
 pressurised versus unpressurised filling, 750–755
 pyrolysis, 147
 residual stress, 614–615, 625f
 'rosettes' (cells), 212
 runner, 683–690
 salt cores, 799
 silica, 154
 silicon, 275
 steel inserts, 185
 stress relief, 628–630
 uniform contraction, 914–915
Great Paul bell, 507–508
Greensand moulds:
 accuracy, 904–905
 binders, 807–808
 high-pressure moulding, 153
 inert property, 797
 iron castings, 151–153
 mould penetration, 153
 pyrolysis, 147
 residual stress, 614
 steels, 144
 structure, 136, 138f
 vapour zones, 140, 140f
Greensand spelling, 807–808
Griffin Process (steel railway stock):
 carbon, 806
 graphite dies, 909
 gravity dies, 825
 inert moulds and cores, 797–802
 low-pressure casting, 842, 856
 production, 790
Griffith cracks, 89, 444, 527
Grit blasting, 499, 927–928
Growth:
 aluminium alloys, 230, 231f–232f, 242, 243f, 248, 248f
 atoms, 198
 cellular front in solids, 202
 dendrites, 201–202, 204–205, 205f–206f, 207–209, 211
 ductile iron, 299f
 morphology of graphite, 290
 solid in alloys, 202
 transparent organic alloys, 204, 206f
Growth restriction parameter (Q), 242, 243f
Gumbel distribution, 461–462, 506
Gunmetal alloys:
 castings, 364f
 description, 270
 grain refinement, 483
 phosphorus and porosity, 410, 411f

H

'H Process' (sequential-filling), 711
H13 hot work steel, 862
Hadfield Manganese steel, 322, 528, 625–626
Hafnium in Ni-base alloys, 331, 882
Hall–Petch relation:
 aluminium alloys, 483f
 dendrite arm spacing, 484–485
 grain refinement, 482
 magnesium alloys, 230, 233, 483f
 residual hardening, 485
 yield strength, 479
Harris, Ken, 802
HAZ. *See* Heat affected zones
Heat diffusivity, 113, 173, 176t, 180, 181f
Heat pipes, 192–193

Heat transfer coefficient (HTC):
 casting–chill interface, 165
 description, 203–204
 die coatings, 172
 fluidity, 114
 light alloys, 172
 metal–mould interface, 170–172
 quenching, 888
Heat transfer and solidification:
 convection, 193
 external chills, 179–182
 flow channel structure, 194–197
 increased heat transfer, 178–193
 remelting, 193–194
 resistances, 163–178
Heat-transfer requirement rule (feeding), 576–577
Heat treatment:
 homogenisation and solution treatments, 929–930
 post-casting processing, 929–935
 reduction and/or elimination, 930–932
 residual stress, 617–618, 626–628
Heat affected zones (HAZ), 934
Hero process (RS), 887–888
Hesitation and reversal (laminar entrainment of surface films), 555–556
Heuvers circles (feeding), 584, 585f
Hexachloroethane (aluminium degassing), 780
Hexagonal close packed structure (hcp)
 grain refinement, 201
 in magnesium alloys, 233
High-temperature alloy, term, 331
High-alloy steel castings and gravity segregation, 222
High-pressure die-casting (HPDC):
 aluminium alloys, 862–863, 907–908
 blister formation, 932
 cold chamber, 863, 863f
 dies, 798
 heat treatment, 866, 932–933
 hot chamber, 863
 impregnation, 942
 leaks, 942
 medium-pressure die-casting, 845
 metal moulds, 907–909
 'pore free process', 866
 post-casting processing, 927
 salt cores, 799–800
 solid mixtures, 791
 squeeze pins, 866
 vacuum processes, 866
 zinc alloys, 223, 226–227
High-strength low-alloy (HSLA) steels, 321
High temperature tensile properties, 509–511
High volume Al alloys and pouring basin, 660

HIPping. *See* Hot isostatic pressing
Hipping of solid castings, 78–80
Hiratsuka, S. et al, 130f, 131
Hitchener Company, 840–841
Holder failure (melting), 776
Holding, transfer and distribution (melting):
 description, 774–775
 holder failure, 776
 transfer and distribution, 776–777
Holding furnaces, 784
Hollow castings, 618–619, 619f
Hollow ceramic spheres, 805–806
Homogeneous nucleation temperature, 198
Homogenisation and solution treatments, 929–930
Homogenisation treatment, term, 930
'Horizontal' stack moulding, 711
Horizontal stack moulding (H process), 826–827
Horizontal stream flow, 553–555
Horizontal transfer ('level pour'), 646
Horizontal transfer casting:
 controlled tilt, 829–834
 level pour, 828–829
 roll-over after casting, 836–838
Horizontal velocity in the mould, 695–696
Horizontal vortex well, 721, 721f
Horn gates, 703
Hot isostatic pressing (HIPping):
 bi-Weibull analysis, 461, 938
 centrifugal casting, 857
 post-casting, 935–938
 rapid solidification, 887
Hot tearing:
 characteristics, 417–420
 control
  alloying, 440
  brackets, 439
  castings, 436–438, 440
  chilling, 438
  contracting length, 440
  grain refinement, 439
  improved mould filling, 436
  reduced constraint, 438–439
 cracks, 433f
 CSC model, 432
 filling system, 417
 grain boundary wetting by liquid, 420–421
 initiation, 425–426
 mechanisms of, 423
 opposed cones test, 436
 porosity summary, 440–442
 pre-tear extension, 421–423
 predictive techniques, 442
 ring die test, 431f, 434

stainless steels, 442
strain concentration, 423–424
stress concentration, 424–425
susceptibility prediction, 429–433
tear growth, 426–428
tear initiation, 425–426
testing methods, 433–436
Hot box process, 811, 906
Hoult, Fred, 812, 826
Hydraulic lock, 741
Hydrogen:
 aluminium, 779, 782
 aluminium alloys, 545, 782
  description, 267–269
  pick up, 269
  rotary degassing, 268–269
  solubility, 6, 6f
 bifilms, 618
 binders, 808–809
 embrittlement, 398, 541
 magnesium, 783
 magnesium alloys, 783
 porosity, 406–409
 steel melts, 541
 steels, 788
Hydrostatic tensions in liquids, 345
Hypereutectic alloys, 244–247, 289
Hypoeutectic alloys, 247–248, 289

I

I-beam ('dog-bone') test for hot tearing, 433
Ice cubes and freezing, 343
IDECO test (Germany), 83–84
'If in doubt, visualise water' concept, 646
Impregnation of castings, 942
Incipient melting, 933–934
'Inclusion', term, 724
Inclusion control: filters and traps:
 dross trap, 724–725
 slag pockets, 725–726
 swirl trap, 726–728
'Inclusion shape control', 81
Inclusions:
 aluminium alloys, 3–5, 4t, 239, 537
 electrical conductivity, 537
 liquid aluminium, 3
 steels, 317–319, 324
Increased heat transfer:
 chills and fins, 178–179
 external chills, 179–182
  fins, 178–179, 185–193
   pins, 193
 internal chills, 182–185

Indirect gates, 690
Indirect gating (into up-runner/riser), 707–710
Indirect squeeze casting, 868
Induction melting, 772
Industrial revolution and grey iron, 275
Inert gas cloud (pouring basin), 660
Inert gas shroud (steels), 788, 789f
Inert moulds and cores:
 ceramic moulds and cores, 800–801
 description, 797
 permanent metal moulds, 797–799
 salt cores, 799–800
'Ingates'. See Gates
Interdendritic feeding, 355–357, 440
Inter-granular corrosion, 518
Intermetallics:
 Ag_3Sn, 223
 Al-Si alloys, 71
 brittle behaviour, 510–511
 $CuAl_2$, 263
 Fe_3Al, 510–511
 FeSi, 292
 fluidity, 103, 105f, 106
 iron-rich aluminium alloys, 259–262
 $Mg_{17}Al_{12}$, 232f, 234–235
 Ti-rich, 263
 TiAl, 336–337, 339–340, 403, 857
 $TiAl_3$, 201, 240–241, 482, 785
Internal chills, 182–185
Internal porosity by surface initiation, 365–367
Internal running system, 685f
International Magnesium Association, 228–229
International Meehanite Metal Company, 291
International Nickel Company, New York, 792
International Standards Organisation. See ISO
International Zinc Association, 224
'Interrupted pouring', 712–713, 826
Interstitial diffusion, 9
'Inverse chill' in cast iron, 178
Inverse segregation. See Dendritic segregation
Invest term, 800
Investment mould, 874
Investment shell (lost wax) moulds, 800
Investment 2-part block moulds (Shaw Process), 801
Invocast method (roll-over casting), 835
Iron:
 bifilms, 790
 binders, 807–809, 811–813
 casting, 822
 cupola furnace, 772–773
 diffusion coefficients, 10f
 flux and slag inclusions, 30–33
 'inverse chill', 178

Iron: (*Continued*)
    α-iron, 197, 197f, 259
    β-iron:
        Al-7Si-0.4Mg alloy, 487, 488f, 490, 492f
        bifilms, 71
        iron-rich intermetallics, 259–262
    δ-iron, 326, 329–330
    liquid, 9, 326
    residual stress, 614–615, 625f
    structure, 197, 197f
    surface films, 13
    type A/C, 296–297
    type D/E, 300–301. *See also* Cast iron; Ductile iron; Grey iron
Iron alloys:
    crack growth in bcc structure, 445
    ferro (liquid), 31, 33f
    Fe-3Si, TiB$_2$, 201
    Fe-C
        eutectics, 247
        eutectics and interdendritic feeding, 356
        fluidity, 106, 108f
        freezing, 343, 344f
        micrographs, 302
        phase diagram, 289f
        white iron, 312
    Fe-C-Mn porosity, 217f
    filters, 734. *See also* Steels
Iron Bridge, UK, 642
Iron carbide, 290–291, 312–313
Iron-rich intermetallics (aluminium alloys), 259–262
ISO (International Standards Organisation)
    accuracy, 893, 897, 925

J

Jewellery and centrifuge casting, 857
Jorstad, John, 119
Junction effect and gates (filling system), 696–700
Junction requirement rule (feeding), 582–583
Just-in-time slurry production, 792

K

K-mould test (evaluation of melt cleanness), 450
Killed steels, 215f, 216, 221f
Kirksite alloy, 223
Kish graphite, 285–288, 297–298
Kiss gate. *See* Touch gates
Krypton gas and leaks, 523

L

Ladle (flotation of oxides), 82, 550, 552, 788
Laid, Erik, 828
Lap-type defects and rolling waves, 92
Large bubbles and entrainment, 25
Latent heat (H) and solidification time, 115–119, 118f

Layer porosity:
    Al-Mg alloys, 356–357
    ductility, 492
    gas porosity, 465, 466f, 467
    shrinkage pore structure, 382–387
    yield strength, 479
Layout and filling system design, 748
Lead alloys:
    binders, 808–809
    brasses/bronzes, 270
    copper, 270
    Pb-Sb resistance and heat transfer, 165
    Pb-Sn alloys
        fluidity, 101, 101f
        shrinkage pores, 381
    Sb-Pb alloys, fluidity of, 103f
Lead distribution, 776
Lead-bronze and vapour transport, 155
'Leaded gunmetal', 270
Leak testing of castings, description, 943
Leak tightness, 519–523
Leakage control and filters, 735–739
Ledeburite (iron carbide eutectic), 313
LEFM. *See* Linear elastic fracture mechanics
Lego and net shape concept, 897
Level pour (side pour) casting, 646, 828–829
Liberty Bell. *See* 'Freedom Bell'
Light alloys:
    dimensional accuracy, 900
    machining, 939
    woven cloth or mesh filters, 729–730
Light Alloys (2006), 336
LiMCA (Liquid Metal Cleanness Analyser), 3–4, 538, 537
Linear elastic fracture mechanics (LEFM), 495
Liquid aluminium:
    alloys, 238–239, 769
    bubble trail, 67
    critical velocity, 39, 40f, 44, 49
    density, 805
    furnaces, 238–239
    gravity dies, 824
    holding, 776
    hydrogen, 267–268
    mould gas explosions, 145
    'no-fall' requirement, 547
    oxides, 23t, 237
    oxygen, 8
    packed beds, 786
    strength-related defects, 3
    surface films, 19
    thermal conductivity, 770–771
    velocity, 39
    zircon sand, 805

Liquid basalt, 821
Liquid cast iron, 277–288
Liquid copper packed beds, 787
Liquid ferro-alloys granulation, 31, 33f
Liquid ferromanganese granulation, 31, 33f
Liquid gold, 3
Liquid iron, 9, 326, 769
Liquid magnesium, 155, 552
Liquid metals:
    ablation casting, 887–888
    advancement, 550
    air, 545
    aluminium, 770–771
    aluminium alloys, 769
    batch melting, 769–772
    buoyancy forces, 899
    cast material, 790–791
    Cosworth Process, 481
    critical velocities, 544
    dry hearth furnaces, 774–775
    expansion, 550
    filling systems design, 642, 646
    gas precipitated from solution, 467
    greensand, 808–809
    maximum velocity requirement, 544
    melts, 536–537
    microjetting, 58
    models, 3
    primary inclusions, 323
    pumps
        centrifugal, 847
        Cosworth Process, 851
        electro magnetic, 847–849
        pneumatic pumps, 852–853
    sand cores, 389
    slot gates, 710
    surface turbulence, 37–38
    tensile strength, 346–347
Liquid silver, 13
Liquid steel, 31–32, 185, 326
Liquid titanium, 335
Liquids:
    bubble detachment, 391–392, 395–396
    critical heights/velocities, 41t
    feeding (shrinkage porosity), 352–365
    flow behaviour, 46
    fluxes, 32–33, 82
    fracture pressure, 348, 348t
    maximum heights supported by one atmosphere, 592
    penetration: surface tension and pressure, 151–153
    surface and entrainment, 544
    velocities, 35
L-junctions:
    feeding, 583

    fins, 188, 188f
    gates, 699
Loam, 702, 812
Location points:
    clamping points, 635f
    cylindrical systems, 636
    datums, 631–633
    description, 633
    integrated manufacture, 638
    jigs, 637
    metrology, 925
    rectilinear systems, 634–635
    six-point system, 633–634, 635f
    term, 633–637
    thin-walled boxes, 636–637
    triangular systems, 636
Long freezing range alloys:
    85Cu-5Sn-5Zn-5Pb, 356
    bubble damage, 562f
    freezing distance, 588
    layer porosity, 383
    porosity, 359
Longden, E., 920
Loss on ignition (LOI), 578f, 804, 815–817
'Lost air' castings, 877–878
Lost crucible technique, 853–854
Lost foam casting:
    description, 875–878
    exterior shapes, 898
    Replicast, 878
Lost foam castings, 145, 285
Lost wax casting, 800, 898, 914
Lost wax process, 800
Lost wax/ceramic mould casting processes:
    description, 871
    plaster investment, 874
    Shaw process, 872–873
Low alloy steels:
    cooling, 329–330
    crack growth, 445
    DAS and bifilms, 489
    entrained inclusions, 319–320
    growth, 204, 205f
    microsegregation, 216
    superheat/fluidity, 114, 115f
Low-carbon steels:
    carburisation, 156
    cooling, 329–330
    crack growth, 445
    grain evolution, 330f
'Low nitrogen' binders, 412–413
Low-pressure casting:
    counter-gravity, 840f, 842
    failure modes, 856

Low-pressure die casting, mould accuracy, 908
Low-pressure filling, 545–546
LPS (low-pressure sand) system, 851
Lustrous carbon. *See* Carbon films

M

Machined-to-form mould process, 813
Macrosegregation, 224–225, 610–611
Magnesium:
   aluminium alloys, 237–238
   critical velocity, 41
   hydrogen, 783
   liquid, 155, 552
   melting, 8
   metal moulds, 907–909
   oxidation, 155
   oxide bifilms, 302, 306–310, 310f
   plaster moulds, 154
   sulphur hexafluoride, 155
Magnesium alloys:
   ablation casting, 889–891, 890f–891f
   aided sedimentation, 785
   aluminium, 230, 233
   aluminium carbide, 233
   argon, 229
   AZ91E counter-gravity filling, 878
   bifilms, 234–235
   binders, 808
   calcium, 238
   carbon black, 125
   carbon dioxide, 229
   chills, 179
   'Chinese Script', 235
   detrainment, 783
   dies, 805–806
   elastic modulus, 508
   films, 228–230
   fins, 883
   flow channel structure, 707
   flux inclusions, 235
   grain refinement, 201, 233, 482
   Hall–Petch relation, 233, 483f
   hexagonal close packed structure, 233
   high-pressure die-casting, 798–799
   HPDC (hot chamber), 863
   hydrogen, 783
   lost crucible technique, 853–854
   magnesium aluminate, 230
   manganese, 237
   melting, 769
   metal surface reactions, 155
   metallurgy, 228

Mg-9Al-1Zn corrosion resistance, 228
Mg-Al, 233–235, 430
Mg-Li carbon, 806
Mg-Zn
   hot tearing, 430, 467
   shrinkage porosity, 356–357, 384
   TS and elongation, 492
Mg-Zr description, 233–234
   microstructure, 234–235
   multiple sprue, 675
   oxidation, 228–229
   potassium borofluoride, 413
   protective atmospheres, 228–230
   recycling, 228
   sand inclusions, 235–236
   slurry production from granules, 792
   strengthening, 230–234
   sulphur hexafluoride, 228
   superplastic forming, 510
   surface films, 13
   surface turbulence, 230
   susceptibility prediction of hot tearing, 430
   T-Mag process, 846
   uniform contraction, 912
   vapourisation, 14
   ZE41A-T5, 508
   zinc, 230. *See also* Casting alloys
Magnesium aluminate, 230, 936–937
Magnesium oxide, 228–229, 237, 302
Magnesium silicate (dross 'stringers'), 282
Magnetic moulding, 802
Major, Fred, 379
Mandl and Berger, Austria, 835
Manganese:
   evaporation, 15, 150
   grain refinement, 201
   magnesium alloys, 230
   steels, 15, 314, 623
Mass feeding (shrinkage porosity), 353–355
Master brake cylinders and leakage, 945
Matrix structure:
   differentiation of solid, 209–212
   general, 197–198
   growth of solid, 201–209
   nucleation of solid, 198–201
Maximum fluidity (unrestricted flow):
   extended fluidity, 128–131
   fluidity tests, 125–128
   running system, 95
   solidification, 97–107
   solidification time t_f, 112–121
   surface tension, 121–124
   unstable substrates, 124–125

INDEX 1015

velocity, 110
viscosity, 110–112
Maximum fluidity length, 96, 132, 132f
Maximum velocity requirement, 41, 42f, 543–546, 642
Medium-pressure (MP) die-casting, 845
Meehan, Gus, 291
'Meehanite', 291
'Meehanite Metal', 291
Meehanite Metal Corporation, 291
Melt treatments:
  additions, 787–788
  cast material, 790–793
  degassing, 777–783
  detrainment, 783–787
  pouring, 788–790
  re-melting processes, 794–796
Melting:
  automatic bottom pouring, 772
  continuous, 772–774
  crucible melting, 771
  induction, 772
  melt treatments, 777–790
  reverberatory furnaces, 770–771
  titanium alloys, 339–340
Melting process:
  batch, 769–772
  cast material, 790–793
  continuous melting, 772–774
  holding, transfer and distribution, 774–777
  melt treatments, 777–790
  remelting, 794–796
Melts:
  aluminium, 3
  bifilms, 398
  cleaning and grain refinement, 243–244
  combustion, 5
  description, 3–5
  environment, 5–9, 6f–8f
  oxide films, 11–13, 12f, 19, 25, 559
  pressurised filtration tests, 5
  Reduced pressure test, 5, 538–540, 539f–540f
  surface films, 11–13, 12f
  tilt casting, 832–833
  transfer, 547, 547f
  transport of gases, 9, 10f
  ultrasonic reflections, 4
  vapourisation, 14–15, 14f. *See also* Rule 1 for good casting
Meniscus (laminar entrainment), 550–551
Mercury Marine Company, US, 799
Mercury tensile strength, 346
Metal moulds (dies), 797
  accuracy, 907–909
  die life, 908t

Metal surface reactions:
  boron, 159
  carbon (pickup and loss), 155–156
  grain refinement, 158–159
  nitrogen, 156–157
  oxidation, 155
  phosphorus, 157
  sulphur, 157
  surface alloying, 158
  tellurium, 159
  water vapour, 155
Metal/matrix composites (MMCs):
  air exclusion from sprue, 707
  aluminium alloys, Al-Si, 35, 238, 793
  Al-Mg$_2$Si, 263f, 264
  Al-TiB$_2$, 27f
  cast material, 792–793
  description, 19–20, 20f
  in situ, 793
  viscosity, 111
  vortex technique, 33–34
Metal–mould interface:
  heat transfer coefficient, 170–172
  resistance and heat transfer
    air gap, 167–170
    description, 163
    heat transfer coefficient, 170–172
  sequential gating, 711
Metal–mould reactions:
  magnesium alloys, 413
  nitrogen, 412
Metals:
  bifilms, 537–541, 539f–540f
  chills, 160–161
  defects, 536–537
  dendrite arm spacing, 3
  fins and thermal conductivity, 191–192
  growth of solid, 201–209
  hydrogen solubility, 6, 6f
  liquids, metal moulds, 797
  shrinkage rates, 341–387
  solidification shrinkage, 343t
  strength-related properties, 3, 537. *See also* Liquid metals
Methane and combustion of melts, 5
'Metallic inclusions' term, 724
Methoding approach (filling system design), 748
Metrology:
  description, 924
  digital laser-reading systems, 925
  gauges, 925–926, 926f
Mg$_2$Si ('Chinese script') in magnesium alloys, 235
Mg$_{17}$Al$_{12}$ and magnesium alloys, 234–235
Micro-blows, 397–398

Microjetting:
  entrainment, 57–58, 548
  flow, 91
  mechanical properties of casts, 58
  mechanism, 58
  plaster investment moulds, 874–875
  surface tension controlled filling, 655–656
  touch gates, 700
Microsegregation, 216–218, 224–225, 610
  single crystal solidification, 886–887
Microstructure and bifilms, 481–482
Mild steel liquid and freezing, 185
Miller, Glen, 291
Mitsubishi aluminium alloys, 226
MMCs. *See* Metal/matrix composites
Modification of Al-Si eutectic, 250, 793
Modulus:
  of common shapes, 176t
  dimensions, 576
  feeder, 576–577
  gradient technique, 584–585
  ratio, 574, 576, 586f
  solidification, 584–585
  solidification time, 112–121
Mould accuracy:
  aggregate moulds, 904–907
  ceramic moulds, 907
  metal moulds (dies), 907–909
  process routes, 894–895, 895f
Mould atmosphere:
  composition, 144–145
  explosions, 145–147
Mould cavity:
  conical basin, 655, 668–669
  filling system design, 648
  gates, 644f, 690, 711
  sprue, 671–672
Mould coatings:
  aggregate moulds, 159–160
  dry coatings, 161
  grain refinement, 201
  permanent moulds and metal chills, 160–161
Mould constraint, 915–917
Mould contamination:
  gravity die casters, 150
  iron carbonyl, 150
  manganese evaporation, 150
Mould design:
  assembly methods, 902–904
  'cake core', 900, 901f
  cope-to-drag location, 904
  dimensional accuracy, description, 900
  infringement of 10 rules, 902

Mould penetration:
  chemical interactions, 154
  liquids penetration: surface tension and pressure, 151–153
  penetration barriers, 153–154
  temperature and time dependence, 154
  vapour transport, 154–155
Mould processes:
  machined-to-form, 813
  precision core assembly, 813
  unbonded aggregate moulds, 814
Mould surface reactions:
  lustrous carbon film, 148–149
  mould contamination, 149–150
  mould penetration, 151–155
  pyrolysis, 147–148
  sand reactions, 149
Moulding:
  aggregate moulding materials, 802–806
  binders, 807–813
  casting industry, 797
  inert moulds and cores, 797–802
  reclamation and re-cycling of aggregates, 814–819
Moulds:
  blows, 563–564
  filling, 542–543
  gating, 547–548
  investment, 874
  and metal constants, 175t
  resistance and heat transfer, 172–178
  rubber, 814
  temperature and solidification time, 119–121
  wall movement, 167, 168f
Moulds and cores:
  computer simulation, 571–573
  evaporation and condensation zones, 139–144
  gas evolution rates, 135, 136f
  greensand mould structure, 136, 138f
  metal surface reactions, 155–159
  moulds
    atmosphere, 144–147
    coatings, 159–161
    gases composition, 137f, 139
    inert or reactive, 135–136
    surface reactions, 147–155
    transformation zones, 136–139
Multiple gates for filling system, 692, 693f–694f
Multiple sprue, 675–677

N

Naturally pressurised system concept, 683, 753
'Navy Gun Metal', 270
NDT. *See* Nondestructive testing
'Near-net shape' concept, 897, 906

Negatively tapered sprue, 674
Net shape concept:
  casting process, 897, 900
Ni-base alloys:
  air melting and casting, 331–332
  automatic bottom pouring, 772
  bifilms, 333, 934
  ceramic moulds and cores, 800
  cobalt aluminate, 158
  convection, 605
  description, 330–331
  dye penetrant inspection, 332, 335
  furling and unfurling, 74, 75f
  gravity segregation, 222
  hafnium, 882
  heat affected zones, 934
  incipient melting, 934
  layer porosity, 362f, 384
  melts, 542
  modulus, 577
  oxidisable elements, 331
  porosity, 361–362, 362f
  production, 542
  steel inserts, 185
  turbine blades, 332, 542, 605, 614f
  ultrasonic reflections, 4
  vacuum melting and casting, 332–335, 881–882
  Waspalloy, 794
Ni-base superalloys:
  aerofoil test, 119–121, 120f
  bifilm crack, 335f
  ceramic moulds and cores, 800
  dendrites, 211
  directional solidification (DS), 885
  flow channels, 194–197
  flowability and fillabilty, 120f, 123
  'freckles', 197
  HIPping, 937
  machining, 938
  mould temperature and solidification time, 119–121
  nitrogen, 333
  nucleation, 198–199
  oxide flow tubes, 57
  oxides in melts, 542
  single crystal solidification, 528, 886–887
  vacuum casting, 67
  waterfall flow, 552–553
  white iron, 312
Nitride bifilms, 283, 306
Nitrogen:
  binders and pyrolysis, 148
  cast iron, 276–277
  metal surface reactions, 156–157
  nucleation, 412
  porosity, 412–413
Noble metals (gold, platinum and iridium) and surface films, 13
Nodular iron. See Ductile iron
'No-Fall' requirement (casting), 547–550, 708–709
Non-classical initiation of pores:
  high-energy radiation, 375–376
  pore initiation on bifilms, 378–381
  pre-existing suspension of bubbles, 376–378
Non-destructive testing (NDT):
  bifilms, 89
  dye penetrant inspection (DPI), 943, 945
  leak testing, 943–944
  resonant frequency testing, 944–945
  X-ray radiography, 942–943
Non-ferrous casting, 773–774, 809
Non-metallic inclusion term, 724
Non-uniform contraction (distortion):
  casting constraint, 917–922
  ductile iron, 920, 928
  elastic–plastic model, 920
  grey iron, 920, 921f
  mould constraint, 915–917
  plate shaped castings, 919–920
Non-uniform elongation, 477
Norite (aggregate moulding materials), 805
Nozzle, 31–32, 34f, 49, 645, 660–661, 666, 713–714, 789f, 790, 823, 826, 957
Nucleation:
  austenite matrix, 299–300
  burst feeding and shrinkage porosity, 358, 359f
  cast iron, 299–300, 326–329
  cold cracking, 443–444
  graphite, 290, 309–310, 313
  heterogeneous, 370–372
  homogeneous, 368–369
  nitrogen, 412
  oxygen, 410
  primary inclusions, 323
  shrinkage pores, 372–373
  silicon, 248–250
  solids and grain refinement, 201, 239–244
  term, 198

O

Offset sprue, 722–723
Offset step (weir) basin, 449, 661–662, 745
'Oil-can' distortion, 919–920
Olivine sand (moulding aggregate), 803–805, 818–819
'One pass filling' (OPF) designs, 549, 651–652
Ounce metal, 270

Outgassing:
    loss of ignition, 568f
    pressure in cores, 564–568
Oxidation:
    magnesium, 155
    magnesium alloys, 228–229
    metal surface reactions, 155
Oxidation and corrosion resistance:
    corrosion, 514
    filiform corrosion, 517–518
    inter-granular corrosion, 518
    internal oxidation, 512
    pitting corrosion, 515–517
    stress corrosion cracking, 519
'Oxidation index', 412
Oxide bifilms:
    Al-12Si alloys, 260f–261f
    aluminium, 782
    aluminium alloys, 236, 784
    carbides, 314
    crack initiation, 444
    direct chill aluminium alloys, 328
    graphite, 297–298
    grey iron, 280
    magnesium, 302, 306–310, 310f
    populations of, 771
    stainless steels, 316–317
'Oxide crack', 23
Oxide films:
    aluminium, 25
    aluminium alloys, 237–239
    cast iron, 280
    down-runner, 650
    entrainment, 19–25, 33–35, 549
    ferrous alloys, 546
    flow, 99
    flux and slag inclusions, 30–33
    furling and unfurling, 74
    inclusion control, 724
    iron-rich intermetalics, 260
    leaks, 944f
    melts, 11–13, 12f, 19–20, 27f, 556
    microjetting, 58
    old oxide, 33–35
    sand inclusions, 35–37
Oxide flow tubes, 47, 56–57
    double bifilms, 57
    laminar entrainment of surface films, 552–553
    pouring, 47
    waterfall flow, 552–553
Oxide lap defects:
    confluence welds, 52–56
    entrainment, 50–56
    surface films, 555–557
    surface flooding, 50–52
Oxides:
    aluminium alloys, 230, 231f–232f
    machining, 938–939
    sedimentation, 784–785
    single crystal solidification, 886
    skins
        bubble damage, 60–61
        melt charge materials, 46–47
Oxygen:
    liquid aluminium, 8
    nucleation, 410
    steels, 784

P

Packed beds and detrainment, 786–787
Padding and feeding, 585, 595f
Painting of castings, 940
Paris–Erdogan equation, 506
Partially solid mixtures (casting), 791–793
Particle-on-particle ('POP') attrition, 815–817, 819
Passive degassing, 777
Pasty zones:
    freezing (mode of solidification), 97
    hot tears, 435f, 437f
    shrinkage porosity, 385
Patternmaker's contraction allowance, 342, 899
Patterns and accuracy, 895f
PCRT. See Process-compensated resonance testing
PDPC. See Counter-pressure casting
Peeling process (heat transfer), 179
Pegasus engine, Harrier Jump Jet, 937
Pencil gates, 702–703
Permanent metal moulds (dies):
    description, 797
    gravity/low pressure die-casting dies, 798
    high-pressure die-casting dies, 798–799
Permanent mould casting (US), 160–161, 646
Phenol-urethane no-bake (PUNB) binders, 805, 809
Phenol-urethane (PU) binders, 809–810, 815–817
Phenolic-urethane (PF) binders:
    aluminium alloys, 397
    carbon (pickup and loss), 156
    isocyanate, 413
    pyrolysis, 145
    resins, 407, 409f
    strength, 42f
Phosphoric acid and mould surface reactions, 148
Phosphorus:
    Al-Si alloys, 250
    copper alloys, 272–273
    grey iron, 106, 109f, 275

gunmetal, 410, 411f
metal surface reactions, 157
Pilling–Bedworth ratio, 228
Pitting corrosion, 515–517
Pitting resistance equivalent (PRE) for stainless steels, 316
Planar front:
  growth in solids, 201–209
  hypoeutectic Al-Si alloys, 244–247
  segregation, 213–216, 610–613
Plaster investment (integral block moulds), 874
Plaster investment block moulds, 801
Plaster of Paris, 801
Plastic working (forging, rolling, extrusion), 941–942
Plate fracture defect in ductile iron, 282–283
Plate shaped castings distortion, 919–920
Pneumatic pumps, 852–853
PoDFA (porous disc filtration analysis) tests, 3, 537, 782
Poisson's ratio, 945
Polymer quenching, 620–623
'Pore free process' (HPDC), 866
Pore initiation on bifilms, 378–381
Porosity:
  ablation casting, 888, 890
  aluminium alloys, 538
  diagnosis, 414t
    core blows, 387
    entrained, 388–389
    gas porosity, 413, 414t
    shrinkage, 414–415
  Fe-C-Mn alloys, 217f
  fin length, 188, 190f
  gas, 387–413
  hot tearing summary, 440–442
  leak testing, 943
  nitrogen, 412–413
  shrinkage, 341–387
  steels, 412–413
  surface pinhole, 411
Positive segregation, 620f
Post-casting processing:
  blister formation, 932–933
  description, 927
  fluid beds, 934–935
  heat treatment, 929–935
    reduction and/or elimination, 929–930
  hot isostatic pressing, 935–938
  impregnation, 942
  incipient melting, 933–934
  machining, 938–940
  non-destructive testing, 942–945
  painting, 940
  plastic working, 941–942
  surface cleaning, 927–929

Potassium borofluoride (Mg alloys), 413
Pour rate of steels, 957–958
Pouring:
  contact, 660–661, 823
  critical fall height, 48–50
  defects, 642
  entrained inclusions, 326
  entrainment, 47–50
  filling systems
    speed, 645
    time, 755–757
  oxide films, 50
  oxide flow tubes, 47
  shrouded, 822
  sprue, 789–790
  steels, 773, 789–790
Pouring basin:
  conical basin, 657–660
  contact pouring, 660–661
  design, 759
  inert gas shroud, 660
  offset step (weir) basin, 662–667, 681–682
  sharp cornered/undercut basin, 667–669
  stoppers, 669
PRC. See Pressure riserless casting
PRE. See Pitting resistance equivalent
Precision core assembly mould process, 813
Prefil tests, 3, 537
Premature filling problem via early gates, 692–695
Press casting, 871
Pressure assisted casting:
  description, 861
  high-pressure die casting, 862–868
  squeeze casting, 868–871
Pressure die casting:
  dimensional accuracy, 900
  magnesium and uniform contraction, 912
  oxide flow tubes, 553
  precautions, 865
  rotary degassing and leakage, 782
  surface finish, 524–525
  surface centred control filling, 656
  zinc alloys, 66f, 122–123
Pressure requirement (feeding), 590–593
  cast iron, 596
Pressure riserless casting (PRC), 846
Pressurised filtration tests, 5, 538
Pressurised versus unpressurised (filling system design):
  choke, 752–753
  gating ratios, 753–755
Pre-tear extension, 421–423
Primary inclusions, 322–324
Priming of gates (filling system), 712–714

Principles of Magnesium Technology, 228, 234–235
Process comparison (casting accuracy), 923–924
Process-compensated resonance testing (PCRT), 945
Process routes for castings manufacture, 894–895, 895f, 923–924
'Product Design and Development for Magnesium Die Castings', 228
Programmable control for countergravity casting, 855
Properties of castings:
    defects, 463–470
    Elastic (Young's modulus) and damping capacity, 507–508
    fatigue, 498–507
    fracture toughness, 493–498
    gas porosity, 465
    high temperature tensile properties, 509–511
    leak tightness, 519–523
    oxidation and corrosion resistance, 511–519
    quality indices, 526–527
    residual stress, 507–508
    statistics of failure, 450–462
    surface finish, 523–526
    tears, cracks and bifilms, 467–470
    tensile properties, 470–493
    test bars, 447–450
PU. See Phenol-urethane
Pumps, 30–31, 82, 770–771, 775, 790, 839, 846–853, 880
Pyrolysis:
    coal, 147–148
    description, 147
    nitrogen containing binders, 148

Q

'Quality assurance' (QA), 89
'Quality control' (QC), 89
Quality indices, 526–527
Quartz transition, 803
Quenching:
    polymers and other quenchants, 620–623
    quench factor analysis, 626
    residual stress, 617–620
    stress relief, 628–630

R

RA. See Reduction in area
Radial choke for sprue, 681
Radius of sprue/runner junction, 681–683
Railroad wheels, 797, 806
Rapid solidification (RS) casting:
    ablation casting, 888–891
    Sophia and Hero processes, 887–888
Rappaz, M., 417, 420, 433, 441–442
Recalescence, 201
Reclamation and recycling of aggregates:
    aluminium foundry, 814–816
    ductile iron, 816–817
    facing and backing sands, 818–819
    impact attrition, 815
    POP attrition and classification, 815–816
    soluble inorganic binders, 817–818
    thermal reclamation, 815–816
Rectilinear systems and location points, 634–635
'Reduce residual stress' rule:
    beneficial residual stress, 628
    controlled quenching, 620–623
    distortion, 626
    epilogue, 631
    heat treatment developments, 626–628
    quenching, air, 623–624
    strength reduction by heat treatment, 624–626
    stress relief, 628–630
'Reduce segregation damage' rule:
    macrosegregation, 610–611
    microsegregation, 610
    planar front segregation, 610–613
Reduced pressure test (RPT):
    bifilms, 69f, 72–73, 72f, 83–84, 85f
    copper alloys, 274
    feeding of castings, 349
    gas porosity in situ, 402–403, 402f
    inclusions, 538, 540f
    melts, 5, 538–540, 539f–540f, 782–783
Reduction in area (RA) and ductility, 477
Relative diffusivities and chills, 181f
Re-melting process:
    electro-slag, 794–795
    heat transfer, 193–194
    vacuum arc, 795–796
Re-oxidation inclusions, 325
Replicast, lost foam casting, 878
Residual stress:
    beneficial, 628
    castings, 613–617
    description, 508
    distortion, 626
    quenching, 617–624
'Resin concretes', 821
Resistance and heat transfer:
    castings, 165
    heat transfer coefficient, 170–172
    metal mould interface, 165–167
    moulds, 172–178
Resonant frequency testing, 944–945
Resonant ultrasound spectrometry, 945
Reverberatory furnaces, melting, 770–771
Reverse tapered feeder, 579
Reverse tapered sprue, 681
Reynold's number (Re), 45–46, 650

Rheocasting (cast material), 791
Ribs and liquid distribution, 550
Riley, Percy, 401–402
Rimming steel ingots and segregation, 214–216, 215f
Ring die test for hot tearing, 431f, 434
Riser (up-runner):
    flow channel structure, 195f, 197
    indirect gating, 707–710
    rotary surge, 719
'Rock candy structure', 322
'Rodding' of castings, 359
Rohatgi, Pradeep, 792
Roll-over:
    after casting process, 836–838
    convection, 610
    Cosworth Process, 851
Roll-over after casting (inversion):
    Cosworth Process, 449, 838
    description, 449, 836
Rolled steels, 31–32, 34f
Rolling waves, 91–92
Room temperature operation and mould accuracy, 905–906
Rosettes (cells):
    flake graphite, 300, 301f
    grey cast iron, 212
Ross Meehan Foundry, Chattanooga, US, 291
Rotocast process (roll-over casting), 835
Rotary degassing:
    aluminium alloys, 268–269, 780–783
    detrainment, 82
    grain refinement, 785
Rotary furnaces, 771
Rotary surge risers, 719
Rover (UK), 851
RS. See Rapid solidification
Rubber moulds, 814
Rules for good castings:
    1: 'use a good-quality melt', 536–542, 724
    2: 'avoid turbulent entrainment', 542–550, 642, 649–650, 655–656, 822
    3: 'avoid laminar entrainment of the surface film', 550–557
    4: 'avoid bubble damage', 557–561, 648
    5: 'avoid core blows', 561–573
    6: 'avoid shrinkage damage', 573–603
    7: 'avoid convection damage', 603–610, 845
    8: 'reduce segregation damage', 610–613
    9: 'reduce residual stress', 613–631
    10: 'provide location points', 631–638, 904
Runner:
    bubble damage, 686–687
    description, 683
    'diffuser', 688
    expanded, 706
    external system, 685f
    filters, 686
    internal system, 685f
    slot, 686–687, 687f, 761–762
    speed, 684–686
    tapered, 687–688
    vortex, 722–723
Running system. See Filling system

S

St. Paul's Cathedral, London, 507–508
Salt cores, 799–800
Sand:
    casting, 285, 607, 825, 898
    cores composition, 568
    inclusions, 35–37
    machining, 938–939
    moulds
        10 test bar, 448
        magnesium alloys, 230
        penetration, 153
        resin binders, 135, 137f
    reactions, 149
    silica, 803–804
Sand blasting, 927
Sb-Cd/Pb alloys and fluidity, 101, 103f
Scanning electron microscopy (SEM):
    bifilms, 83, 86–87
    bubble trails, 60, 62f
    compact graphite iron, 308–310, 313, 313f
    ductile iron, 512
    gas porosity in situ, 400
    graphite spherule, 304f
    hot tears, 418f
    intermetallics, 262
    lustrous carbon film, 285, 287f, 288
    Mg-Zr alloys, 233–234
    Ni-base superalloy, 333, 334f
    stress corrosion cracking, 519
    superplastic forming, 510
    vapour transport in moulds, 154
    zinc alloys, 66f
Schoefer diagram for stainless steels, 316, 317f
Sealing of castings (impregnation), 942
Seawater corrosion, 516f, 518
Secondary dendrite arm spacing. See Dendrite arm spacing
Sedimentation and detrainment, 82–83
Segregation:
    definition, 213
    dendritic, 218–219, 869
    gravity, 219–222
    microsegregation, 216–218
    planar front, 213–216

Segregation: (*Continued*)
  positive, 620f
  ratio, 216
  solidification, 213–222. *See also* Rule 8 for good castings
Semi-killed steels, 215–216, 215f
Sequential gates, 711
Sessile (sitting) drops:
  critical fall height, 48–49, 642
  critical velocity, 542–544, 546, 642
  entrainment, 544–546
  surface flooding, 50
  vapour transport, 154
Seven feeding rules. *See* Feeding
Sharp cornered/undercut basin and pouring, 667–669
Shaw, Clifford, 801
Shaw, Noel, 801
Shaw process (two-part block mould), 872–873
Shear modulus, 945
Shell process, 811–812
Short-freezing-range alloys (shrinkage cavity), 383f
Shot blasting, 927–928
Shower nucleation, 211
Shrinkage cavity or pipe, 382–383
Shrinkage damage. *See also* Rule 6 for good castings
Shrinkage porosity:
  computer simulated, 366, 367f
  defects in castings
    macroshrinkage, 467
    microshrinkage, 466
  description, 341
  diagnosis, 413–415
  feeding criteria, 348–352, 590–593
  feeding mechanisms
    burst feeding, 357–360
    interdendritic feeding, 355–357
    liquid feeding, 352–353
    mass feeding, 353–355
    solid feeding, 360–365
  initiation
    external porosity (surface links), 367–368
    growth of shrinkage pores, 381
    internal porosity by surface initiation, 365–367
    non-classical initiation of pores, 375–381
    nucleation of internal porosity, 368–375
  pore structure
    cavity or pipe, 382–383
    layer porosity, 383–387
  reduction, 346
  solidification shrinkage, 342–348
  types, 345, 345f
Shrouded pouring, 822
Siemens' open hearth furnace, 770
Sievert's law (gas concentration and pressure), 6

Si/Fe competition for sites (intermetallics), 262
Silica:
  moulds and cores, 138–139
  sand, 803–804, 906–907
  thermal shock, 803
Silicate, sodium, 810–811, 817–818, 906, 942
Silicic acid (binder), 800
Silicon:
  bifilms, 250
  grey iron, 275
  latent heat, 107f, 115–119, 115f
  nucleation, 248–250
Silicon carbide, 292–293
Silicosis, 803
SIMA. *See* Strain-induced melt activation
Single crystal solidification, 885–887
Siphons and aluminium, 776
Siting:
  filters, 738f
  gates, 690
Six-point location system, 633–634, 635f
'Skeins of geese' (bronzes), 63–64
Skin-freezing, 97–98
Slag:
  inclusions, 30–33
  pockets, 725–726
Slot:
  gates, 699, 707–711, 708f, 723f
  runner, 688–690, 689f, 761–762
  sprue, 689f, 761–762
Small bubbles, 25, 60–61
Smalley, Oliver, 291
Smith, C.S., 420
Snorkel ladle, 713–714, 826
Sodium:
  aluminium alloys, 117, 118f, 238–239
  latent heat, 117
  vapourisation, 14, 14f
Sodium polyphosphate glass binder, 810
Sodium silicate, impregnation, 942
Sodium silicate binder, 810–811, 817–818, 906
  inorganic, 811
Soldering, 862
Solid feeding:
  dangers, 600–601
  shrinkage porosity, 360–365
Solid surface film and surface finish, 525–526
Solidification:
  alloys, 211
  castings and computer simulation, 576
  controlled, 883
  fatigue in steel, 499
  fluidity, 97–107

heat transfer, 163–197
matrix structure, 197–212
modulus, 584–585
segregation, 213–222
times, 565f
Solidification shrinkage:
 hydrostatic tensions in liquids, 346–348
 metals, 341–342
 pores, 345, 345f
 sphere model, 345
Solidification time (t_f):
 DAS and grain size, 207–209, 208f
 heat-transfer coefficient, 114
 latent heat, 115–119, 118f
 maximum fluidity, 112–121
 modulus, 113–114
 mould temperature, 119–121
 superheat, 114
Soluble, transient films, 80–81
Soluble inorganic binders and aggregate reclamation, 817–818
Solution treatment term, 930
Sophia Process, 887–888
Southern bentonite (clay), 144
Sparks, Robert, 802
SPC. See Statistical process control
Speed of filling systems, 646, 650, 684–686
Spherical wax castings and feeding behaviour, 362, 363f
Spheroidal graphite iron (SGI) (ductile iron):
 description, 302
 exploded nodular graphite, 307
 mis-shapen spheroids, 306–307. See also Ductile iron
'Spiking' in fracture surfaces, 283
Spinels, 29f, 237–238, 785
Spiral tests for fluidity, 125–128
Spraying time, 741
Sprue (down-runner):
 base, 678–679
 design, 669
 division, 677–678
 expanding runner, 688–690
 filling systems design, 644f, 645–646, 650, 759–763
 filters, 737f
 gravity filled systems, 548
 and metal fall, 643–645
 multiple, 675–677
 negatively tapered, 674
 'no fall requirement', 547–548
 offset, 722–723
 pouring, 789–790
 priming of gates, 712–714
 radial choke, 681
 radius of runner junction, 681–683
 reverse tapered, 681
 slot, 688–690, 689f, 761–762
 stoppers, 669
 straight tapered, 671, 674f
 surge control, 719
 tapered, 449, 671
 tilt casting, 832
 vortex, 683, 720–721
 well, 679–681
 zero tapered, 673–674
Sprue/runner junction, radius of, 681–683
Squeeze casting:
 ablation casting, 888
 cast preforms for forging, 871
 direct, 868
 indirect, 868
 press casting, 871
 term, 869
'Squeeze forming' term, 869–871
Squeeze pins (HPDC), 866
Stainless steels:
 ferrite, 316, 317f
 hot tearing, 442
 martensitic, 315
 oxide bifilms, 316
 pitting resistance equivalent, 316
 pouring, 47
 radiography, 64–65
 Schoefer diagram, 316, 317f, 328
 shot, 927–928
 σ phase, 317
 superheat and fluidity, 114, 115f
Statistical process control (SPC), 602
Statistics and casting dimensions, 894f
Statistics of failure:
 accuracy limits, 461
 bimodal distribution, 456f, 459
 extreme value distributions, 461–462
 Weibull analysis, 453–461
Steam reaction in copper alloys, 271
Steel boxes and core blows, 564
Steel Foundry Society of America (SFSA), 788
Steels:
 1.5Cr-1C, segregation damage, 216
 20Cr-20Ni-6Mo, 316f
 aluminium nitride, 322
 austenitic, 179, 327
 automatic bottom pouring, 772
 bcc structure, 483
 bifilms, 329, 790
 binders, 808–809
 and boron, 528
 calcium cyanamide, 327
 carbon, 315, 319–320, 433f, 445

Steels: (*Continued*)
    carbon-based filters, 744
    casting into ingots, 215
    chaplets, 899
    chromite sand, 817–818
    cold mild and freezing, 184f
    columnar grains, 211
    contraction, 911, 913f
    cooling, 921
    cracks, 443–444
    creep, 509
    CSC model, 432
    C-Cr microsegregation, 216, 217f
    dimensional accuracy, 893
    directional solidification (DS), 884f, 885
    dross defects, 391, 391f
    entrained inclusions, 319–322
    expansion and contraction, 899
    fatigue and solidification, 499
    feeders, 579, 586, 599f
    feeding distance, 587–590
    Fe-3%25Si, 329
    films and deoxidation, 52
    filters, 734
    fins, 883
    fluidity, 106–107
    flux and slag inclusions, 31
    freezing times of plate shaped castings, 173, 174f
    gases, 412
    grain refinement, 328–329, 483
    gravity dies, 824–825
    gravity segregation, 219–220, 221f, 613
    greensand moulds, 144
    heat pipes, 192–193
    high-alloy castings and gravity segregation, 222
    high-strength low-alloy, 321
    hot work H14, 862
    hydrogen, 779
    hydrogen in melts, 540
    inclusions, 317–319, 324–326
    inert gas shroud, 788, 789f
    ingots and gravity segregation, 219–222, 220f–221f, 612
    inserts and grey iron, 185
    investment casting, 156
    killed, 215f, 216
    layer porosity, 383–387
    liquid, 31–32, 326
    low-alloy, 114, 115f, 204, 205f, 216
    low carbon, 154–156
    magnetic moulding, 802
    manganese, 15
    melting, 315
    melts, 783
    microsegregation, 216–218
    nucleation, 198–201, 326–329
    olivine, 803–805, 818–819
    oxygen, 777, 779
    permanent moulds, 160
    phase changes in cooling, 211
    pitting corrosion, 515–517
    plasticity of inclusions, 942
    porosity, 341–416
    pour rate, 957–958
    pouring, 755, 788–790
    primary inclusions, 322–324
    railway stock. *See* Griffin Process
    residual stress from quenching, 618–619
    rimming ingots and segregation, 214–216, 215f
    rolled, 31–32, 34f
    rolling waves, 92
    secondary inclusions and second phases, 324–326
    semi-killed, 215–216, 215f
    shot, 799, 802, 805
    Siemens' open hearth furnace, 770
    strength reduction by heat treatment, 624–626
    structure development in solid, 329–330
    surface, 933f
        films, 4, 13
        finish, 523–526
        tension, 652, 654–655, 654f
    ultrasonics, 4, 78, 86
    unzipping waves, 91–95
    welding, 330
    yield point, 477–478
    zircon sand, 805. *See also* Stainless steels
Stokes velocity, 81–82, 280
Stoppers for pouring basin, 669
Straight-tapered sprue, 671, 674f
Strain concentration (hot tearing), 423–424
Strain-induced melt activation (SIMA), 792
Strainers (filling system), 729
Straube-Pfeiffer test, 83–84, 402
Street, Arthur, 223, 862
Stress concentration (hot tearing), 424–425
Stress corrosion cracking (SCC), 519
Stress relief:
    alloys, 629–630
    elongation failure and quenching, 629–630
    subresonance, 630
Stringers:
    alumina, 31–32, 34f
    dross, 280–282
    ductile iron, 280–282
Strontium (Sr):
    aluminium, 788
    aluminium alloys, 914

Al-Si alloys, 238–239, 253–254, 929
zinc alloys, 227
Structure development in solid (steels), 329–330
Substitutional diffusion, 9
Subsurface:
pinholes in copper alloys, 274
pores, 411
porosity, 824
Sulphides, 324–325, 552–553, 552f
Sulphur and metal surface reactions, 157
Sulphur hexafluoride, 155, 228–229
Sulphonic acids pyrolysis, 148
Sump (oil pan) casting:
datums, 632, 632f
gravity filling, 51, 51f
leak tightness, 519–523
leakage, 943–944
runner system, 683, 685f
thin-walled boxes, 636–637
Superalloys, 335f, 885
term, 331
Supercooling/undercooling, 170, 198–199, 201, 202f–203f, 203–204, 240, 250–251, 265–266, 266f, 292, 300, 326, 327f, 886
Superheat
defined, 114
and fluidity, 114, 115f–117f
Superplastic forming, 510
Surface alloying, 158
Surface cleaning (castings), 927–929
Surface films:
aluminium, 19
aluminium alloys, 13, 23, 31
confluence welds, 52–56
copper alloys, 270, 273
entrainment, 13, 19, 31
free energy of formation, 11, 12f, 13
gates, 710
laminar entrainment, 550–557
liquid cast iron, 277–288
bifilms, 280–283
carbon films, 284–288
oxide films, 280
plate fracture defect, 282–283
magnesium alloys, 13
melts, 11–13, 12f
noble metals, 13
oxides, 11–13, 12f
speed of growth, 19
steels, 13
surface finish, 525–526
titanium alloys, 339–340
Surface finish, 523–526, 925

Surface flooding, 50–52
Surface forms on filling:
rolling waves, 91–92
surface tension, 91–92
unzipping waves, 91–95
Surface lap defects in confluence welds, 52–56
Surface pinhole porosity, 411
Surface tension:
core blows, 561
flow, 91–92
maximum fluidity, 121–124
surface finish, 523–525
thin-walled castings, 549
Surface turbulence:
counter-gravity casting, 648
critical heights/velocities, 41
elimination in filling systems, 648
Froude number (Fr), 44–45
liquid metal, 37–38
machining, 938–939
magnesium alloys, 230
oxide lap defects, 556–557
physics, 38–39
Reynolds number (Re), 45–46
Weber number, 41–44
Surface tension controlled filling:
aluminium, 654–655
filling systems design, 651–656
steels, 652, 654–655, 654f
Surface venting (of chills), 561
Surge control system (filling), 716–719
Susceptibility prediction of hot tearing, 429–433
Swirl traps, 726–728
Synthetic aggregates, 806
Synthetic grains, 135–136

T

Tangential filter gate, 714, 714f–715f
Tangential placement of filters, 741–742
Taper and sprue, 674
Tapered runner, 687–688
Tear. *See* Hot tearing
Tears, cracks and bifilms (defects in castings), 467–470
Tellurium and metal surface reactions, 159
Temperature and tooling, 909–910
Template gauges, 925
Tensile properties:
ductility, 473–477
micro-structural failure, 470–473
strength in liquids, 346–347
ultimate tensile strength, 490–493
yield strength, 477–490

Test bars for castings:
  10 test bar mould, 448
  history, 447
  K-mould test, 450
'The breaking of the surface oxide' concept, 21
'The breaking of the surface tension' concept, 21
'The Diecasting Book', 223
Thermal analysis of aluminium alloys, 264–267, 266f
Thermal conductivity and chills/fins, 191–192
Thermal diffusivity, 165, 176t, 180, 619, 625, 629, 801, 805
Thermal fatigue, 506–507
Thick section castings, 602
Thin section castings, 602, 758
Thin-walled castings:
  carbon black, 125
  feeders, 575–576
  surface tension, 549
Thixocasting (cast material), 791
Thread-forming fasteners (TTFs), 939
Three-bar frame castings and residual stress, 613
Tilt casting and convection, 610
'Tilt pouring' concept, 831f
Time (pouring), 755–757
Tin and iron castings, 185
Tin (Sn) addition to aluminium alloys, 939
Tin bronze:
  dendritic segregation, 218–219
  vapour transport, 155
Ti-rich intermetallics, 263
Tiryakioğlu, Ergin, 596–598
Tiryakioğlu, M.
  bi-Weibull distributions, 459–461
  fatigue, 501–504, 506
  feeder rules, 348–351
  feeding logic, 564
  freezing time of casting, 182f
  quality indices, 526–527
  tensile properties, 470–473
Tital, Germany, 888
Titanium:
  aluminium alloys, 239, 241–242, 242f, 482, 875
  Al-Si alloys, 243, 263
  binders, 809
  ferrite, 328
  grain refinement of aluminium, 239, 241–243, 242f
  graphite, 825
  liquid, 335
  metal surface reactions, 155
  nickel alloys, 331
  properties, 337
  $TiAl_3$, 201, 240–241, 482, 785
Titanium alloys:
  'α case', 155, 797
  aluminides, 200f, 239, 241–242, 242f, 336–337, 339–340, 403
  bifilms, 339–340
  CaO crucible, 337
  carbon (pickup and loss), 155
  description, 336
  hipping, 336
  inert moulds and cores, 797
  melting and casting, 336–337
  metal surface reactions, 155
  mould materials, 337
  soluble, transient films, 81
  surface films, 339–340
  Ti-5Al-2.5Sn, 336
  Ti-6Al-4V, 336
  Ti-22Al-26Nb, 339–340
    tungsten droplets, 724
  Ti-45Al, 338f
Titanium aluminide (TiAl, Ti_3Al), 336–337, 339–340, 403, 857
Titanium diboride (TiB_2), 33, 35, 200f, 201, 241–242, 785
Titanium oxide and vertical filling of casts, 51–52
T-junctions:
  feeding, 583, 594f
  fins, 185–187, 187f
  gates, 696, 697f–698f, 698–699
T-Mag process for countergravity casting, 846
Tolerances (dimensional accuracy), 896f, 923–924
Tooling, 909–910
Tooling point (TP), 633
Top-gating, 645, 702–703
Touch gates, 700
Tower (shaft) furnaces, 772–773
Transfer and distribution (melting), 776–777
Transformation induced plasticity (TRIP), 921
Transmission electron microscopy (TEM), 241–242, 249, 249f
Transparent organic alloys (growth), 204, 206f
Transport of gases in melts, 9, 10f
Transverse wave effects, 93f
Triangular systems and location points, 636
Trident Gate, 715–716, 715f, 764
True centrifugal casting, 857
'True gas porosity', 398
TS. See Ultimate tensile strength
TTFs. See Thread forming fasteners
Turbine blades:
  ceramic moulds and cores, 800
  conventional shaped castings, 883
  lost wax assembly, 605, 614f
  machining, 938
  Pegasus engine, Harrier Jump Jet, 937
  single crystal solidification, 885–886
  titanium, 335
  vacuum melting and casting, 332
  waterfall flow, 552–553

Turbulence and filling systems, 646–647
Two-stage filling (priming) for gravity casting, 825–826
'Two-stage pour', 712–713, 826
Type A/C iron, 296–297
Type D/E iron, 300–301

U

Ultimate tensile strength (UTS), 490–493
Ultra-clean aluminium alloys, 541
Ultrasonics:
    aluminium alloys, 4, 86, 481, 537–538
    bifilms, 4, 86, 537–538
    reflections, 4, 537–538
Unbonded aggregate moulds process, 814
Undercooling/supercooling, 170, 198–199, 201, 202f–203f, 203–204, 240, 250–251, 265–266, 266f, 292, 300, 326, 327f, 886
Underfeeder shrinkage porosity, 350, 582
Underside shrinkage in zinc alloys, 227
Uniform contraction and casting accuracy, 910–915
Uniform elongation, 477
Unimpeded flow, 96
Unstable substrates and maximum fluidity, 124–125
Unzipping waves, 91–95
Up-runner. *See* Riser
'Use a good-quality melt' rule, 536–542

V

Vacuum arc remelting (VAR), 179, 519, 795–796
Vacuum castings, 57, 67, 403, 653
Vacuum degassing of melts, 783
Vacuum lock, 741
Vacuum melting and casting, 332–335, 823, 881–882
Vacuum moulding (V process), 878–880, 906
Vacuum processes (HPDC), 866
Vacuum riserless casting (VRC), 846
Vacuum-assisted casting, 880–881
Vacuum-assisted filling, 653–654, 654f
Vapourisation:
    alloys melted in vacuum, 15
    copper alloys with lead, 15
    ductile iron, 14
    magnesium alloys, 15
    manganese steels, 15
    melts, 14–15, 14f
    sodium, 14
    zinc, 14
Vapour pressure, 271–272, 272f
Vapour zones:
    greensand moulds, 140, 140f
    transport, 141–142, 142f, 154–155
    water content, 141f
VAR. *See* Vacuum arc remelting

Velocity:
    flow channels, 196
    maximum fluidity, 110
Venting (of cores), 523, 570, 581f, 900
Vertical filling of casts, 51, 52f
Vertical gates, 703–706, 710, 741
Vertical stack moulding, 826
V-Grooves (for chills), 191, 450
Vibration, 128, 204–205, 208f, 211
Viscosity and maximum fluidity, 110–112
VK (Voya Kondic) fluidity strip test, 122f, 125–127
Volatile organic compounds (VOCs), 160
Volume criterion. *See* Feeding, rule 3
Vortex:
    base, 683
    gates, 723
    MMCs, 33–34
    runner, 683
    runner (offset sprue), 722–723
    sprue, 683, 685f, 719–723
    well, 721–722
VRC. *See* Vacuum riserless casting

W

Wall plaques, 821
Warm box processes (binders), 811
Waspalloy (Ni-base), 794
Water:
    flow, 646
    hammer (momentum effect) test piece, 152f
    tensile strength, 347
Water vapour:
    aluminium, 8
    cores, 561
    metal surface reactions, 155
    zinc alloys, 8
Waterfall effect, 56f, 552–553
Weber number (We):
    critical velocities, 41–44
    HPDC, 866
Wedge trap, 717f, 725
Weibull analysis:
    background, 453–455
    bi-Weibull, 459–461
    description, 451
    extreme value distributions, 462
    runner system, 722
    sample number, 464
    three-parameter, 459
    tilt casting, 832
    two-parameter, 455–459
    two-stage filling, 826
Weight and volume estimate (filling system), 748–750

Welders in foundries, 320
Welding steels, 330
Wells for sprue, 679–681
Western bentonite (clay), 144
White iron (iron carbide), 312–313
Wilkinson, John, 812
'Will the metal fill the mould'? 95
Wood and tooling, 909
Woolley, J.W. and Woodbury, K.A.
    (heat transfer coefficients), 172
Woven cloth or mesh filters, 729–733
Wright, George, 53

X

X-ray studies:
    bubble traps, 735
    filling systems, 641
    horizontal transfer casting, 832
    lost foam casting, 877
    low-pressure die casting, 844
    nondestructive testing, 942–943, 945
    pressurised/unpressurised systems, 750, 751f, 753
    radiography of turbine blades, 332
    runner, 683
    sprue, 674, 680
    tangential placement of filters, 741–742
    tomography, 893–894
    video technique for bubble trails in aluminium castings, 63

Y

Yield strength (σ_y)
    dendrite arm spacing, 481, 484–490
    grain size, 481–484
    Hall–Petch relation, 479, 485
    layer porosity, 479
    magnesium alloys, 233
Yielding fracture mechanics (YFM), 495
Young's modulus (elastic stiffness), 226, 507–508, 614, 945

Z

Zero tapered sprue, 673–674
Zildjian cymbals, 508
Zinc:
    boiling point, 271–272, 272f
    copper alloys, 271–272
    distribution, 776
    Mg-Zn alloys, 230, 356–357, 384
    mould coats, 160
    post-casting processing, 927
    vapourisation, 14
    vapour, 410
    vapour pressure, 272f
Zinc alloys:
    accuracy, 900
    aging, 226
    alumina, 226
    aluminium, 224–227
    bifilms, 225–227
    binders, 808
    blisters, 226–227
    carbon-based dies, 797
    casting, 224t, 225f
    creep resistance, 225
    dendrite arm spacing, 227
    dimensional accuracy, 900
    failure, 226
    grain refinement, 201
    graphite dies, 797, 909
    gravity dies, 825
    gunmetal, 270
    heavy metal castings, 227
    high-pressure die-casting, 223, 226–227
    HPDC (hot chamber), 863
    macrosegregation, 224–225
    microsegregation, 224–225
    oxides, 226
    pressure die casting, 65–67, 119
    rubber moulds, 814
    strontium, 227
    underside shrinkage, 227
    water vapour, 8
    ZA series, 119, 226–227, 806
    ZA27, 125, 226–227, 914
    ZAMAK series, 224
    zinc flare, 271
    Zn-4Al, 178, 915f
    Zn-27Al, 122f, 864f
    Zn-Al phase diagram, 225f
    Zn-Al-Cu, 119. *See also* Casting alloys
Zip fasteners (zinc alloys), 226
Zircon sand (zirconium silicate), 800, 803, 805, 811, 818–819
Zirconium:
    binders, 809
    grain refinement, 233–234, 483
    Zr-Cu alloys and vapour transport, 154–155